Praise for LIFESPAN

"In this insightful and provocative book that asks questions about how we age, and whether humans can overcome decay and degeneration, Sinclair grapples with some of the most fundamental questions around the science of aging. The result is an elegant and exciting book that deserves to be read broadly and deeply."

—Siddhartha Mukherjee, Pulitzer Prize–winning
and #1 *New York Times* bestselling author

"If you ever wondered how we age, if we can slow or even reverse aging, and if we can live a healthy 100-plus years, then David Sinclair's new book, *Lifespan* . . . will guide you through the science and the practical strategies to make your health span equal your lifespan, and make your lifespan long and vibrant."

—Mark Hyman, MD Director, Cleveland Clinic Center
for Functional Medicine and #1 *New York Times* bestselling author

"This is the most visionary book about aging I have ever read. Seize the day—and seize this book!"

—Dean Ornish, MD, founder & president, Preventive Medicine
Research Institute, and *New York Times* bestselling author of *UnDo It!*

"In *Lifespan*, David Sinclair eloquently tells us the secret everyone wants to know: how to live longer and age slower. Sinclair convinces us that it is not only possible to live beyond one hundred years, it is inevitable that we will be able to one day do so. If you are someone who wants to know how to beat aging, *Lifespan* is a must-read."

—William W. Li, MD, *New York Times*
bestselling author of *Eat to Beat Disease*

"[I]nsightful, inspiring, and informative. [Sinclair] has translated a wealth of molecular detail into a program that we can all use to live longer and healthier. For anyone interested in understanding the aging process, living longer, and avoiding the diseases of aging, this is *the* book to read."

—Dale Bredesen, MD, *New York Times* bestselling author of *The End of Alzheimer's*

"A visionary book from one of the most masterful longevity scientists of our time. *Lifespan* empowers us to change our health today while revealing a potential future when we live younger for longer."

—Sara Gottfried, MD, *New York Times* bestselling author of *The Hormone Cure*

"Prepare to have your mind blown. You are holding in your hands the precious results of decades of work, as shared by Dr. David Sinclair, the rock star of aging and human longevity."

—Dave Asprey, founder and CEO of Bulletproof and *New York Times* bestselling author of *The Bulletproof Diet*

"Imagine a world in which we can live long enough to meet not just our grandchildren, but our great-grandchildren. This is Sinclair's vision for the future of humankind, a vision that looks to science, nature, history, and even politics to make the case that it is possible to live well into our hundreds. *Lifespan* is boldly leading the way."

—Jason Fung, MD, author of *The Diabetes Code* and *The Obesity Code*

"In *Lifespan*, Dr. David Sinclair . . . provides us with the everyday tools that we can all use to stop what he now calls 'the disease of aging.'. . . You owe it to yourself and your loved ones to read and follow his advice, as I have for the last 15 years!"

—Steven R. Gundry, MD, *New York Times* bestselling author of *The Longevity Paradox* and medical director of the International Heart and Lung Institute

"*Lifespan* . . . transcends everything we know about aging and longevity—a combination of brilliant scientific work, a pioneering mind, and the dream for a longer, healthier and happier life. *Lifespan* provides a vision for our future and the road map on how to get there, merging scientific breakthroughs and simple lifestyle changes to not only help us feel younger, but actually become younger."

—Naomi Whittel, *New York Times* bestselling author of *Glow15*

"David Sinclair masterfully presents a bold vision of the future in which humanity is able to slow or reverse the aging process and live younger, healthier lives for longer."

—Victor J. Dzau, MD, president of the US National Academy of Medicine and CEO of Duke University Medical Center

"There are few books that have ever made me think about science in a fundamentally new way. David Sinclair's book did that for me on aging. This is a book that anyone who ages must read."

—Leroy Hood, PhD, professor at the California Institute of Technology, inventor, entrepreneur, member of all three US National Academies, and coauthor of *Code of Codes*

"In *Lifespan*, the full force of [Sinclair's] optimism, humor, and soft-spoken eloquence as a storyteller-scientist come through. I'm hoping we have David Sinclair with us and doing his science and writing books for another 500 years, give or take a century."

—David Ewing Duncan, award-winning journalist, bestselling author, and curator of *Arc Fusion*

"*Lifespan* gives us hope for an extraordinary life. As the brilliant Dr. David Sinclair explains, aging is a disease, and that disease is treatable. This eye-opening book takes you to front lines of incredible breakthroughs. Enjoy this must-read masterpiece!"

—Peter H. Diamandis, MD, *New York Times* bestselling author of *Abundance* and *Bold*

"[D]escribes real science that will question the foundation of everything we assume about our life and society."

—Salman Khan, founder of Khan Academy

"David is a pioneer poised to change how we think about and understand aging."

—Stephanie Lederman, CEO of the American Federation
for Aging Research (AFAR), New York

"The most important message and priority of our time. For years to come, humanity will reflect on this book with awe and respect. Read it. . . . Your life depends on it."

—Marc Hodosh, former owner & cocreator of TEDMED

"A tour de force. Sinclair's book, and his life's work, ranks with humanity's greatest contributions to helping enhance the joy and happiness of life, ranking with the works of Jenner, Pasteur, Salk, Locke, Gandhi, and Edison. A masterpiece."

—Martine Rothblatt, founder, Chairwoman of the Board, and
CEO of United Therapeutics and creator
of SiriusXM Satellite Radio

"Stepping on the moon changed humanity. In *Lifespan*, Sinclair takes the ultimate step for humanity that will transform our lives beyond anything we could ever have imagined. The author is bold, the science is profound, and our future is here."

—Henry Markram, PhD, professor at EPFL, Switzerland,
director of the Blue Brain Project, and
founder of Frontiers open-access journals

"An intellectually fascinating book with tantalizing insights on the most important issue about yours and everyone's future."

—Andrew Scott, PhD, professor of economics at
the London Business School and author of *The 100-Year Life*

LIFESPAN

WHY WE AGE—

AND

WHY WE DON'T HAVE TO

David A. Sinclair, PhD, AO,

with Matthew D. LaPlante
Illustrations by Catherine L. Delphia

ATRIA BOOKS

NEW YORK   LONDON   TORONTO   SYDNEY   NEW DELHI

An Imprint of Simon & Schuster, Inc.
1230 Avenue of the Americas
New York, NY 10020

First Atria Books hardcover edition September 2019

ATRIA B O O K S and colophon are trademarks of Simon & Schuster, Inc.

For information about special discounts for bulk purchases, please contact Simon & Schuster Special Sales at 1-866-506-1949 or business@simonandschuster.com.

The Simon & Schuster Speakers Bureau can bring authors to your live event. For more information, or to book an event, contact the Simon & Schuster Speakers Bureau at 1-866-248-3049 or visit our website at www.simonspeakers.com.

Illustrations and Sinclairfont by Catherine L. Delphia.
Illustrations in Cast of Characters by David A. Sinclair, PhD.
For reproductions, contact David A. Sinclair, PhD.

Interior design by Ruth Lee-Mui

Manufactured in the United States of America

1   3   5   7   9   10   8   6   4   2

Library of Congress Cataloging-in-Publication Data
Names: Sinclair, David A., 1969– author. | LaPlante, Matthew D., author.
Title: Lifespan : the revolutionary science of why we age—and why we don't have to / David A. Sinclair, Ph.D., A.O. with Matthew D. LaPlante ; illustrations and Sinclairfont by Catherine L. Delphia.
Description: First Atria Books hardcover edition. | New York : Atria Books, 2019. | Includes bibliographical references and index.
Identifiers: LCCN 2019007196 (print) | LCCN 2019009229 (ebook) | ISBN 9781501191992 (eBook) | ISBN 9781501191978 (hardback)
Subjects: LCSH: Life spans (Biology) | Longevity. | BISAC: SCIENCE / Life Sciences / Genetics & Genomics. | HEALTH & FITNESS / Diseases / Genetic.
Classification: LCC QH528.5 (ebook) | LCC QH528.5 .S56 2019 (print) | DDC 570—dc23
LC record available at https://lccn.loc.gov/2019007196

ISBN 978-1-5011-9197-8
ISBN 978-1-5011-9199-2 (ebook)

Thanks to Christine Liu and her team at the Innovative Genomics Institute (IGI) for permission to use the Glossary icons. Thanks to Wikipedia for biographical facts in the cast of characters.

To my grandmother Vera,
who taught me to see the world the way it could be.

To my mother, Diana,
who cared more about her children than herself.

To my wife, Sandra,
my bedrock.

And to my great-great-grandchildren;
I am looking forward to meeting you.

Contents

THE BUSH. In the wild and wonderful world of the Garigal clan, waterfalls and saltwater estuaries wind through ancient sandstone escarpments, under shadowy canopies of charred bloodwoods, angophoras, and scribbly gums that kookaburras, currawongs, and wallabies call home.

INTRODUCTION

A GRANDMOTHER'S PRAYER

I GREW UP ON THE EDGE OF THE BUSH. IN FIGURATIVE TERMS, MY BACKYARD was a hundred-acre wood. In literal terms, it was much bigger than that. It went on as far as my young eyes could see, and I never grew tired of exploring it. I would hike and hike, stopping to study the birds, the insects, the reptiles. I pulled things apart. I rubbed the dirt between my fingers. I listened to the sounds of the wild and tried to connect them to their sources.

And I played. I made swords from sticks and forts from rocks. I climbed trees and swung on branches and dangled my legs over steep precipices and jumped off of things that I probably shouldn't have jumped off. I imagined myself as an astronaut on a distant planet. I pretended to be a hunter on safari. I lifted my voice for the animals as though they were an audience at the opera house.

"Coooeey!" I would holler, which means "Come here" in the language of the Garigal people, the original inhabitants.

I wasn't unique in any of this, of course. There were lots of kids in the northern suburbs of Sydney who shared my love of adventure and exploration and imagination. We expect this of children. We *want* them to play this way.

Until, of course, they're "too old" for that sort of thing. Then we want them to go to school. Then we want them to go to work. To find a partner. To save up. To buy a house.

Because, you know, the clock is ticking.

My grandmother was the first person to tell me that it didn't have to be that way. Or, I guess, she didn't tell me so much as show me.

She had grown up in Hungary, where she spent Bohemian summers swimming in the cool waters of Lake Balaton and hiking in the mountains of its northern shore at a holiday resort that catered to actors, painters, and poets. In the winter months, she helped run a hotel in the Buda Hills before the Nazis took it over and converted it to the central command of the Schutzstaffel, or "SS."

A decade after the war, in the early days of the Soviet occupation, the Communists began to shut down the borders. When her mother tried to cross illegally into Austria, she was caught, arrested, and sentenced to two years in jail and died shortly after. During the Hungarian Uprising in 1956, my grandmother wrote and distributed anti-Communist newsletters in the streets of Budapest. After the revolution was crushed, the Soviets began arresting tens of thousands of dissidents, and she fled to Australia with her son, my father, reasoning that it was the furthest they could get from Europe.

She never set foot in Europe again, but she brought every bit of Bohemia with her. She was, I have been told, one of the first women to sport a bikini in Australia and got chased off Bondi Beach because of it. She spent years living in New Guinea—which even today is one of the most intensely rugged places on our planet—all by herself.

Though her bloodline was Ashkenazi Jew and she had been raised a Lutheran, my grandmother was a very secular person. Our equivalent of the Lord's Prayer was the English author Alan Alexander Milne's poem "Now We Are Six," which ends:

> *But now I am six,*
> *I'm as clever as clever.*
> *So I think I'll be six now*
> *for ever and ever.*

She read that poem to my brother and me again and again. Six, she told us, was the very best age, and she did her damnedest to live life with the spirit and awe of a child of that age.

Even when we were very young, my grandmother didn't want us to call her "grandmother." Nor did she like the Hungarian term, "nagy-mama," or any of the other warm terms of endearment such as "bubbie," "grandma," and "nana."

To us boys, and everyone else, she was simply Vera.

Vera taught me to drive, swerving and swaying across all of the lanes, "dancing" to whatever music was on the car's radio. She told me to enjoy my youth, to savor the feeling of being young. Adults, she said, always ruined things. Don't grow up, she said. Never grow up.

Well into her 60s and 70s, she was still what we call "young at heart," drinking wine with friends and family, eating good food, telling great stories, helping the poor, sick, and less fortunate, pretending to conduct symphonies, laughing late into the night. By just about anyone's standard, that's the mark of a "life well lived."

But yes, the clock was ticking.

By her mid-80s, Vera was a shell of her former self, and the final decade of her life was hard to watch. She was frail and sick. She still had enough wisdom left to insist that I marry my fiancée, Sandra, but by then music gave her no joy and she hardly got out of her chair; the vibrancy that had defined her was gone.

Toward the end, she gave up hope. "This is just the way it goes," she told me.

She died at the age of 92. And, in the way we've been taught to think about these things, she'd had a good, long life. But the more I have thought about it, the more I have come to believe that the person she *truly* was had been dead many years at that point.

Growing old may seem a distant event, but every one of us will experience the end of life. After we draw our last breath, our cells will scream for oxygen, toxins will accumulate, chemical energy will be exhausted, and cellular structures will disintegrate. A few minutes later, all of the education, wisdom, and memories that we cherished, and all of our future potential, will be irreversibly erased.

I learned this firsthand when my mother, Diana, passed away. My father, my brother, and I were there. It was a quick death, thankfully, caused by a buildup of liquid in her remaining lung. We had just been laughing together about the eulogy I'd written on the trip from the United States to Australia, and then suddenly she was writhing on the bed, sucking for air that couldn't satisfy her body's demand for oxygen, staring at us with desperation in her eyes.

I leaned in and whispered into her ear that she was the best mom I could have wished for. Within a few minutes, her neurons were dying, erasing not just the memory of my final words to her but all of her memories. I know some people die peacefully. But that's not what happened to my mother. In those moments she was transformed from the person who had raised me into a twitching, choking mass of cells, all fighting over the last residues of energy being created at the atomic level of her being.

All I could think was "No one tells you what it is like to die. Why doesn't anyone tell you?"

There are few people who have studied death as intimately as the Holocaust documentary filmmaker Claude Lanzmann. And his assessment—indeed, his warning—is chilling. "Every death is violent," he said in 2010. "There is no natural death, unlike the picture we like to paint of the father who dies quietly in his sleep, surrounded by his loved ones. I don't believe in that."[1]

Even if they don't recognize its violence, children come to understand

A "GOOD, LONG LIFE." My grandmother "Vera" sheltered Jews in World War II, lived in primitive New Guinea, and was removed from Bondi Beach for wearing a bikini. The end of her life was hard to watch. "This is just the way it goes," she said. But the person she truly was had been dead many years at that point.

the tragedy of death surprisingly early in their lives. By the age of four or five, they know that death occurs and is irreversible.[2] It is a shocking thought for them, a nightmare that is real.

At first, because it's calming, most children prefer to think that there are certain groups of people who are protected from death: parents, teachers, and themselves. Between 5 and 7, however, all children come to understand the universality of death. Every family member will die. Every pet. Every plant. Everything they love. Themselves, too. I can remember first learning this. I can also very well remember our oldest child, Alex, learning it.

"Dad, you won't *always* be around?"

"Sadly, no," I said.

Alex cried on and off for a few days, then stopped, and never asked me about it again. And I've never again mentioned it, either.

It doesn't take long for the tragic thought to be buried deep in the recesses of our subconscious. When asked if they worry about death, children tend to say that they don't think about it. If asked what they do think about it, they say it is not a concern because it will occur only in the remote future, when they get old.

That's a view most of us maintain until well into our fifties. Death is simply too sad and paralyzing to dwell on each day. Often, we realize it too late. When it comes knocking, and we are not prepared, it can be devastating.

For Robin Marantz Henig, a columnist at the *New York Times*, the "bitter truth" about mortality came late in life, after she became a grandparent. "Beneath all the wonderful moments you may be lucky enough to share in and enjoy," she wrote, "your grandchild's life will be a long string of birthdays you will not live to see."[3]

It takes courage to consciously think about your loved ones' mortality before it actually happens. It takes even more courage to deeply ponder your own.

It was the comedian and actor Robin Williams who first demanded this courage from me through his portrayal of John Keating, the teacher and hero in the film *Dead Poets Society*, who challenges his teenage students to stare into the faces of the long-dead boys in a fading photo.[4]

"They are not that different from you, are they?" Keating says. "Invincible, just like you feel. . . . Their eyes are full of hope . . . But you see, gentlemen, these boys are now fertilizing daffodils."

Keating encourages the boys to lean in closer to listen for a message from the grave. Standing behind them, in a quiet, ghostly voice, he whispers, "*Carpe. Carpe diem.* Seize the day, boys. Make your lives extraordinary."

That scene had an enormous impact on me. It is likely that I would not have had the motivation to become a Harvard professor if it hadn't been for that movie. At the age of 20, I had finally heard someone else say what my grandmother had taught me at an early age: Do your part to make humanity be the best it can be. Don't waste a moment. Embrace your youth; hold on to it for as long as you can. Fight for it. Fight for it. Never stop fighting for it.

But instead of fighting for youth, we fight for life. Or, more specifically, we fight against death.

As a species, we are living much longer than ever. But not much better. Not at all. Over the past century we have gained additional years, but not additional life—not life worth living anyway.[5]

And so most of us, when we think about living to 100, still think "God forbid," because we've seen what those final decades look like, and for most people, most of the time, they don't look appealing at all. Ventilators and drug cocktails. Broken hips and diapers. Chemotherapy and radiation. Surgery after surgery after surgery. And hospital bills; my God, the hospital bills.

We're dying slowly and painfully. People in rich countries often spend a decade or more suffering through illness after illness at the ends of their lives. We think this is normal. As lifespans continue to increase in poorer nations, this will become the fate of billions of additional people. Our successes in extending life, the surgeon and doctor Atul Gawande has noted, have had the effect of "making mortality a medical experience."[6]

But what if it didn't have to be that way? What if we could be younger longer? Not years longer but decades longer. What if those final years didn't look so terribly different from the years that came before them? And what if, by saving ourselves, we could also save the world?

Maybe we can never be six again—but how about twenty-six or thirty-six?

What if we could play as children do, deeper into our lives, without worrying about moving on to the things adults *have to do* so soon? What if all of the things we need to compress into our teenage years didn't need to be so compressed after all? What if we weren't so stressed in our 20s? What if we weren't feeling middle-aged in our 30s and 40s? What if, in our 50s, we wanted to reinvent ourselves and couldn't think of a single reason why we shouldn't? What if, in our 60s, we weren't fretting about leaving a legacy but *beginning* one?

What if we didn't have to worry that the clock was ticking? And what if I told you that soon—very soon, in fact—we won't?

Well, that's what I'm telling you.

I'm fortunate that after thirty years of searching for truths about human biology, I find myself in a unique position. If you were to visit me in Boston, you'd most likely find me hanging out in my lab at Harvard Medical School, where I'm a professor in the Department of Genetics and codirector of the Paul F. Glenn Center for the Biological Mechanisms of Aging. I also run a sister lab at my alma mater, the University of New South Wales in Sydney. In my labs, teams of brilliant students and PhDs have both accelerated and reversed aging in model organisms and have been responsible for some of the most cited research in the field, published in some of the world's top scientific journals. I am also a cofounder of a journal, *Aging*, that provides space to other scientists to publish their research on one of the most challenging and exciting questions of our time, and a cofounder of the Academy for Health and Lifespan Research, a group of the top twenty researchers in aging worldwide.

In trying to make practical use of my discoveries, I've helped start a number of biotechnology companies and sit as a chair of the scientific boards of advisers of several others. These companies work with hundreds of leading academics in scientific areas ranging from the origin of life to genomics to pharmaceuticals.[7] I am, of course, aware of my own labs' discoveries years before they are made public, but through these associations, I'm also aware of many other transformational discoveries ahead of time, sometimes a decade ahead. The coming pages will serve as your backstage pass and your front-row seat.

Having received the equivalent of a knighthood in Australia and taken on the role of an ambassador, I've been spending quite a bit of my time briefing political and business leaders around the world about the ways our understanding of aging is changing—and what that means for humanity going forward.[8]

I've applied many of my scientific findings to my own life, as have many of my family members, friends, and colleagues. The results—which, it should be noted, are completely anecdotal—are encouraging. I'm now 50, and I feel like a kid. My wife and kids will tell you I act like one, too.

That includes being a *stickybeak,* the Australian term for someone who is overly inquisitive, perhaps derived from the currawong crows that used to punch through the foil lids of the milk bottles delivered to our homes and drink the milk out of them. My old high school friends still like to tease me about how, whenever they came over to my parents' house, they would find me pulling something apart: a pet moth's cocoon, a spider's curled-up leaf shelter, an old computer, my father's tools, a car. I became quite good at it. I just wasn't very good at putting these things back together.

I couldn't bear *not* knowing how something worked or where it came from. I still can't—but at least now I get paid for it.

My childhood home is perched on a rocky mountainside. Below is a river that runs into Sydney Harbor. Arthur Phillip, the first governor of New South Wales, explored these valleys in April 1788, only a few months after he and his First Fleet of marines, prisoners, and their families established a colony on the shores of what he called the "finest and most extensive harbor in the universe." The person most responsible for him being there was the botanist Sir Joseph Banks, who eight years earlier had sailed up the Australian coastline with Captain James Cook on his "voyage round the world."[9]

After returning to London with hundreds of plant specimens to impress his colleagues, Banks lobbied King George III to start a British penal colony on the continent, the best site for which, he argued, not coincidentally, would be a bay called "Botany" on "Cape Banks."[10] The First Fleet settlers soon discovered that Botany Bay, despite its most excellent name, had no source of water, so they sailed up to Sydney Harbor and found one of the world's largest "rias," a highly branched, deep waterway that formed when the Hawkesbury River system had been flooded by rising sea levels after the last ice age.

At the age of 10, I had already discovered through exploration that the river in my backyard flowed down into Middle Harbor, a branch of Sydney Harbor. But I could no longer stand not knowing where the river originated. I needed to know what the *beginning* of a river looked like.

I followed it upstream, left the first time it forked and right the time after that, wending into and out of several suburbs. By nightfall I was miles from home, beyond the last mountain on the horizon. I had to ask a stranger to let me call my mother to beg her to come pick me up. A few times after that I tried searching upstream, but never did get anywhere close to the fount. Like Juan Ponce de León, the Spanish explorer of Florida known for his apocryphal quest to find the Fountain of Youth, I failed.[11]

Ever since I can remember, I have wanted to understand why we grow old. But finding the source of a complex biological process is like searching for the spring at the source of a river: it's not easy.

On my quest, I've wound my way left and right and had days when I wanted to give up. But I've persevered. Along the way, I have seen a lot of tributaries, but I've also found what may be the spring. In the coming pages, I will present a new idea about why aging evolved and how it fits into what I call the Information Theory of Aging. I will also tell you why I have come to see aging as a disease—the most common disease—one that not only can but *should* be aggressively treated. That's part I.

In part II, I will introduce you to the steps that can be taken right now—and new therapies in development—that may slow, stop, or reverse aging, bringing an end to aging as we know it.

And yes, I fully recognize the implications of the words "bringing an end to aging as we know it," so, in part III, I will acknowledge the many possible futures these actions could create and propose a path to a future that we can look forward to, a world in which the way we can get to an increased lifespan is through an ever-rising *healthspan*, the portion of our lives spent without disease or disability.

There are plenty of people who will tell you that's a fairy tale—closer to the works of H. G. Wells than those of C. R. Darwin. Some of them are very smart. A few are even people who understand human biology quite well and whom I respect.

Those people will tell you that our modern lifestyles have cursed us with shortening lifespans. They'll say you're unlikely to see 100 years of age and that your children aren't likely to get to the century mark, either. They'll say they've looked at the science of it all and done the projections, and it sure doesn't seem likely that your grandchildren will get to their 100th birthdays, either. And they'll say that if you *do* get to 100, you probably won't get there healthy and you definitely won't be there for long. And if they grant you that people will live longer, they'll tell you that it's the worst thing for this planet. Humans are the enemy!

They've got good evidence for all of this—the entire history of humanity, in fact.

Sure, little by little, millennia by millennia, we've been adding years to the *average* human life, they will say. Most of us didn't get to 40, and then we did. Most of us didn't get to 50, and then we did. Most of us didn't get to 60, and then we did.[12] By and large, these increases in life expectancy came as more of us gained access to stable food sources and clean water. And largely the average was pushed upward from the bottom; deaths during infancy and childhood fell, and life expectancy rose. This is the simple math of human mortality.

But although the *average* kept moving up, the *limit* did not. As long as we've been recording history, we have known of people who have reached their 100th year and who might have lived a few years beyond that mark. But very few reach 110. Almost no one reaches 115.

Our planet has been home to more than 100 billion humans so far. We know of just one, Jeanne Calment of France, who ostensibly lived past the age of 120. Most scientists believe she died in 1997 at the age of 122, although it's also possible that her daughter replaced her to avoid paying taxes.[13] Whether or not she actually made it to that age really doesn't matter; others have come within a few years of that age but most of us, 99.98 percent to be precise, are dead before 100.

So it certainly makes sense when people say that we might continue to chip away at the average, but we're not likely to move the limit. They say it's easy to extend the maximum lifespan of mice or of dogs, but we humans are different. We simply live too long already.

They are wrong.

There's also a difference between extending life and prolonging vitality. We're capable of both, but simply keeping people alive—decades after their lives have become defined by pain, disease, frailty, and immobility—is no virtue.

Prolonged vitality—meaning not just more years of life but more active, healthy, and happy ones—is coming. It is coming sooner than most people expect. By the time the children who are born today have reached middle age, Jeanne Calment may not even be on the list of the top 100 oldest people of all time. And by the turn of the next century, a person who is 122 on the day of his or her death may be said to have lived a full, though not particularly long, life. One hundred and twenty years might be not an outlier but an expectation, so much so that we won't even call it longevity; we will simply call it "life," and we will look back with sadness on the time in our history in which it was not so.

What's the upward limit? I don't think there is one. Many of my colleagues agree.[14] There is no biological law that says we must age.[15] Those who say there is don't know what they're talking about. We're probably still a long way off from a world in which death is a rarity, but we're not far from pushing it ever farther into the future.

All of this, in fact, is inevitable. Prolonged healthy lifespans are in sight. Yes, the entire history of humanity suggests otherwise. But the science of lifespan extension in this particular century says that the previous dead ends are poor guides.

It takes radical thinking to even begin to approach what this will mean for our species. Nothing in our billions of years of evolution has prepared us for this, which is why it's so easy, and even alluring, to believe that it simply cannot be done.

But that's what people thought about human flight, too—up until the moment someone did it.

Today the Wright brothers are back in their workshop, having successfully flown their gliders down the sand dunes of Kitty Hawk. The world is about to change.

And just as was the case in the days leading up to December 17, 1903, the majority of humanity is oblivious. There was simply no context with which to construct the idea of controlled, powered flight back then, so the idea was fanciful, magical, the stuff of speculative fiction.[16]

Then: liftoff. And nothing was ever the same again.

We are at another point of historical inflection. What hitherto seemed magical will become real. It is a time in which humanity will redefine what is possible; a time of ending the inevitable.

Indeed, it is a time in which we will redefine what it means to be human, for this is not just the start of a revolution, it is the start of an evolution.

PART I

WHAT

WE

KNOW

(THE PAST)

ONE

VIVA PRIMORDIUM

IMAGINE A PLANET ABOUT THE SIZE OF OUR OWN, ABOUT AS FAR FROM ITS STAR, rotating around its axis a bit faster, such that a day lasts about twenty hours. It is covered with a shallow ocean of salty water and has no continents to speak of—just some sporadic chains of basaltic black islands peeking up above the waterline. Its atmosphere does not have the same mix of gases as ours. It is a humid, toxic blanket of nitrogen, methane, and carbon dioxide.

There is no oxygen. There is no life.

Because this planet, our planet as it was 4 billion years ago, is a ruthlessly unforgiving place. Hot and volcanic. Electric. Tumultuous.

But that is about to change. Water is pooling next to warm thermal vents that litter one of the larger islands. Organic molecules cover all surfaces, having ridden in on the backs of meteorites and comets. Sitting on dry, volcanic rock, these molecules will remain just molecules, but when dissolved in pools of warm water, through cycles of wetting and drying at the pools' edges, a special chemistry takes place.[1] As the nucleic acids concentrate, they grow into polymers, the way salt crystals form when a seaside puddle evaporates. These are the world's first RNA molecules, the predecessors to DNA. When the pond refills, the primitive genetic material becomes encapsulated by fatty acids to form microscopic soap bubbles—the first cell membranes.[2]

It doesn't take long, a week perhaps, before the shallow ponds are covered with a yellow froth of trillions of tiny precursor cells filled with short strands of nucleic acids, which today we call genes.

Most of the protocells are recycled, but some survive and begin to evolve primitive metabolic pathways, until finally the RNA begins to copy itself. That point marks the origin of life. Now that life has formed—as fatty-acid soap bubbles filled with genetic material—they begin to compete for dominance. There simply aren't enough resources to go around. May the best scum win.

Day in and day out, the microscopic, fragile life-forms begin to evolve into more advanced forms, spreading into rivers and lakes.

Along comes a new threat: a prolonged dry season. The level of the scum-covered lakes has dropped by a few feet during the dry season, but the lakes have always filled up again as the rains returned. But this year, thanks to unusually intense volcanic activity on the other side of the planet, the annual rains don't fall as they usually do and the clouds pass on by. The lakes dry up completely.

What remains is a thick, yellow crust covering the lake beds. It is an ecosystem defined not by the annual waxing and waning of the waters but by a brutal struggle for survival. And more than that: it is a fight for the future—because the organisms that survive will be the progenitors of every living thing to come: archaea, bacteria, fungi, plants, and animals.

Within this dying mass of cells, each scrapping for and scraping by on the merest minimums of nutrients and moisture, each one doing whatever it can to answer the primal call to reproduce, there is a unique species. Let's call it *Magna superstes*. That's Latin for "great survivor."

It does not look very different from the other organisms of the day, but *M. superstes* has a distinct advantage: it has evolved a genetic survival mechanism.

There will be far more complicated evolutionary steps in the eons to come, changes so extreme that entire branches of life will emerge. These changes—the products of mutations, insertions, gene rearrangements, and the horizontal transfer of genes from one species to another—will create organisms with bilateral symmetry, stereoscopic vision, and even consciousness.

By comparison, this early evolutionary step looks, at first, to be rather simple. It is a circuit. A gene circuit.

The circuit begins with gene A, a caretaker that stops cells from re-producing when times are tough. This is key, because on early planet Earth, *most* times are tough. The circuit also has a gene B, which encodes for a "silencing" protein. This silencing protein shuts gene A off when times are good, so the cell can make copies of itself when, and only when, it and its offspring will likely survive.

The genes themselves aren't novel. All life in the lake has these two genes. But what makes *M. superstes* unique is that the gene B silencer has mutated to give it a second function: it helps repair DNA. When the cell's DNA breaks, the silencing protein encoded by gene B moves from gene A to help with DNA repair, which turns on gene A. This temporar-ily stops all sex and reproduction until the DNA repair is complete.

This makes sense, because while DNA is broken, sex and reproduc-tion are the last things an organism should be doing. In future multicel-lular organisms, for instance, cells that fail to pause while fixing a DNA break will almost certainly lose genetic material. This is because DNA is pulled apart prior to cell division from only one attachment site on the DNA, dragging the rest of the DNA with it. If DNA is broken, part of a chromosome will be lost or duplicated. The cells will likely die or multi-ply uncontrollably into a tumor.

With a new type of gene silencer that repairs DNA, too, *M. superstes* has an edge. It hunkers down when its DNA is damaged, then revives. It is superprimed for survival.

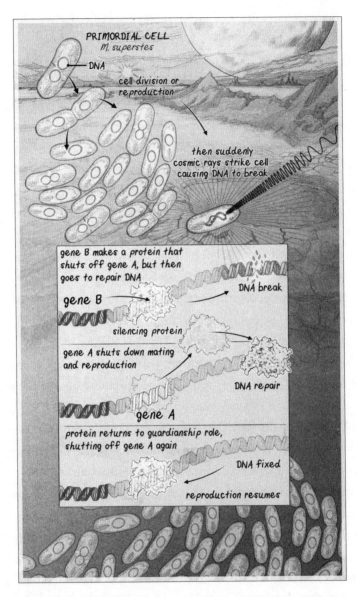

THE EVOLUTION OF AGING. A 4-billion-year-old gene circuit in the first life-forms would have turned off reproduction while DNA was being repaired, providing a survival advantage. Gene A turns off reproduction, and gene B makes a protein that turns off gene A when it is safe to reproduce. When DNA breaks, however, the protein made by gene B leaves to go repair DNA. As a result, gene A is turned on to halt reproduction until repair is complete. We have inherited an advanced version of this survival circuit.

And that's good, because now comes yet another assault on life. Powerful cosmic rays from a distant solar eruption are bathing the Earth, shredding the DNA of all the microbes in the dying lakes. The vast majority of them carry on dividing as if nothing has happened, unaware that their genomes have been broken and that reproducing will kill them. Unequal amounts of DNA are shared between mother and daughter cells, causing both to malfunction. Ultimately, the endeavor is hopeless. The cells all die, and nothing is left.

Nothing, that is, but *M. superstes*. For as the rays wreak their havoc, *M. superstes* does something unusual: thanks to the movement of protein B away from gene A to help repair the DNA breaks, gene A switches on and the cells stop almost everything else they are doing, turning their limited energy toward fixing the DNA that has been broken. By virtue of its defiance of the ancient imperative to reproduce, *M. superstes* has survived.

When the latest dry period ends and the lakes refill, *M. superstes* wakes up. *Now* it can reproduce. Again and again it does so. Multiplying. Moving into new biomes. Evolving. Creating generations upon generations of new descendants.

They are our Adam and Eve.

Like Adam and Eve, we don't know if *M. superstes* ever existed. But my research over the past twenty-five years suggests that every living thing we see around us today is a product of this great survivor, or at least a primitive organism very much like it. The fossil record in our genes goes a long way to proving that every living thing that shares this planet with us still carries this ancient genetic survival circuit, in more or less the same basic form. It is there in every plant. It is there in every fungus. It is there in every animal.

It is there in us.

I propose the reason this gene circuit is conserved is that it is a rather simple and elegant solution to the challenges of a sometimes brutish and sometimes bounteous world that better ensures the survival of the organisms that carry it. It is, in essence, a primordial survival kit that diverts energy to the area of greatest need, fixing what exists in times when the stresses of the world are conspiring to wreak havoc on the genome, while permitting reproduction only when more favorable times prevail.

And it is so simple and so robust that not only did it ensure life's continued existence on the planet, it ensured that Earth's chemical survival circuit was passed on from parent to offspring, mutating and steadily improving, helping life continue for billions of years, no matter what the cosmos brought, and in many cases allowing individuals' lives to continue for far longer than they actually needed to.

The human body, though far from perfect and still evolving, carries an advanced version of the survival circuit that allows it to last for decades past the age of reproduction. While it is interesting to speculate why our long lifespans first evolved—the need for grandparents to educate the tribe is one appealing theory—given the chaos that exists at the molecular scale, it's a wonder we survive thirty seconds, let alone make it to our reproductive years, let alone reach 80 more often than not.

But we do. Marvelously we do. Miraculously we do. For we are the progeny of a very long lineage of great survivors. Ergo, we are great survivors.

But there is a trade-off. For this circuit within us, the descendant of a series of mutations in our most distant ancestors, is also the reason we age.

And yes, that definite singular article is correct: it is *the* reason.

TO EVERYTHING THERE IS A REASON

If you are taken aback by the notion that there is a singular cause of aging, you are not alone. If you haven't given any thought at all as to why we age, that's perfectly normal, too. A lot of biologists haven't given it much thought, either. Even gerontologists, doctors who specialize in aging, often don't ask why we age—they simply seek to treat the consequences.

This isn't a myopia specific to aging. As recently as the late 1960s, for example, the fight against cancer was a fight against its symptoms. There was no unified explanation for why cancer happens, so doctors removed tumors as best they could and spent a lot of time telling patients to get their affairs in order. Cancer was "just the way it goes," because that's what we say when we can't explain something.

Then, in the 1970s, genes that cause cancer when mutated were discovered by the molecular biologists Peter Vogt and Peter Duesberg. These so-called oncogenes shifted the entire paradigm of cancer research. Pharmaceutical developers now had targets to go after: the tumor-inducing proteins encoded by genes, such as *BRAF*, *HER2*, and *BCR-ABL*. By inventing chemicals that specifically block the tumor-promoting proteins, we could finally begin to move away from using radiation and toxic chemotherapeutic agents to attack cancers at their genetic source, while leaving normal cells untouched. We certainly haven't cured all types of cancer in the decades since then, but we no longer believe it's impossible to do so.

Indeed, among an increasing number of cancer researchers, optimism abounds. And that hopefulness was at the heart of what was arguably the most memorable part of President Barack Obama's final State of the Union address in 2016.

"For the loved ones we've all lost, for the family we can still save, let's make America the country that cures cancer once and for all," Obama said as he stood in the House of Representatives chamber and called for a "cancer moon shot." When he placed then Vice President Joe Biden— whose son Beau had died of brain cancer a year earlier—in charge of the effort, even some of the Democrats' staunch political enemies had trouble holding back the tears.

In the days and weeks that followed, many cancer experts noted that it would take far more than the year remaining to the Obama-Biden administration to end cancer. Very few of those experts, however, said it absolutely couldn't be done. And that's because, in the span of just a few decades, we had completely changed the way we think about cancer. We no longer submit ourselves to its inevitability as part of the human condition.

One of the most promising breakthroughs in the past decade has been immune checkpoint therapy, or simply "immunotherapy." Immune T-cells continually patrol our body, looking for rogue cells to identify and kill before they can multiply into a tumor. If it weren't for T-cells, we'd all develop cancer in our twenties. But rogue cancer cells evolve ways to fool cancer-detecting T-cells so they can go on happily multiplying. The latest and most effective immunotherapies bind to proteins on the cancer cells' surface. It is the equivalent of taking the invisible cloak off cancer cells so T-cells can recognize and kill them. Although fewer than 10 percent of all cancer patients currently benefit from immunotherapy, that number should increase thanks to the hundreds of trials currently in progress.

We continue to rail against a disease we once accepted as fate, pouring billions of dollars into research each year, and the effort is paying off. Survival rates for once lethal cancers are increasing dramatically. Thanks to a combination of a BRAF inhibitor and immunotherapy, survival of melanoma brain metastases, one of the deadliest types of cancer, has increased by 91 percent since 2011. Between 1991 and 2016, overall deaths from cancer in the United States declined by 27 percent and continue to fall.[3] That's a victory measured in millions of lives.

Aging research today is at a similar stage as cancer research was in the 1960s. We have a robust understanding of what aging looks like and what it does to us and an emerging agreement about what causes it and what keeps it at bay. From the looks of it, aging is not going to be that hard to treat, far easier than curing cancer.

Up until the second half of the twentieth century, it was generally accepted that organisms grow old and die "for the good of the species"—an idea that dates back to Aristotle, if not further. This idea feels quite intuitive. It is the explanation proffered by most people at parties.[4] But it is dead wrong. We do not die to make way for the next generation.

In the 1950s, the concept of "group selection" in evolution was going out of style, prompting three evolutionary biologists, J. B. S. Haldane, Peter B. Medawar, and George C. Williams, to propose some important ideas about why we age. When it comes to longevity, they agreed, individuals look out for themselves. Driven by their selfish genes, they press on and try to breed for as long and as fast as they can, so long as it doesn't kill them. (In some cases, however, they press on too much, as my great-grandfather Miklós Vitéz, a Hungarian screenwriter, proved to his bride forty-five years his junior on their wedding night.)

If our genes don't ever want to die, why don't we live forever? The trio of biologists argued that we experience aging because the forces of natural selection required to build a robust body may be strong when we are 18 but decline rapidly once we hit 40 because by then we've likely replicated our selfish genes in sufficient measure to ensure their survival. Eventually, the forces of natural selection hit zero. The genes get to move on. We don't.

Medawar, who had a penchant for verbiage, expounded on a nuanced theory called "antagonistic pleiotropy." Put simply, it says genes that help us reproduce when we are young don't just become less helpful as we age, they can actually come back to bite us when we are old.

Twenty years later, Thomas Kirkwood at Newcastle University framed the question of why we age in terms of an organism's available resources. Known as the "Disposable Soma Hypothesis," it is based on the fact that there are always limited resources available to species—energy, nutrients, water. They therefore evolve to a point that lies somewhere between two very different lifestyles: breed fast and die young, or breed slowly and maintain your *soma*, or body. Kirkwood reasoned that organisms can't breed fast *and* maintain a robust, healthy body—there simply isn't enough energy to do both. Stated another way, in the history of life, any line of creature with a mutation that caused it to live fast and attempt to die old soon ran out of resources and was thus deleted from the gene pool.

Kirkwood's theory is best illustrated by fictitious but potentially real-life examples. Imagine you are a small rodent that is likely to be picked off by a bird of prey. Because of this, you'll need to pass down your genetic material quickly, as did your parents and their parents before them. Gene combinations that would have provided a longer-lasting body were not enriched in your species because your ancestors likely didn't escape predation for long (and you won't, either).

Now consider instead that you are a bird of prey at the top of the food chain. Because of this, your genes—well, actually, your ancestors' genes—benefited from building a robust, longer-lasting body that could breed for decades. But in return, they could afford to raise only a couple of fledglings a year.

Kirkwood's hypothesis explains why a mouse lives 3 years while some birds can live to 100.[5] It also quite elegantly explains why the American chameleon lizard, *Anolis carolinensis*, is evolving a longer lifespan as we speak, having found itself a few decades ago on remote Japanese islands without predators.[6]

These theories fit with observations and are generally accepted. Individuals don't live forever because natural selection doesn't select for immortality in a world where an existing body plan works perfectly well to pass along a body's selfish genes. And because all species are resource limited, they have evolved to allocate the available energy either to reproduction or to longevity, but not to both. That was as true for *M. superstes* as it was and still is for all species that have ever lived on this planet.

All, that is, except one: *Homo sapiens*.

Having capitalized on its relatively large brain and a thriving civilization to overcome the unfortunate hand that evolution dealt it—weak limbs, sensitivity to cold, poor sense of smell, and eyes that see well only in daylight and in the visible spectrum—this highly unusual species continues to innovate. It has already provided itself with an abundance of food, nutrients, and water while reducing deaths from predation, exposure, infectious diseases, and warfare. These were all once limits to its evolving a longer lifespan. With them removed, a few million years of evolution might double its lifespan, bringing it closer to the lifespans of some other species at the top of their game. But it won't have to wait that long, nowhere near that. Because this species is diligently working to invent medicines and technologies to give it the robustness of a much longer lived one, literally overcoming what evolution failed to provide.

CRISIS MODE

Wilbur and Orville Wright could never have built a flying machine without a knowledge of airflow and negative pressure and a wind tunnel. Nor could the United States have put men on the moon without an understanding of metallurgy, liquid combustion, computers, and some measure of confidence that the moon is not made of green cheese.[7]

In the same way, if we are to make real progress in the effort to alleviate the suffering that comes with aging, what is needed is a unified explanation for why we age, not just at the evolutionary level but at the fundamental level.

But explaining aging at a fundamental level is no easy task. It will have to satisfy all known laws of physics and all rules of chemistry and be consistent with centuries of biological observations. It will need to span the least understood world between the size of a molecule and the size of a grain of sand,[8] and it should explain simultaneously the simplest and the most complex living machines that have ever existed.

It should, therefore, come as no surprise that there has never been a unified theory of aging, at least not one that has held up—though not for lack of trying.

One hypothesis, proposed independently by Peter Medawar and Leo Szilard, was that aging is caused by DNA damage and a resulting loss of genetic information. Unlike Medawar, who was always a biologist, who built a Nobel Prize–winning career in immunology, Szilard had come to study biology in a roundabout way. The Budapest-born polymath and inventor lived a nomadic life with no permanent job or address, preferring to spend his time staying with colleagues who satisfied his mental curiosities about the big questions facing humanity. Early in his career, he was a pioneering nuclear physicist and a founding collaborator on the

Manhattan Project, which ushered in the age of atomic warfare. Horrified by the countless lives his work had helped end, he turned his tortured mind toward making life maximally long.[9]

The idea that mutation accumulation causes aging was embraced by scientists and the public alike in the 1950s and 1960s, at a time when the effects of radiation on human DNA were on a lot of people's minds. But although we know with great certainty that radiation can cause all sorts of problems in our cells, it causes only a subset of the signs and symptoms we observe during aging,[10] so it cannot serve as a universal theory.

In 1963, the British biologist Leslie Orgel threw his hat into the ring with his "Error Catastrophe Hypothesis," which postulated that mistakes made during the DNA-copying process lead to mutations in genes, including those needed to make the protein machinery that copies DNA. The process increasingly disrupts those same processes, multiplying upon themselves until a person's genome has been incorrectly copied into oblivion.[11]

Around the same time that Szilard was focusing on radiation, Denham Harman, a chemist at Shell Oil, was also thinking atomically, albeit in a different way. After taking time off to finish medical school at Stanford University, he came up with the "Free Radical Theory of Aging," which blames aging on unpaired electrons that whiz around within cells, damaging DNA through oxidation, especially in mitochondria, because that is where most free radicals are generated.[12] Harman spent the better part of his life testing the theory.

I had the pleasure of meeting the Harman family in 2013. His wife told me that Professor Harman had been taking high doses of alpha-lipoic acid for most of his life to quench free radicals. Considering that he worked tirelessly on his research well into his 90s, I suppose, at the very least, it didn't hurt.

Through the 1970s and 1980s, Harman and hundreds of other researchers tested whether antioxidants would extend the lifespan of animals. The results overall were disappointing. Although Harman had some success increasing the average lifespan of rodents, such as with the food additive butylated hydroxytoluene, none showed an increase in *maximum* lifespan. In other words, a cohort of study animals might live a few weeks longer, on average, but none of the animals was setting records for individual longevity. Science has since demonstrated that the positive health effects attainable from an antioxidant-rich diet are more likely caused by stimulating the body's natural defenses against aging, including boosting the production of the body's enzymes that eliminate free radicals, not as a result of the antioxidant activity itself.

If old habits die hard, the free-radical idea is heroin. The theory was overturned by scientists within the cloisters of my field more than a decade ago, yet it is still widely perpetuated by purveyors of pills and drinks, who fuel a $3 billion global industry.[13] With all that advertising, it is not surprising that more than 60 percent of US consumers still look for foods and beverages that are good sources of antioxidants.[14]

Free radicals do cause mutations. Of course they do. You can find mutations in abundance, particularly in cells that are exposed to the outside world[15] and in the mitochondrial genomes of old individuals. Mitochondrial decline is certainly a hallmark of aging and can lead to organ dysfunction. But mutations alone, especially mutations in the nuclear genome, conflict with an ever-increasing amount of evidence to the contrary.

Arlan Richardson and Holly Van Remmen spent about a decade at the University of Texas at San Antonio testing if increasing free-radical damage or mutations in mice led to aging; it didn't.[16] In my lab and others, it has proven surprisingly simple to restore the function of mitochondria in old mice, indicating that a large part of aging is not due to mutations in mitochondrial DNA, either, at least not until late in life.[17]

Although the discussion about the role of nuclear DNA mutations in aging continues, there is one fact that contradicts all these theories, one that is difficult to refute.

Ironically, it was Szilard, in 1960, who initiated the demise of his own theory by figuring out to how to clone a human cell.[18] Cloning gives us the answer as to whether or not mutations cause aging. If old cells had indeed lost crucial genetic information and this was the cause of aging, we shouldn't be able to clone new animals from older individuals. Clones would be born old.

It's a misconception that cloned animals age prematurely. It has been widely perpetuated in the media and even the National Institutes of Health website says so.[19] Yes, it's true that Dolly, the first cloned sheep, created by Keith Campbell and Ian Wilmut at the Roslin Institute at the University of Edinburgh, lived only half a normal lifespan and died of a progressive lung disease. But extensive analysis of her remains showed no sign of premature aging.[20] Meanwhile, the list of animal species that have been cloned and proven to live a normal, healthy lifespan now includes goats, sheep, mice, and cows.[21]

Because of the fact that nuclear transfer works in cloning, we can say with a high degree of confidence that aging isn't caused by mutations in nuclear DNA. Sure, it's possible that some cells in the body don't mutate and those are the ones that end up making successful clones, but that seems highly unlikely. The simplest explanation is that old animals retain all the requisite genetic information to generate an entirely new, healthy animal and that mutations are not the primary cause of aging.[22]

It's certainly no dishonor to those brilliant researchers that their theories haven't withstood the test of time. That's what happens to most science, and perhaps all of it eventually. In *The Structure of Scientific Revolutions*, Thomas Kuhn noted that scientific discovery is never complete; it goes through predictable stages of evolution. When a theory succeeds at explaining previously unexplainable observations about the world, it becomes a tool that scientists can use to discover even more.

Inevitably, however, new discoveries lead to new questions that are not entirely answerable by the theory, and those questions beget more questions. Soon the model enters crisis mode and begins to drift as scientists seek to adjust it, as little as possible, to account for that which it cannot explain.

Crisis mode is always a fascinating time in science but one that is not for the faint of heart, as doubts about the views of previous generations continue to grow against the old guard's protestations. But the chaos is ultimately replaced by a paradigm shift, one in which a new consensus model emerges that can explain more than the previous model.

That's what happened about a decade ago, as the ideas of leading scientists in the aging field began to coalesce around a new model—one that suggested that the reason so many brilliant people had struggled to identify a single cause of aging was that there wasn't one.

In this more nuanced view, aging and the diseases that come with it are the result of multiple "hallmarks" of aging:

- Genomic instability caused by DNA damage
- Attrition of the protective chromosomal endcaps, the telomeres
- Alterations to the epigenome that controls which genes are turned on and off
- Loss of healthy protein maintenance, known as proteostasis
- Deregulated nutrient sensing caused by metabolic changes
- Mitochondrial dysfunction
- Accumulation of senescent zombielike cells that inflame healthy cells
- Exhaustion of stem cells
- Altered intercellular communication and the production of inflammatory molecules

Researchers began to cautiously agree: address these hallmarks, and you can slow down aging. Slow down aging, and you can forestall disease. Forestall disease, and you can push back death.

Take stem cells, which have the potential to develop into many other kinds of cells: if we can keep these undifferentiated cells from tiring out, they can continue to generate all the differentiated cells necessary to heal damaged tissues and battle all kinds of diseases.

Meanwhile, we're improving the rates of acceptance of bone marrow transplants, which are the most common form of stem cell therapy, and using stem cells for the treatment of arthritic joints, type 1 diabetes, loss of vision, and neurodegenerative diseases such as Alzheimer's and Parkinson's. These stem cell–based interventions are adding years to people's lives.

Or take senescent cells, which have reached the end of their ability to divide but refuse to die, continuing to spit out panic signals that inflame surrounding cells: if we can kill off senescent cells or keep them from accumulating in the first place, we can keep our tissues much healthier for longer.

The same can be said for combating telomere loss, the decline in proteostasis, and all of the other hallmarks. Each can be addressed one by one, a little at a time, in ways that can help us extend human healthspans.

Over the past quarter century, researchers have increasingly homed their efforts in on addressing each of these hallmarks. A broad consensus formed that this would be the best way to alleviate the pain and suffering of those who are aging.

There is little doubt that the list of hallmarks, though incomplete, comprises the beginnings of a rather strong tactical manual for living longer and healthier lives. Interventions aimed at slowing any one of these hallmarks may add a few years of wellness to our lives. If we can address all of them, the reward could be vastly increased *average* lifespans.

As for pushing way past the *maximum* limit? Addressing these hallmarks might not be enough.

But the science is moving fast, faster now than ever before, thanks to the accumulation of many centuries of knowledge, robots that analyze tens of thousands of potential drugs each day, sequencing machines that read millions of genes a day, and computing power that processes trillions of bytes of data at speeds that were unimaginable just a decade ago. Theories on aging, which were slowly chipped away for decades, are now more easily testable and refutable.

Although it is in its early days, a new shift in thinking is again under way. Once again we find ourselves in a period of chaos—still quite

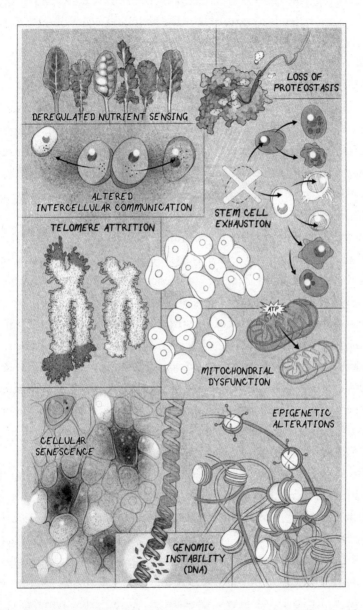

THE HALLMARKS OF AGING. Scientists have settled on eight or nine hallmarks of aging. Address one of these, and you can slow down aging. Address all of them, and you might not age.

confident that the hallmarks are accurate indicators of aging and its myriad symptoms but unable to explain why the hallmarks occur in the first place.

It is time for an answer to this very old question.

Now, finding a universal explanation for anything—let alone something as complicated as aging—doesn't happen overnight. Any theory that seeks to explain aging must not just stand up to scientific scrutiny but provide a rational explanation for every one of the pillars of aging. A universal hypothesis that seems to provide a reason for cellular senescence but not stem cell exhaustion would, for example, explain neither.

Yet I believe that such an answer exists—a cause of aging that exists upstream of all the hallmarks. Yes, a *singular* reason why we age.

Aging, quite simply, is a loss of information.

You might recognize that loss of information was a big part of the ideas that Szilard and Medawar independently espoused, but it was wrong because it focused on a loss of *genetic* information.

But there are two types of information in biology, and they are encoded entirely differently. The first type of information—the type my esteemed predecessors understood—is *digital*. Digital information, as you likely know, is based on a finite set of possible values—in this instance, not in base 2 or binary, coded as 0s and 1s, but the sort that is quaternary or base 4, coded as adenine, thymine, cytosine, and guanine, the nucleotides A, T, C, G of DNA.

Because DNA is digital, it is a reliable way to store and copy information. Indeed, it can be copied again and again with tremendous accuracy, no different in principle from digital information stored in computer memory or on a DVD.

DNA is also robust. When I first worked in a lab, I was shocked by how this "molecule of life" could survive for hours in boiling water and thrilled that it was recoverable from Neanderthal remains at least 40,000 years old.[23] The advantages of digital storage explain why chains of nucleic acids have remained the go-to biological storage molecule for the past 4 billion years.

The other type of information in the body is *analog*.

We don't hear as much about analog information in the body. That's in part because it's newer to science, and in part because it's rarely described in terms of information, even though that's how it was first described when geneticists noticed strange nongenetic effects in plants they were breeding.

Today, analog information is more commonly referred to as the *epigenome*, meaning traits that are heritable that aren't transmitted by genetic means.

The term *epigenetics* was first coined in 1942 by Conrad H. Waddington, a British developmental biologist, while working at Cambridge University. In the past decade, the meaning of the word epigenetics has expanded into other areas of biology that have less to do with heredity—including embryonic development, gene switch networks, and chemical modifications of DNA-packaging proteins—much to the chagrin of orthodox geneticists in my department at Harvard Medical School.

In the same way that genetic information is stored as DNA, epigenetic information is stored in a structure called chromatin. DNA in the cell isn't flailing around disorganized, it is wrapped around tiny balls of protein called histones. These beads on a string self-assemble to form loops, as when you tidy your garden hose on your driveway by looping it into a pile. If you were to play tug-of-war using both ends of a chromosome, you'd end up with a six foot-long string of DNA punctuated by thousands of histone proteins. If you could somehow plug one end of the DNA into a power socket and make the histones flash on and off, a few cells could do you for holiday lights.

In simple species, like ancient *M. superstes* and fungi today, epigenetic information storage and transfer is important for survival. For complex life, it is essential. By complex life, I mean anything made up of more than a couple of cells: slime molds, jellyfish, worms, fruit flies, and of course mammals like us. Epigenetic information is what orchestrates the assembly of a human newborn made up of 26 billion cells from a single fertilized egg and what allows the genetically identical cells in our bodies to assume thousands of different modalities.[24]

If the genome were a computer, the epigenome would be the software.

It instructs the newly divided cells on what type of cells they should be and what they should remain, sometimes for decades, as in the case of individual brain neurons and certain immune cells.

That's why a neuron doesn't one day behave like a skin cell and a dividing kidney cell doesn't give rise to two liver cells. Without epigenetic information, cells would quickly lose their identity and new cells would lose their identity, too. If they did, tissues and organs would eventually become less and less functional until they failed.

In the warm ponds of the primordial Earth, a digital chemical system was the best way to store long-term genetic data. But information storage was also needed to record and respond to environmental conditions, and this was best stored in analog format. Analog data are superior for this job because they can be changed back and forth with relative ease whenever the environment within or outside the cell demands it, and they can store an almost unlimited number of possible values, even in response to conditions that have never been encountered before.[25]

The unlimited number of possible values is why many audiophiles still prefer the rich sounds of analog storage systems. But even though analog devices have their advantages, they have a major disadvantage. In fact, it's the reason we've moved from analog to digital. Unlike digital, analog information degrades over time—falling victim to the conspiring forces of magnetic fields, gravity, cosmic rays, and oxygen. Worse still, information is lost as it's copied.

No one was more acutely disturbed by the problem of information loss than Claude Shannon, an electrical engineer from the Massachusetts Institute of Technology (MIT) in Boston. Having lived through World War II, Shannon knew firsthand how the introduction of "noise" into analog radio transmissions could cost lives. After the war, he wrote a short but profound scientific paper called "The Mathematical Theory of Communication" on how to preserve information, which many consider the foundation of Information Theory. If there is one paper that propelled us into the digital, wireless world in which we now live, that would be it.[26]

Shannon's primary intention, of course, was to improve the robustness of electronic and radio communications between two points. His work may ultimately prove to be even more important than that, for what he discovered about preserving and restoring information, I believe, can be applied to aging.

Don't be disheartened by my claim that we are the biological equivalent of an old DVD player. This is actually good news. If Szilard had turned out to be right about mutations causing aging, we would not be able to easily address it, because when information is lost without a backup, it is lost for good. Ask anyone who's tried to play or restore content from a DVD that's had an edge broken off: what is gone is gone.

But we can usually recover information from a *scratched* DVD. And if I am right, the same kind of process is what it will take to reverse aging.

As cloning beautifully proves, our cells retain their youthful digital information even when we are old. To become young again, we just need to find some polish to remove the scratches.

This, I believe, is possible.

A TIME TO EVERY PURPOSE

The Information Theory of Aging starts with the primordial survival circuit we inherited from our distant ancestors.

Over time, as you might expect, the circuit has evolved. Mammals, for instance, don't have just a couple of genes that create a survival circuit, such as those that first appeared in *M. superstes*. Scientists have found more than two dozen of them within our genome. Most of my colleagues call these "longevity genes" because they have demonstrated the ability to extend both average and maximum lifespans in many organisms. But these genes don't just make life longer, they make it healthier, which is why they can also be thought of as "vitality genes."

Together, these genes form a surveillance network within our bodies, communicating with one another between cells and between organs by releasing proteins and chemicals into the bloodstream, monitoring and responding to what we eat, how much we exercise, and what time of day it is. They tell us to hunker down when the going gets tough, and they tell us to grow fast and reproduce fast when the going gets easier.

And now that we know these genes are there and what many of them do, scientific discovery has given us an opportunity to explore and exploit them; to imagine their potential; to push them to work for us in different ways. Using molecules both natural and novel, using technology both simple and complex, using wisdom both new and old, we can read them, turn them up and down, and even change them altogether.

The longevity genes I work on are called "sirtuins," named after the yeast *SIR2* gene, the first one to be discovered. There are seven sirtuins in mammals, *SIRT1* to *SIRT7*, and they are made by almost every cell in the body. When I started my research, sirtuins were barely on the scientific radar. Now this family of genes is at the forefront of medical research and drug development.

Descended from gene B in *M. superstes*, sirtuins are enzymes that remove acetyl tags from histones and other proteins and, by doing so, change the packaging of the DNA, turning genes off and on when needed. These critical epigenetic regulators sit at the very top of cellular control systems, controlling our reproduction and our DNA repair. After a few billion years of advancement since the days of yeast, they have evolved to control our health, our fitness, and our very survival. They have also evolved to require a molecule called nicotinamide adenine dinucleotide, or NAD. As we will see later, the loss of NAD as we age, and the resulting decline in sirtuin activity, is thought to be a primary reason our bodies develop diseases when we are old but not when we are young.

Trading reproduction for repair, the sirtuins order our bodies to "buckle down" in times of stress and protect us against the major diseases of aging: diabetes and heart disease, Alzheimer's disease and osteoporosis, even cancer. They mute the chronic, overactive inflammation that drives diseases such as atherosclerosis, metabolic disorders, ulcerative colitis, arthritis, and asthma. They prevent cell death and boost mitochondria, the power packs of the cell. They go to battle with muscle wasting, osteoporosis, and macular degeneration. In studies on mice, activating the sirtuins can improve DNA repair, boost memory, increase exercise endurance, and help the mice stay thin, regardless of what they eat. These are not wild guesses as to their power; scientists have established all of this in peer-reviewed studies published in journals such as *Nature, Cell,* and *Science.*

And in no small measure, because sirtuins do all of this based on a rather simple program—the wondrous gene B in the survival circuit—they're turning out to be more amenable to manipulation than many other longevity genes. They are, it would appear, one of the first dominos in the magnificent Rube Goldberg machine of life, the key to understanding how our genetic material protects itself during times of adversity, allowing life to persist and thrive for billions of years.

Sirtuins aren't the only longevity genes. Two other very well studied sets of genes perform similar roles, which also have been proven to be manipulable in ways that can offer longer and healthier lives.

One of these is called target of rapamycin, or TOR, a complex of proteins that regulates growth and metabolism. Like sirtuins, scientists have found TOR—called mTOR in mammals—in every organism in which they've looked for it. Like that of sirtuins, mTOR activity is exquisitely regulated by nutrients. And like the sirtuins, mTOR can signal cells in stress to hunker down and improve survival by boosting such activities as DNA repair, reducing inflammation caused by senescent cells, and, perhaps its most important function, digesting old proteins.[27]

When all is well and fine, TOR is a master driver of cell growth. It senses the amount of amino acids that is available and dictates how much protein is created in response. When it is inhibited, though, it forces cells to hunker down, dividing less and reusing old cellular components to maintain energy and extend survival—sort of like going to the junkyard to find parts with which to fix up an old car rather than buying a new one, a process called autophagy. When our ancestors were unsuccessful in bringing down a woolly mammoth and had to survive on meager rations of protein, it was the shutting down of mTOR that permitted them to survive.

The other pathway is a metabolic control enzyme known as AMPK, which evolved to respond to low energy levels. It has also been highly conserved among species and, as with sirtuins and TOR, we have learned a lot about how to control it.

These defense systems are all activated in response to biological stress. Clearly, some stresses are simply too great to overcome—step on a snail, and its days are over. Acute trauma and uncontrollable infections will kill an organism without *aging* that organism. Sometimes the stress inside a cell, such as a multitude of DNA breaks, is too much to handle. Even if the cell is able to repair the breaks in the short term without leaving mutations, there is information loss at the epigenetic level.

Here's the important point: there are plenty of stressors that will activate longevity genes without damaging the cell, including certain types of exercise, intermittent fasting, low-protein diets, and exposure to hot and cold temperatures (I discuss this in chapter 4). That's called hormesis.[28] Hormesis is generally good for organisms, especially when it can be induced without causing any lasting damage. When hormesis happens, all is well. And, in fact, all is *better* than well, because the little bit of stress that occurs when the genes are activated prompts the rest of the system to hunker down, to conserve, to survive a little longer. That's the start of longevity.

Complementing these approaches are hormesis-mimicking molecules. Drugs in development and at least two drugs on the market can turn on the body's defenses without creating *any* damage. It's like making a prank call to the Pentagon. The troops and the Army Corps of Engineers are sent out, but there's no war. In this way, we can mimic the benefits of exercise and intermittent fasting with a single pill (I discuss this in chapter 5).

Our ability to control all of these genetic pathways will fundamentally transform medicine and the shape of our everyday lives. Indeed, it will change the way we define our species.

And yes, I realize how that sounds. So let me explain why.

TWO

THE DEMENTED PIANIST

ON APRIL 15, 2003, NEWSPAPERS, TELEVISION PROGRAMS, AND WEBSITES around the world carried the story: the mapping of the human genome was complete.

There was just one pesky problem: it really wasn't. There were, in fact, huge gaps in the sequence.

This wasn't a case of the mainstream news media blowing things out of proportion. Highly respected scientific journals such as *Science* and *Nature* told pretty much the same story. It also wasn't a case of scientists overstating their work. The truth is simply that, at the time, most researchers involved in the thirteen-year, $1 billion project agreed that we'd come as close as we possibly could—given the technology of the time—to identifying each of the 3 billion base pairs in our DNA.

The parts of the genome that were missing, generally overlapping sections of repetitive nucleotides, were just not considered important. These were areas of the code of life that were once derided as "junk DNA" and that are now a little better respected but still generally disregarded as "noncoding." From the perspective of many of the best minds in science at the time, those regions were little more than the ghosts of genomes past, mostly remnants of dead hitchhiking viruses that had integrated into the genome hundreds of thousands of years ago. The stuff that makes us who we are, it was thought, had largely been identified, and we had what we needed to propel forward our understanding of what makes us human.

Yet by some estimates, that genetic dark matter accounts for as much as 69 percent of the total genome,[1] and even within the regions generally regarded as "coding," some scientists believe, up to 10 percent has yet to be decoded, including regions that impact aging.[2]

In the relatively short time that has come and gone since 2003, we have come to find out that within the famous double helix, there were sequences that were not just unmapped but essential to our lives. Indeed, many thousands of sequences had gone undetected because the original algorithms to detect genes were written to disregard any gene less than 300 base pairs long. In fact, genes can be as short as 21 base pairs, and today we're discovering hundreds of them all over the genome.

These genes tell our cells to create specific proteins, and these proteins are the building blocks of the processes and traits that constitute human biology and lived experiences. And as we get closer to identifying a complete sequence of our DNA, we've come closer to having a "map" of the genes that control so much of our existence.

Even once we have a complete code, though, there's something we still won't be able to find.

We won't be able to find an aging gene.

We have found genes that impact the *symptoms* of aging. We've found longevity genes that control the body's defenses against aging and thus offer a path to slowing aging through natural, pharmaceutical, and technological interventions. But unlike the oncogenes that were discovered in the 1970s and that have given us a good target for going to battle against cancer, we haven't identified a singular gene that causes aging. And we won't.

Because our genes did not evolve to *cause* aging.

YEAST OF EDEN

My journey toward formulating the Information Theory of Aging was a long one. And in no small part, it can be traced to the work of a scientist who toiled without fame but whose work helped set the stage for a lot of the longevity research being done around the world today.

His name was Robert Mortimer, and if there was one adjective that seemed to come up more than any other about him after he passed away, it was "kind."

"Visionary" was another. "Brilliant," "inquisitive," and "hardworking," too. But I've long been inspired by the example Mortimer set for his fellow scientists. Mortimer, who died in 2007, had played a tremendously important role in elevating *Saccharomyces cerevisiae* from a seemingly lowly, single-celled yeast with a sweet tooth (its name means "sugar-loving") to its rightful place as one of the world's most important research organisms.

Mortimer collected thousands of mutant yeast strains in his lab, many of which had been developed right there at the University of California, Berkeley. He could have paid for his research, and then some, by charging the thousands of scientists he supplied through the university's Yeast Genetic Stock Center. But anyone, from impecunious undergraduates to tenured professors at the world's best-funded research institutions, could browse the center's catalog, request any strain, and have it promptly delivered for the cost of postage.[3]

And because he made it so easy and so inexpensive, yeast research bloomed.

When Mortimer began working on *S. cerevisiae* alongside fellow biologist John Johnston[4] in the 1950s, hardly anyone was interested in yeast. To most, it didn't seem we could learn much about our complex selves by studying a tiny fungus. It was a struggle to convince the scientific community that yeast could be useful for something more than baking bread, brewing beer, and vinting wine.

What Mortimer and Johnston recognized, and what many others

began to realize in the years to come, was that those tiny yeast cells are not so different from ourselves. For their size, their genetic and biochemical makeup is extraordinarily complex, making them an exceptionally good model for understanding the biological processes that sustain life and control lifespans in large complex organisms such as ourselves. If you are skeptical that a yeast cell can tell us anything about cancer, Alzheimer's disease, rare diseases, or aging, consider that there have been five Nobel Prizes in Physiology or Medicine awarded for genetic studies in yeast, including the 2009 prize for discovering how cells counteract telomere shortening, one of the hallmarks of aging.[5]

The work Mortimer and Johnston did—and, in particular, a seminal paper in 1959 that demonstrated that mother and daughter yeast cells can have vastly different lifespans—would set the stage for a world-shattering change in the way we view the limits of life. And by the time of Mortimer's death in 2007, there were some 10,000 researchers studying yeast around the globe.

Yes, humans are separated from yeast by a billion years of evolution, but we still have a lot in common. *S. cerevisiae* shares some 70 percent of our genes. And what it does with those genes isn't so different from what we do with them. Like a whole lot of humans, yeast cells are almost always trying to do one of two things: either they're trying to eat, or they're trying to reproduce. They're hungry or they're horny. As they age, much like humans, they slow down and grow larger, rounder, and less fertile. But whereas humans go through this process over the course of many decades, yeast cells experience it in a week. That makes them a pretty good place to start in the quest to understand aging.

Indeed, the potential for a humble yeast to tell us so much about ourselves—and do so quite quickly relative to other research organisms— was a big part of the reason I decided to begin my career by studying *S. cerevisiae*. They also smell like fresh bread.

I met Mortimer in Vienna in 1992, when I was in my early 20s and attending the International Yeast Conference—yes, there is such a thing—with my two PhD supervisors, Professor Ian Dawes, a rule-avoiding Australian from the University of New South Wales,[6] and Professor Richard Dickinson, a rule-abiding Briton from the University of Cardiff, Wales.

Mortimer was in Vienna to discuss a momentous scientific endeavor: the sequencing of the yeast genome. I was there to be inspired. And I was.[7] If I'd harbored any doubts about my decision to dedicate the opening years of my scientific career to a single-celled fungus, they all went away when I came face to face with people who were building great knowledge in a field that had hardly existed a few decades before.

It was shortly after that conference that one of the world's top scientists in the yeast field, Leonard Guarente of the Massachusetts Institute of Technology, came to Sydney on holiday to visit Ian Dawes. Guarente and I ended up at a dinner together, and I made sure I was sitting opposite him.

I was then a graduate student using yeast to understand an inherited condition called maple syrup urine disease. As you might imagine from its name, the disease is not something most polite people discuss over dinner. Guarente, though, engaged me in a scientific discussion with a curiosity and enthusiasm that was nothing short of enchanting. The conversation soon turned to his latest project—he had begun studying aging in yeast the past few months—work that had its roots in the workable genetic map that Mortimer had completed in the mid-1970s.

That was it. I had a passion for understanding aging, and I knew something about wrangling a yeast cell with a microscope and micromanipulator. Those were essential skills needed to figure out why yeast age. That night, Guarente and I agreed on one thing: if we couldn't solve the problem of aging in yeast, we had no chance in humans.

I didn't just *want* to work with him. I *had* to work with him.

Dawes wrote him to tell him that I was keen to join his lab and I was "skilled at the bench."

"It would be a pleasure to work with David," he replied a few weeks later, the same way he probably did to so many other enthusiastic applicants. "But he's got to come with his own funding." Later I learned he had been excited only because he'd thought I was the *other* student he'd met at dinner.

I had a foot in the door, but my chances were slim. At the time, foreigners weren't considered for prestigious postdoctoral awards in the United States, but I insisted I be interviewed and paid for a flight to Boston myself. I was interviewed by a giant in the stem cell field, Douglas Melton, for a Helen Hay Whitney Foundation Fellowship, which has been providing research support to postdoctoral biomedical students since 1947. After waiting in line outside his office with the other four candidates, I had my chance. This was my moment. I don't remember being nervous. I figured I probably wouldn't get the award anyway. So I went for it.

I told Melton about my lifelong quest to understand aging and find "life-giving genes," then sketched out on his whiteboard how the genes work and what I'd be doing for the next three years if I got the money. To show my gratitude, I gave him a bottle of red wine that I'd brought from Australia.

Afterward, two things became clear. One, don't bring wine to an interview because it can be seen as a bribe. And two, Melton must have liked what I said and how I said it, because I flew home, got the fellowship, and then got onto a plane back to Boston. It was, without a doubt, the most life-changing meeting of my life.[8]

At the time of my arrival, in 1995, I had expected to build our understanding of aging by studying Werner syndrome, a terrible disease that occurs in less than 1 in 100,000 live births, with symptoms that include a loss of body strength, wrinkles, gray hair, hair loss, cataracts, osteoporosis, heart problems, and many other telltale signs of aging—not among folks in their 70s and 80s but rather among people in their 30s and 40s. Life expectancy for someone with Werner is 46 years.

Within two weeks of my arrival in the United States, though, a research team at the University of Washington, headed by the wise and supportive grandfather of aging research, George Martin, announced that they had found the gene that, when mutated, causes Werner syndrome.[9]

It was deflating at the time to have been "scooped," but the discovery allowed me to take a bigger first step toward my ultimate objective. Indeed, it became the key to formulating the Information Theory of Aging.

Now that the Werner gene, known as WRN, had been identified in humans, the next step was to test if the similar gene in yeast had the same function. If so, we could use yeast to more rapidly determine the cause of Werner syndrome and perhaps help us better understand aging in general. I marched into Guarente's office to tell him I was now studying Werner's syndrome in yeast and that's how we would solve aging.

In yeast, the equivalent of the WRN gene is Slow Growth Suppressor 1, or *SGS1*. The gene was already suspected to code for a type of enzyme called a DNA helicase that untangles tangled strands of DNA before they break. Helicases are especially important in repetitive DNA sequences that are inherently prone to tangling and breaking. Functionality of proteins, such as the ones coded for by the Werner gene, is therefore vital, since more than half of our genome is, in fact, repetitive.

Through a gene-swapping process in which cells are tricked into picking up extra pieces of DNA, we swapped out the functional *SGS1* gene with a mutant version. In effect, we were testing to see if it was possible to give the yeast Werner syndrome.

After the swap, the yeast cells' lifespan was cut in half. Ordinarily, this would not have been news. Many events unrelated to aging—such as being eaten by a mite, drying out on a grape, or being placed in an oven—can and do shorten the lifespan of yeast cells. And here we'd messed with their DNA, which could have short-circuited the cells in a thousand different ways to cause early death.

But those cells weren't *just* dying. They were dying after a precipitous decline in health and function. As the *SGS1* mutants became older, they slowed down in their cell cycle. They grew larger. Both male and female "mating-type" genes (descendants of gene A) were switched on at the same time, so they were sterile and couldn't mate. These were all known hallmarks of aging in yeast. And it was happening more quickly in the mutants we'd made. It certainly looked like a yeast version of Werner's.

Using specialized stains, we colored the DNA blue and used red for the nucleolus, which sits inside the nucleus of all eukaryotic cells. That made it easier to see under the microscope what was happening at a cellular level.

And what was happening was fascinating.

The nucleolus is a part of the nucleus in which ribosomal DNA, or rDNA, resides. rDNA is copied into ribosomal RNA, which is used by ribosome enzymes to stitch amino acids together to make every new protein.

In the aged *SGS1* cells, the nucleolus looked as if it had exploded. Instead of a single red crescent swimming in a blue ocean, the nucleolus was scattered into half a dozen small islands. It was tragic and beautiful. The picture, which would later appear in the August 1997 issue of the prestigious journal *Science*, still hangs in my office.

What happened next was both enchanting and illuminating. In response to the damage, like rats to the call of the Pied Piper, the protein called Sir2—the first known sirtuin, which is encoded by the gene *SIR2*[10] and descended from gene B—had moved away from the mating genes that control fertility and into the nucleolus.

That was a beautiful sight to me, but it was a problem for the yeast. Sir2 has an important job: it is an epigenetic factor, an enzyme that sits on genes, bundles up the DNA, and keeps them silent. At the molecular level, Sir2 achieves this via its enzymatic activity, making sure that chemicals called acetyls don't accumulate on the histones and loosen the DNA packaging.

When sirtuins left the mating genes—the ones descended from gene A that controlled fertility and reproduction—the mutant cells turned on both male and female genes, causing them to lose their sexual identity, just as in normal old cells, but much earlier.

I didn't understand at first why the nucleolus was exploding, let alone why the sirtuins were moving toward it as the cells grew older. I agonized over the question for weeks.

And then one night, after a long day in the lab, I woke up with an idea.

It came in the space between sleep-deprived delirium and deep dreaming. The wisps of a concept. A few words jumbled together. A muddled picture of something. That was enough, though, to jolt me awake and pull me from my bed.

I grabbed my notebook and went to the kitchen. There, hunched over the table in the early morning hours of October 28, 1996, I began to write.

A theory on replicative senescence in yeast & other organisms. -

I wrote for about an hour, jotting down ideas, drawing pictures, sketching out graphs, formulating new equations.[11] Scientific observations that had previously made no sense to me were falling perfectly into a larger picture. Broken DNA causes genome instability, I wrote, which distracts the Sir2 protein, which changes the epigenome, causing the cells to lose their identity and become sterile while they fixed the damage. Those were the analog scratches on the digital DVDs. Epigenetic changes cause aging.

There was, I imagined, a singular process that controlled them all. Not a countless number of separate cellular changes or diseases. Not even a set of hallmarks that could be addressed one at a time. There was something bigger—and more singular—than any of that.

This was the foundation for understanding the survival circuit and its role in aging.

The next day I showed Guarente my notes. I was excited; it felt like the biggest idea I'd ever had. But I was nervous, too; afraid he would find a hole in my logic and tear it apart. Instead, he looked over my notebook quietly, asked a few questions, and sent me on my way with six words.

"I like it," he said. "Go prove it."

THE RECITAL

To understand the Information Theory of Aging, we need to pay another visit to the epigenome, the part of the cell that the sirtuins help control.

Up close, the epigenome is more complex and wonderful than anything we humans have invented. It consists of strands of DNA wrapped around spooling proteins called histones, which are bound up into bigger loops called chromatin, which are bound up into even bigger loops called chromosomes.

Sirtuins instruct the histone spooling proteins to bind up DNA tightly, while they leave other regions to flail around. In this way, some genes stay silent, while others can be accessed by DNA-binding transcription factors that turn genes on.[12] Accessible genes are said to be in "euchromatin," while silent genes are in "heterochromatin." By removing chemical tags on histones, sirtuins help prevent transcription factors from binding to genes, converting euchromatin into heterochromatin.

Every one of our cells has the same DNA, of course, so what differentiates a nerve cell from a skin cell is the epigenome, the collective term for the control systems and cellular structures that tell the cell which genes should be turned on and which should remain off. And this, far more than our genes, is what actually controls much of our lives.

One of the best ways to visualize this is to think of our genome as a grand piano.[13] Each gene is a key. Each key produces a note. And from instrument to instrument, depending on the maker, the materials, and the circumstances of manufacturing, each will sound a bit different, even if played the exact same way. These are our genes. We have about 20,000 of them, give or take a few thousand.[14]

Each key can also be played *pianissimo* (soft) or *forte* (with force). The notes can be *tenuto* (held) or *allegretto* (played quickly). For master pianists, there are hundreds of ways to play each individual key and endless ways to play the keys together, in chords and combinations that create music we know as jazz, ragtime, rock, reggae, waltzes, whatever.

The pianist that makes this happen is the epigenome. Through a process of revealing our DNA or bundling it up in tight protein packages, and by marking genes with chemical tags called methyls and acetyls composed of carbon, oxygen, and hydrogen, the epigenome uses our genome to make the music of our lives.

Yes, sometimes the size, shape, and condition of a piano dictate what a pianist can do with it. It's tough to play a concerto on an eighteen-key toy piano, and it's mighty hard to make beautiful music on an instrument that hasn't been tuned in fifty years. Likewise, the genome certainly dictates what the epigenome can do. A caterpillar can't become a human being, but it can become a butterfly by virtue of changes in epigenetic expression that occur during metamorphosis, even though its genome never changes. Similarly, the child of two parents from a long line of people with black hair and brown eyes isn't likely to develop blond hair and blue eyes, but twin agouti mice in the lab can turn out brown or golden, depending on how much the *Agouti* gene is turned on during gestation by environmental influences on the epigenome, such as folic acid, vitamin B$_{12}$, genistein from soy, or the toxin bisphenol A.[15]

Similarly, among monozygotic human twins, epigenetic forces can drive two people with the same genome in vastly different directions. It can even cause them to age differently. You can see this clearly in side-by-side photographs of the faces of smoking and nonsmoking twins; their DNA is still largely the same, but the smokers have bigger bags under their eyes, deeper jowls below their chins, and more wrinkles around their eyes and mouths. They are not older, but they've clearly aged faster. Studies of identical twins place the genetic influences on longevity at between 10 and 25 percent which, by any estimation, is surprisingly low.[16]

Our DNA is not our destiny.

Now imagine you're in a concert hall. A virtuoso pianist is seated at a gorgeously polished Steinway grand. The concerto begins. The music is beautiful, breathtaking. Everything is perfect.

But then, a few minutes into the piece, the pianist misses a key. The first time it happens, it's almost unnoticeable—an extra D, perhaps, in a chord that doesn't need that note. Embedded in so many perfectly played notes, hidden among an otherwise flawless chord in an otherwise perfect melody, it's nothing to worry about. But then, a few minutes later, it happens again. And then, with increasing frequency, again and again and again.

It's important to remember that there is nothing wrong with the piano. And the pianist is playing most of the notes prescribed by the composer. She's just also playing some extra notes. Initially, this is just annoying. Over time it becomes unsettling. Eventually it ruins the concerto. Indeed, we'd assume that there was something wrong with the pianist. Someone might even rush onto the stage to make sure she is all right.

Epigenetic noise causes the same kind of chaos. It is driven in large part by highly disruptive insults to the cell, such as broken DNA, as it was in the original survival circuit of *M. superstes* and in the old yeast cells that lost their fertility. And this, according to the Information Theory of Aging, is why we age. It's why our hair grays. It's why our skin wrinkles. It's why our joints begin to ache. Moreover, it's why each one of the hallmarks of aging occurs, from stem cell exhaustion and cellular senescence to mitochondrial dysfunction and rapid telomere shortening.

This is, I acknowledge, a bold theory. And the strength of a theory is based on how well it predicts the results of rigorous experiments, often millions of them, the number of phenomena it can explain, and its simplicity. The theory was simple, and it explained a lot. As good scientists, what we had left to do was to try our best to disprove it and see how long it survived.

To get started, Guarente and I had to get our eyes on some yeast DNA.

We used a technique called a Southern blot, a method of separating DNA based on its size and conformation and lighting it up with a radioactive DNA probe. In the first experiment, we noticed something spectacular. Normally, the rDNA of a yeast cell that is made visible by a Southern blot is tightly packed, like a new spool of rope, with a few

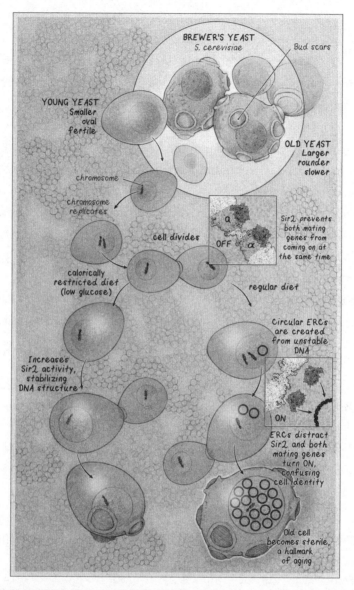

LESSONS FROM YEAST CELLS ABOUT WHY WE AGE. In young yeast cells, male and female "mating-type information" (gene A) is kept in the "off" position by the Sir2 enzyme, the first sirtuin (encoded by a descendant of gene B). The highly repetitive ribosomal DNA (rDNA) is unstable, and toxic DNA circles form; these recombine and eventually accumulate to toxic levels in old cells, killing them. In response to DNA circles and the perceived genome instability, Sir2 moves away from silent mating-type genes to help stabilize the genome. Both male and female genes turn on, causing infertility, the main hallmark of yeast aging.

barely visible wispy loops of supercoiled DNA. But the rDNA of the yeast cells we'd created in our lab—the Werner mutants that seemed to be aging rapidly—were madly unpacking, like a vacuum-sealed bag of yarn that had been ripped open.

The rDNA was in a state of chaos. The genome, it seemed, was fragmenting. DNA was recombining and amplifying, showing up on the Southern blot as dark spots and wispy circles, depending on how coiled up and twisted they were. We called those loops extrachromosomal ribosomal DNA circles, or ERCs, and they were accumulating as the mutants aged.

If we had indeed induced aging, then we would see this same pattern emerge in yeast cells that had aged normally.

We don't count the age of a single yeast cell with birthday candles. They simply don't last that long. Instead, aging in yeast is measured by the number of times a mother cell divides to produce daughter cells. In most cases, a yeast cell gets to about 25 divisions before it dies. That, however, makes obtaining old yeast cells an exceptionally challenging task. Because by the time an average yeast cell expires, it is surrounded by 2^{25}, or 33 million, of its descendants.

It took a week of work, a lot of sleepless nights, and a whole lot of caffeinated beverages to collect enough regular old cells. The next day, when I developed the film to visualize the rDNA, what I saw astounded me.[17]

Just like the mutants, the normal old yeast cells were packed with ERCs.

That was a "Eureka!" moment. Not proof—a good scientist never has proof of anything—but the first substantial confirmation of a theory, the foundation upon which I and others would build more discoveries in the years to come.

The first testable prediction was if we put an ERC into very young yeast cells—and we devised a genetic trick to do that—the ERCs would multiply and distract the sirtuins, and the yeast cells would age prematurely, go sterile, and die young—and they did. We published that work in December 1997 in the scientific journal *Cell*, and the news broke around the world: "Scientists figured out a cause of aging."

It was there and then that Matt Kaeberlein, a PhD student at the time, arrived at the lab. His first experiment was to insert an extra copy of *SIR2* into the genome of yeast cells to see if it could stabilize the yeast genome and delay aging. When the extra *SIR2* was added, ERCs were prevented, and he saw a 30 percent increase in the yeast cells' lifespan, as we'd been hoping. Our hypothesis seemed to be standing up to scrutiny: the fundamental, upstream cause of sterility and aging in yeast was the inherent instability of the genome.

What emerged from those initial results in yeast, and another decade of pondering and probing mammalian cells, was a completely new way to understand aging, an information theory that would reconcile seemingly disparate factors of aging into one universal model of life and death. It looked like this:

Youth → broken DNA → genome instability → disruption of DNA packaging and gene regulation (the epigenome) → loss of cell identity → cellular senescence → disease → death.

The implications were profound: if we could intervene in any of these steps, we might help people live longer.

But what if we could intervene in all of them? Could we stop aging?

Theories must be tested and tested and tested some more—not just by one scientist but by many. And to that end, I was fortunate to have been put onto a research team that included some of the most brilliant and insightful scientists in the world.

There was Lenny Guarente, our indefatigable mentor. There was also Brian Kennedy, who started the yeast-aging project in Lenny's lab and has since played a tremendously important role in understanding premature aging diseases and the impact of genes and molecules that increase health and longevity in model organisms. There were Monica Gotta and Susan Gasser at the University of Geneva, who are now some of the most influential researchers in the field of gene regulation; Shin-ichiro Imai, now a professor at Washington University, who discovered that sirtuins are NAD-utilizing enzymes and now does research into how the body controls sirtuins; Kevin Mills, who ran a lab in Maine, then became a cofounder of and chief scientific officer at Cyteir Therapeutics, which develops novel ways to fight cancer and autoimmune diseases; Nicanor Austriaco, who started the project with Brian, now teacher of biology and theology at Providence College, a great combo; Tod Smeal, chief scientific officer of cancer biology at the global pharmaceutical company Eli Lilly; David Lombard, who is now a researcher in the field of aging at the University of Michigan; Matt Kaeberlein, a professor at the University of Washington, who is testing molecules on dog longevity; David McNabb, whose lab at the University of Arkansas has made key and lifesaving discoveries about fungal pathogens; Bradley Johnson, an expert on human aging and cancer at the University of Pennsylvania; and Mala Murthy, a prominent neuroscientist now at Princeton.

Again and again I have been greatly privileged in the matter of those who work around me. And that was never truer than it was in Guarente's lab at MIT. It was a dream team, and I often felt humbled by the people with whom I was surrounded.

When I began my career in this field, I dreamt of publishing just one study in a top-tier journal. During those years, our group was publishing one every few months.

We demonstrated that the redistribution of Sir2 to the nucleolus is a response to numerous DNA breakages, which happen as a result of ERCs multiplying and inserting back into the genome or joining together to form superlarge ERCs. When Sir2 moves to combat DNA instability, it causes sterility in old, bloated yeast cells. That was the first step of the survival circuit, though at the time we had no idea that it was as ancient and as essential to our very existence as it turned out to be.

We told the world that we could give yeast a Werner-like syndrome, causing exploded nucleoli.[18] We described the ways in which mutants of *SGS1*, those we'd plagued with the yeast equivalent to the Werner

syndrome mutation, accumulated ERCs more rapidly, leading to premature aging and a shortened lifespan.[19] Crucially, by demonstrating that
if you add an ERC to young cells they age prematurely, we had crucial
evidence that ERCs don't just happen during aging, they cause it. And
by artificially breaking the DNA in the cell and watching the cellular response, we showed why sirtuins move—to help with DNA repair.[20] That
turned out to be the second step of the survival circuit.[21] The DNA damage that gave rise to the ERCs was distracting Sir2 from the mating-type
genes, causing them to become sterile, a hallmark of yeast aging.

It was epigenomic noise in its purest form.

It took another twenty years to learn if those findings in yeast were
relevant to organisms more complex than yeast. We mammals have seven
sirtuin genes that have evolved a variety of functions beyond what simple
SIR2 can do. Three of them, SIRT1, SIRT6, and SIRT7, are critical to the
control of the epigenome and DNA repair. The others, SIRT3, SIRT4,
and SIRT5, reside in mitochondria, where they control energy metabolism, while SIRT2 buzzes around the cytoplasm, where it controls cell
division and healthy egg production.

There had been many clues along the way. Brown University's Stephen Helfand showed that adding extra copies of the dSir2 gene to fruit
flies suppresses epigenetic noise and extends their lifespan. We found
that SIRT1 in mammals moves from silent genes to help repair broken
DNA in mouse and human cells.[22] But the true extent to which the survival circuit is conserved between yeast and humans wasn't fully known
until 2017, when Eva Bober's team at the Max Planck Institute for Heart
and Lung Research in Bad Nauheim, Germany, reported that sirtuins
stabilize human rDNA.[23] Then, in 2018, Katrin Chua at Stanford University found that, by stabilizing human rDNA, sirtuins prevent cellular
senescence—essentially the same antiaging function as we had found for
sirtuins in yeast twenty years earlier.[24]

That was an astonishing revelation: over a billion years of separation
between yeast and us, and, in essence, the circuit hadn't changed.

By the time those findings appeared, though, it was clear to me that epigenomic noise was a likely catalyst of human aging. Two decades of research had already been leading us in that direction.[25]

In 1999, I moved from MIT across the river to Harvard Medical School, where I set up a new lab on aging. There I was hoping to answer a new question that had increasingly been occupying my thoughts.

I had noticed that yeast cells fed with lower amounts of sugar were not just living longer, but their rDNA was exceptionally compact— significantly delaying the inevitable ERC accumulation, catastrophic numbers of DNA breaks, nucleolar explosion, sterility, and death.

Why was that happening?

THE SURVIVAL CIRCUIT COMES OF AGE

Our DNA is constantly under attack. On average, each of our forty-six chromosomes is broken in some way every time a cell copies its DNA, amounting to more than 2 trillion breaks in our bodies per day. And that's just the breaks that occur during replication. Others are caused by natural radiation, chemicals in our environment, and the X-rays and CT scans that we're subjected to.

If we didn't have a way to repair our DNA, we wouldn't last long. That's why, way back in primordium, the ancestors of every living thing on this planet today evolved to sense DNA damage, slow cellular growth, and divert energy to DNA repair until it was fixed—what I call the survival circuit.

Since the yeast work, evidence that yeast aren't so different from us has continued to accumulate. In 2003, Michael McBurney from the University of Ottawa in Canada discovered that mouse embryos manipulated to be unable to produce one of the seven sirtuin enzymes, SIRT1, couldn't last past the fourteenth day of development—about two-thirds of the way into a mouse's gestation period.[26] Among the reasons, the team reported in the journal *Cancer Cell*, was an impaired ability to respond to and repair DNA damage.[27] In 2006, Frederick Alt, Katrin Chua, and Raul Mostovslavsky at Harvard showed that mice engineered to lack

SIRT6 underwent the typical signs of aging faster along with shortened lifespans.[28] When the scientists knocked out a cell's ability to create this vital protein, the cell lost its ability to repair double-strand DNA breaks, just as we had showed in yeast back in 1999.

If you are skeptical, and you should be, you might assume these SIRT mutant mice could just be sick and, therefore, short lived. But adding in more copies of the sirtuin genes *SIRT2* and *SIRT6* does just the opposite: it increases the health and extends the lifespan of mice, just as adding extra copies of the yeast *SIR2* gene does in yeast.[29] Credit for these discoveries goes to two of my previous colleagues, Shin-ichiro Imai, my former drinking buddy at the Guarente lab, and Haim Cohen, my first postdoc at Harvard.

In yeast, we had shown that DNA breaks cause sirtuins to relocalize away from silent mating-type genes, causing old cells to become sterile. That was a simple system, and we'd figured it out in a few years.

But is the survival circuit causing aging in mammals? What parts of the system survived the billion years, and which are yeast specific? Those questions are on the cutting edge of human knowledge right now, but the answers are beginning to reveal themselves.

What I'm suggesting is that the *SIR2* gene in yeast and the *SIRT* genes in mammals are all descendants of gene B, the original gene silencer in *M. superstes*.[30] Its original job was to silence a gene that controlled reproduction.

In mammals, the sirtuins have since taken on a variety of new roles, not just as controllers of fertility (which they still are). They remove acetyls from hundreds of proteins in the cell: histones, yes, but also proteins that control cell division, cell survival, DNA repair, inflammation, glucose metabolism, mitochondria, and many other functions.

I've come to think of sirtuins as the directors of a multifaceted disaster response corps, sending out a variety of specialized emergency teams to address DNA stability, DNA repair, cell survivability, metabolism, and cell-to-cell communication. In a way, this is like the command center for the thousands of utility workers who descended upon Louisiana and

Mississippi in the wake of Hurricane Katrina in 2005. Most of the workers weren't from the Gulf Coast, but they came, did their level best to fix what was broken, and then went home. Some were working in the storm-ravaged communities for a few days and others for a few weeks before returning to their normal lives. And for most, it wasn't the first or last time they had done something like that; anytime there's a mass disaster that impacts utilities, they swoop in to help.

When they're home, those folks take care of the typical business of being at home: paying bills, mowing lawns, coaching baseball, whatever. But when they're away, helping keep places like the Gulf Coast from descending into anarchy—a condition that would have had disastrous results for the rest of the nation—a lot of those things have to be put on hold.

When sirtuins shift from their typical priorities to engage in DNA repair, their epigenetic function at home ends for a bit. Then, when the damage is fixed and they head back to home base, they get back to doing what they usually do: controlling genes and making sure the cell retains its identity and optimal function.

But what happens when there's one emergency after another to tend to? Hurricane after hurricane? Earthquake after earthquake? The repair crews are away from home a lot. The work they normally do piles up. The bills come due, then overdue, and then the folks from collections start calling. The grass grows too long, and soon the president of the neighborhood association is sending nastygrams. The baseball team goes coachless, and the team devolves into the Bad News Bears. And most of all, one of the most important things they do while at home— reproducing—doesn't get done. This form of hormesis, the original survival circuit, works fine to keep organisms alive in the short term. But unlike longevity molecules that simply mimic hormesis by tweaking sirtuins, mTOR, or AMPK, sending out the troops on fake emergencies, these real emergencies create life-threatening damage.

What could cause so many emergencies? DNA damage. And what causes that? Well, over time, *life* does. Malign chemicals. Radiation. Even normal DNA copying. These are the things that we've come to believe are the causes of aging, but there is a subtle but vital shift we have to make in that manner of thinking. It's not so much that the sirtuins are overwhelmed, though they probably are when you are sunburned or get an X-ray; what's happening every day is that the sirtuins and their co-workers that control the epigenome don't always find their way back to their original gene stations after they are called away. It's as if a few emergency workers who came to address the damage done in the Gulf Coast by Katrina had lost their home address. Then disaster strikes again and again, and they must redeploy.

Wherever epigenetic factors leave the genome to address damage, genes that should be off, switch on and vice versa. *Wherever* they stop on the genome, they do the same, altering the epigenome in ways that were never intended when we were born.

Cells lose their identity and malfunction. Chaos ensues. The chaos materializes as aging. This is the epigenetic noise that is at the heart of our unified theory.

How does the *SIR2* gene actually turn off genes? *SIR2* codes for a specialized protein called a histone deacetylase, or HDAC, that enzymatically cleaves the acetyl chemical tags from histones, which, as you'll recall, causes the DNA to bundle up, preventing it from being transcribed into RNA.

When the Sir2 enzyme is sitting on the mating-type genes, they remain silent and the cell continues to mate and reproduce. But when a DNA break occurs, Sir2 is recruited to the break to remove the acetyl tags from the histones at the DNA break. This bundles up the histones to prevent the frayed DNA from being chewed back and to help recruit other repair proteins. Once the DNA repair is complete, most of the Sir2 protein goes back to the mating-type genes to silence them and restore fertility. That is, unless there is another emergency, such as the massive genome instability that occurs when ERCs accumulate in the nucleoli of old yeast cells.

For the survival circuit to work and for it to cause aging, Sir2 and other epigenetic regulators must occur in "limiting amounts." In other words, the cell doesn't make enough Sir2 protein to simultaneously silence the mating-type genes *and* repair broken DNA; it has to shuttle Sir2 between the various places on an "as-needed" basis. This is why adding an extra copy of the *SIR2* gene extends lifespan and delays infertility: cells have enough Sir2 to repair DNA breaks *and* enough Sir2 to silence the mating-type genes.[31]

Over the past billion years, presumably millions of yeast cells have spontaneously mutated to make more Sir2, but they died out because they had no advantage over other yeast cells. Living for 28 divisions was no advantage over those that lived for 24 and, because Sir2 uses up energy, having more of the protein may have even been a disadvantage. In the lab, however, we don't notice any disadvantage because the yeast are given more sugar than they could possibly ever eat. By adding extra copies of the *SIR2* gene, we gave the yeast cells what evolution failed to provide.

If the information theory is correct—that aging is caused by overworked epigenetic signalers responding to cellular insult and damage—it doesn't so much matter *where* the damage occurs. What matters is that it *is* being damaged and that sirtuins are rushing all over the place to address that damage, leaving their typical responsibilities and sometimes returning to other places along the genome where they are silencing genes that aren't supposed to be silenced. This is the cellular equivalent of distracting the cellular pianist.

To prove that, we needed to break some mouse DNA.

It's not hard to intentionally break DNA. You can do it with mechanical shearing. You can do it with chemotherapy. You can do it with X-rays.

But we needed to do it with precision—in a way that wouldn't create mutations or impact regions that affect any cellular function. In essence, we needed to attack the wastelands of the genome. To do that, we got our hands on a gene similar to Cas9, the CRISPR gene-editing tool from bacteria that cuts DNA at precise places.

The enzyme we chose for our experiments comes from a goopy yellow slime mold called *Physarum polycephalum*, which literally means "many-headed slime." Most scientists believe that this gene, called I-*Ppo*I, is a parasite that serves only to copy itself. When it cuts the slime mold genome, another copy of I-*Ppo*I is inserted. It is the epitome of a selfish gene.

That's in a slime mold, its native habitat. But when I-*Ppo*I finds itself in a mouse cell, it doesn't have all the slime mold machinery to copy itself. So it floats around and cuts DNA at just a few places in the mouse genome, and there is no copying process. Instead, the cell has no problem pasting the DNA strands back together, leaving no mutations, which is exactly what we were looking for to engage the survival circuit and distract the sirtuins. DNA-editing genes such as Cas9 and I-*Ppo*I are nature's gifts to science.

To create a mouse to test the information theory, we inserted I-*Ppo*I into a circular DNA molecule called a plasmid, along with all the regulatory DNA elements needed to control the gene, and then inserted that DNA into the genome of a mouse embryonic stem cell line we were culturing in plastic dishes in the lab. We then injected the genetically modified stem cells into a 90-cell mouse embryo called a blastocyst, implanted it into a female mouse's uterus, and waited about twenty days for a baby mouse to show up.

This all sounds complicated, but it's not. After some training, a college student can do it. It's such a commodity these days, you can even order a mouse out of a catalog or pay a company to make you one to your specifications.

The baby mice were born perfectly normal, as expected, since the cutting enzyme was switched off at that stage. We called them affectionately "ICE mice," ICE standing for "Inducible Changes to the Epigenome." The "inducible" part of the acronym is vital—because there's nothing different about these mice until we feed them a low dose of tamoxifen. This is an estrogen blocker that is normally used to treat human cancers, but in this case, we'd engineered the mouse so that tamoxifen would turn on the I-*Ppo*I gene. The enzyme would go to work, cutting the genome and slightly overwhelming the survival circuit, without killing any cells. And since tamoxifen has a half-life of only a couple of days, removing it from the mice's food would turn off the cutting.

The mice might have died. They might have grown tumors. Or they might have been perfectly fine, no worse off than if they'd received a dental X-ray. Nobody had ever done this before in a mouse, so we didn't know. But if our hypothesis about epigenetic instability and aging was correct, the tamoxifen would work like the potion that Fred and George Weasley used to age themselves in *Harry Potter and the Goblet of Fire*.

And it worked. Like wizardry, it did.

During the treatment, the mice were fine, oblivious to the DNA cutting and sirtuin distraction. But a few months later, I got a call from a postdoc who was taking care of our lab's animals while I was on a trip to my lab in Australia.

"One of the mice is really sick," she said. "I think we need to put it down."

I asked her to text me a photo of the mouse she was talking about. When the photo came over my phone, I couldn't help but laugh.

"That's not a sick mouse," I replied. "That's an *old* mouse."

"David," she said, "I think you're mistaken. It says here that it's the sister of these other mice in the cage, and they're perfectly normal."

Her confusion was understandable. At 16 months old, a regular lab mouse still has a thick coat of fur, a sturdy tail, a muscular figure, perky ears, and clear eyes. A tamoxifen-triggered ICE mouse at the same age has thinning, graying hair, a bent spine, paper-thin ears, and cloudy eyes.

Remember, we'd done nothing to change the genome. We'd simply broken the mice's DNA in places where there aren't any genes and forced the cell to paste, or "ligate," them back together. Just to make sure, later we broke the DNA in other places, too, with the same results. Those breaks had induced a sirtuin response. When those fixers went to work, their absence from their normal duties and presence on other parts of the

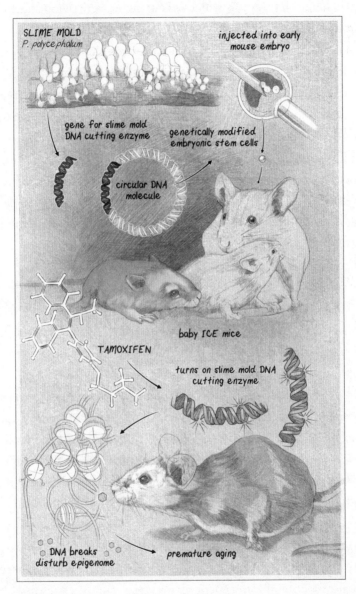

THE MAKING OF THE ICE MOUSE TO TEST IF THE CAUSE OF AGING MIGHT BE INFORMATION LOSS. A gene from a slime mold that encodes an enzyme that cuts DNA at a specific place was inserted into a stem cell and injected into an embryo to generate the ICE mouse. Turning on the slime mold gene cut the DNA and distracted the sirtuins, causing the mouse to undergo aging.

genome altered the ways in which lots of genes were being expressed at the wrong time.

Those findings were aligned to discoveries being made by Trey Ideker and Kang Zhang, at UC San Diego, and Steve Horvath, at UCLA. Steve's name stuck, and today he's the namesake of the Horvath Clock— an accurate way of estimating someone's biological age by measuring thousands of epigenetic marks on the DNA, called methylation. We tend to think of aging as something that begins happening to us at midlife, because that's when we start to see significant changes to our bodies. But Horvath's clock begins ticking the moment we are born. Mice have an epigenetic clock, too. Were the ICE mice older than their siblings? Yes, they were—about 50 percent older.

We'd found life's master clock winder.

In another manner of thinking, we'd scratched up the DVD of life about 50 percent faster than it normally gets scratched. The digital code that is, and was, the basic blueprint for our mice was the same as it had always been. But the analog machine built to read that code was able to pick up only bits and pieces of the data.

Here's the vital takeaway: we could age mice without affecting any of the most commonly assumed causes of aging. We hadn't made their cells mutate. We hadn't touched their telomeres. We hadn't messed with their mitochondria. We hadn't directly exhausted their stem cells. Yet the ICE mice were suffering from a loss of body mass, mitochondria, and muscle strength and an increase in cataracts, arthritis, dementia, bone loss, and frailty.

All of the symptoms of aging—the conditions that push mice, like humans, farther toward the precipice of death—were being caused not by mutation but by the epigenetic changes that come as a result of DNA damage signals.

We hadn't given the mice all of those ailments. We had given them aging.

And if you can give something, you can take it away.

FRUIT OF THE SAME TREE

Like the gnarled hands of giant zombies breaking free of the rocky soil, the ancient bristlecone pine trees of California's White Mountains strike haunting silhouettes against the dewy morning sun.

The oldest of these trees have been here since before the pyramids of Egypt were built, before the construction of Stonehenge, and before the last of the woolly mammoths left our world. They have shared this planet with Moses, Jesus, Muhammad, and the first Buddha. Standing some two miles above sea level, adding fractions of a millimeter of growth to their twisted trunks each year, defying lightning storms and periodic droughts, they are the epitome of perseverance.

It's easy to stand in wonder of these great and ancient things. It's easy to be swept away by their might and majesty. It's easy to simply stare at them in awe. But there's another way to view these antediluvian patriarchs—a harder way, but a way in which we should seek to view every living thing on this planet: as our teachers.

Bristlecones are, after all, our eukaryotic cousins. About half of their genes are close relatives of ours.

Yet they do not age.

Oh, they add years to their lives—thousands upon thousands of them, marked by the nearly microscopic rings hidden in their dense heartwood, which also record in their size, shape, and chemical composition climate events long past, as when the eruption of Krakatoa sent a cloud of ash around the globe in 1883, leaving a fuzzy ring of growth in 1884 and 1885, barely a centimeter from the outer ring of bark that marks our current time.[32]

Yet even over the course of many thousands of years, their cells do not appear to have undergone any decline in function. Scientists call this "negligible senescence." Indeed, when a team from the Institute of Forest Genetics went looking for signs of cellular aging—studying bristlecones from 23 to 4,713 years old—they came up empty-handed. Between young and old trees, their 2001 study found, there were no meaningful differences in the chemical transportation systems, in the rate of shoot growth, in the quality of the pollen they produced, in the size of their seeds, or in the way those seeds germinated.[33]

The researchers also looked for deleterious mutations—the sorts of which many scientists at the time expected to be a primary cause of aging. They found none.[34] I expect that if they were to look for epigenetic changes, they would similarly come up empty-handed.

Bristlecones are outliers in the biological world, but they are not unique in their defiance of aging. The freshwater polyp known as *Hydra vulgaris* has also evolved to defy senescence. Under the right conditions, these tiny cnidarians have demonstrated a remarkable refusal to age. In the wild they might live for only a few months, subject to the powers of predation, disease, and desiccation. But in labs around the world they have been kept alive for upward of 40 years—with no signs that they'll stop there—and indicators of health don't differ significantly between the very young and the very old.

A couple of species of jellyfish can completely regenerate from adult body parts, earning them the nickname "immortal jellies." Only the elegant moon jelly *Aurelia aurita* from the US West Coast and the centimeter-long *Turritopsis dohrnii* from the Mediterranean are currently known to regenerate, but I'm guessing the majority of jellies do. We just need to look. If you separate one of these amazing animals into single cells, the cells jostle around until they form clumps that then assemble back into a complete organism, like the T-1000 cyborg in *Terminator 2*, most likely resetting their aging clock.

Of course, we humans don't want to be mashed into single cells to be immortal. What use is reassembling or spawning if you have no recollection of your present life? We may as well be reincarnated.

What matters is what these biological equivalents of F. Scott Fitzgerald's backward-aging Benjamin Button teach us: that cellular age *can* be fully reset, something I'm convinced we will be able to do one day without losing our wisdom, our memories, or our souls.

Though it's not immortal, the Greenland shark *Somniosus*

microcephalus is still an impressive animal and far more closely related to us. About the size of a great white, it does not even reach sexual maturity until it is 150 years old. Researchers believe the Arctic Ocean could be home to Greenland sharks that were born before Columbus got lost in the New World. Radiocarbon dating estimated that one very large individual may have lived more than 510 years, at least up until it was caught by scientists so they could measure its age. Whether this shark's cells undergo aging is an open scientific question; very few biologists had so much as looked at *S. microcephalus* until the past few years. At the very least, this longest-living vertebrate undergoes the process of aging very, very slowly.

Evolutionarily speaking, all of these life-forms are closer to us than yeast, and just think of what we've learned about human aging from that tiny fungus. But it is certainly forgivable to consider the distances between pine trees, hydrozoans, cartilaginous fish, and mammals like ourselves on the enormous tree of life and say, "No, these things are just too different."

What, then, of another mammal? A warm-blooded, milk-producing, live-birth-giving cousin?

Back in 2007, aboriginal hunters in Alaska caught a bowhead whale that, when butchered, was found to have the head of an old harpoon embedded in its blubber. The weapon, historians would later determine, had been manufactured in the late 1800s, and they estimated the whale's age at about 130. That discovery sparked a new scientific interest in *Balaena mysticetus*, and later research, employing an age-determining method that measures the levels of aspartic acid in the lens of a whale's eye, estimated that one bowhead was 211 years old when it was killed by native whalers.

That bowheads have been selected for exceptional longevity among mammals should perhaps not be surprising. They have few predators and can afford to build a long-lived body and breed slowly. Most likely they maintain their survival program on high alert, repairing cells while maintaining a stable epigenome, thereby making sure the symphony of the cells plays on for centuries.

Can these long-lived species teach us how to live healthier and for longer?

In terms of their looks and habitats, pine trees, jellyfish, and whales are certainly very different from humans. But in other ways, we're very similar. Consider the bowheads. Like us, they are complex, social, communicative, and conscious mammals. We share 12,787 known genes, including some interesting variants in a gene known as *FOXO3*. Also known as *DAF-16*, this gene was first identified as a longevity gene in roundworms by University of California at San Francisco researcher Cynthia Kenyon. She found it to be essential for defects in the insulin hormone pathway to double worm lifespan. Playing an integral role in the survival circuit, *DAF-16* encodes a small transcription factor protein that latches onto the DNA sequence TTGTTTAC and works with sirtuins to increase cellular survival when times are tough.[35]

In mammals, there are four *DAF-16* genes, called *FOXO1*, *FOXO3*, *FOXO4*, and *FOXO6*. If you suspect that we scientists sometimes intentionally complicate matters, you'd be right, but not in this case. Genes in the same "gene family" have ended up with different names because they were named before DNA sequences were easily deciphered. It's similar to the not uncommon situation in which people have their genome analyzed and learn they have a sibling on the other side of town.[36] *DAF-16* is an acronym for *dauer* larvae formation. In German, "dauer" means "long lasting," and this is actually relevant to this story. Turns out, worms become *dauer* when they are starved or crowded, hunkering down until times improve. Mutations that activate *DAF-16* extend lifespan by turning on the worm defense program even when times are good.

I first encountered FOXO/DAF-16 in yeast, where it is known as *MSN2*, which stands for "multicopy suppressor of *SNF1* (AMPK) epigenetic regulator." Like DAF-16, *MSN2*'s job in yeast is to turn on genes that push cells away from cell death and toward stress resistance.[37] We discovered that when calories are restricted *MSN2* extends yeast lifespan by turning up genes that recycle NAD, thereby giving the sirtuins a boost.[38]

Hidden within the sometimes byzantine way scientists talk about science are several repeating themes: low energy sensors (SNF1/AMPK), transcription factors (MSN2/DAF-16/FOXO), NAD and sirtuins, stress resistance, and longevity. This is no coincidence—these are all key parts of the ancient survival circuit.

But what about *FOXO* genes in humans? Certain variants called *FOXO3* have been found in human communities in which people are known to enjoy both longer lifespans and healthspans, such as the people of China's Red River Basin.[39] These *FOXO3* variants likely turn on the body's defenses against diseases and aging, not just when times are tough but throughout life. If you've had your genome analyzed, you can check if you have any of the known variations of *FOXO3* that are associated with a long life.[40] For example, having a C instead of a T variant at position rs2764264 is associated with longer life. Two of our children, Alex and Natalie, inherited two Cs at this position, one from Sandra and one from me, so all other genes being equal, and as long as they don't live terribly negative lifestyles, they should have greater odds of reaching age 95 than I do, with my one C and one T, and substantially greater than someone with two Ts.

It's worth pausing to consider how remarkable it is that we find essentially the same longevity genes in every organism on the planet: trees, yeast, worms, whales, and humans. All living creatures come from the same place in primordium that we do. When we look through a microscope, we're all made of the same stuff. We all share the survival circuit, a protective cellular network that helps us when times are tough. This same network is our downfall. Severe types of damage, such as broken strands of DNA, cannot be avoided. They overwork the survival circuit and change cellular identity. We're all subject to epigenetic noise that should, under the Information Theory of Aging, cause aging.

Yet different organisms age at very different rates. And sometimes, it appears, they do not age at all. What allows a whale to keep the survival circuit on without disrupting the epigenetic symphony? If the piano player's skills are lost, how is it possible for a jellyfish to restore her ability?

These are the questions that have been guiding my thoughts as I have considered where our research is headed. What might seem like fanciful ideas, or concepts straight out of science fiction, are firmly rooted in research. Moreover, they are supported by the knowledge that some of our close relatives have figured out a workaround to aging.

And if they can, we can, too.

THE LANDSCAPE OF OUR LIVES

Before most people could even fathom the idea of mapping our genome, before we had the technology to map a cell's entire epigenome and understand how it bundles DNA to turn genes on and off, the developmental biologist Conrad Waddington was already thinking deeper.

In 1957, the professor of genetics, from the University of Edinburgh, was trying to understand how an early embryo could possibly be transformed from a collection of undifferentiated cells—each one exactly like the next and with the exact same DNA—to the thousands of different cell types in the human body. Perhaps not coincidentally, Waddington's ponderings came in the dawning years of the digital revolution, at the same time that Grace Hopper, the mother of computer programming, was laying the foundation for the first widely used computer language, COBOL. In essence, what Waddington was seeking to ascertain was how cells, all running on the same code, could possibly produce different programs.

There had to be something more than genetics at play: a program that controlled the code.

Waddington conceived of an "epigenetic landscape," a three-dimensional relief map that represents the dynamic world in which our genes exist. More than half a century later, Waddington's landscape remains a useful metaphor to understand why we age.

On the Waddington map, an embryonic stem cell is represented by a marble at the top of a mountain peak. During embryonic development, the marble rolls down the hill and comes to rest in one of hundreds of different valleys, each representing a different possible cell type in the body. This is called "differentiation." The epigenome guides the marbles, but it also acts as gravity after the cells come to rest, ensuring that they don't move back up the slope or hop over into another valley.

The final resting place is known as the cell's "fate." We used to think this was a one-way street, an irreversible path. But in biology there is no such thing as fate. In the last decade, we've learned that the marbles in the Waddington landscape aren't fixed; they have a terrible tendency to move around over time.

At the molecular level, what's really going on as the marble rolls down the slope is that different genes are being switched on and off, guided by transcription factors, sirtuins and other enzymes such as DNA methyltransferases (DNMTs) and histone methyltransferases (HMTs), which mark the DNA and its packing proteins with chemical tags that instruct the cell and its descendants to behave in a certain way.

What's not generally appreciated, even in scientific circles, is how important the stability of this information is for our long-term health. You see, epigenetics was long the purview of scientists who study the very beginnings of life, not folks like me who are studying the other end of things.

Once a marble has settled in Waddington's landscape, it tends to stay there. If all goes well with fertilization, the embryo develops into a fetus, then a baby, then a toddler, then a teenager, then an adult. Things tend to go well in our youth. But the clock is ticking.

Every time there's a radical adjustment to the epigenome, say, after DNA damage from the sun or an X-ray, the marbles are jostled—envision a small earthquake that ever so slightly changes the map. Over time, with repeated earthquakes and erosion of the mountains, the marbles are moved up the sides of the slope, toward a new valley. A cell's identity changes. A skin cell starts behaving differently, turning on genes that were shut off in the womb and were meant to stay off. Now it is 90 percent a skin cell and

10 percent other cell types, all mixed up, with properties of neurons and kidney cells. The cell becomes inept at the things skin cells must do, such as making hair, keeping the skin supple, and healing when injured.

In my lab we say the cell has *ex-differentiated.*

Each cell is succumbing to epigenetic noise. The tissue made up of thousands of cells is becoming a melange, a medley, a miscellaneous set of cells.

As you'll recall, the epigenome is inherently unstable because it is *analog* information—based on an infinite number of possible values— and thus it's difficult to prevent the accumulation of noise and nearly impossible to duplicate without some information loss. The earthquakes are a fact of life. The landscape is always changing.

If the epigenome had evolved to be digital rather than analog, the valley walls would be the equivalent of 100 miles high and vertical, and gravity would be superstrong, so the marbles could never jump over into a new valley. Cells would never lose their identity. If we were built this way, we could be healthy for thousands of years, perhaps longer.

But we are not built this way. Evolution shapes both genomes and epigenomes only enough to ensure sufficient survival to ensure replacement—and perhaps, if we are lucky, just a little bit more—but not immortality. So our valley walls are only slightly sloped, and gravity isn't that strong. A whale that lives two hundred years has probably evolved steeper valley walls and its cells maintain their identity for twice as long as ours do. Yet even whales don't live forever.

I believe the blame lies with *M. superstes* and the survival circuit. The repeated shuffling of sirtuins and other epigenetic factors away from genes to sites of broken DNA, then back again, while helpful in the short term, is ultimately what causes us to age. Over time, the wrong genes come on at the wrong time and in the wrong place.

As we saw in the ICE mice, when you disrupt the epigenome by forcing it to deal with DNA breaks, you introduce noise, leading to an erosion of the epigenetic landscape. The mice's bodies turned into chimeras of misguided, malfunctioning cells.

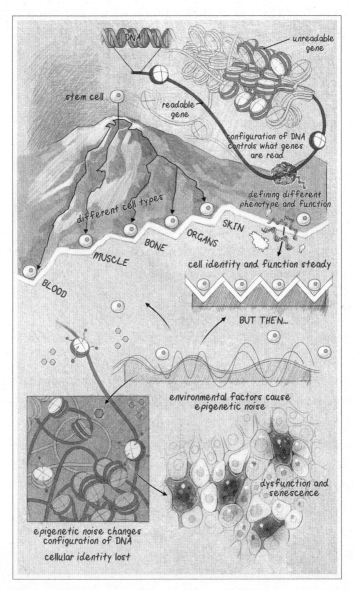

THE CHANGING LANDSCAPE OF OUR LIVES. The Waddington landscape is a metaphor for how cells find their identity. Embryonic cells, often depicted as marbles, roll downhill and land in the right valley that dictates their identity. As we age, threats to survival, such as broken DNA, activate the survival circuit and rejigger the epigenome in small ways. Over time, cells progressively move towards adjacent valleys and lose their original identity, eventually transforming into zombielike senescent cells in old tissues.

That's aging. This loss of information is what leads each of us into a world of heart disease, cancer, pain, frailty, and death.

If the loss of analog information is the singular reason why we age, is there anything we can do about it? Can we stabilize the marbles, keeping the valley walls high and the gravity strong?

Yes. I can say with confidence that there is.

REVERSAL COMES OF AGE

Regular exercise "is a commitment," says Benjamin Levine, a professor at the University of Texas. "But I tell people to think of exercise as part of personal hygiene, like brushing their teeth. It should be something we do as a matter of course to keep ourselves healthy."[41]

I'm sure he's right. Most people *would* exercise a lot more if going to the gym were as easy as brushing their teeth.

Perhaps one day it will be. Experiments in my lab indicate it is possible.

"David, we've got a problem," a postdoctoral researcher named Michael Bonkowski told me one morning in the fall of 2017 when I arrived at the lab.

That's seldom a good way to start the day.

"Okay," I said, taking a deep breath and preparing for the worst. "What is it?"

"The mice," Bonkowski said. "They won't stop running."

The mice he was talking about were 20 months old. That's roughly the equivalent of a 65-year-old human. We had been feeding them a molecule intended to boost the levels of NAD, which we believed would increase the activity of sirtuins. If the mice were developing a running addiction, that would be a very good sign.

"But how can that be a problem?" I said. "That's great news!"

"Well," he said, "it *would* be if not for the fact that they've broken our treadmill."

As it turned out, the treadmill tracking program had been set up to record a mouse running for only up to three kilometers. Once the old mice got to that point, the treadmill shut down. "We're going to have to start the experiment again," Bonkowski said.

It took a few moments for that to sink in.

A thousand meters is a good, long run for a mouse. Two thousand meters—five times around a standard running track—would be a substantial run for a *young* mouse.

But there's a reason why the program was set to three kilometers. Mice simply don't run that far. Yet these *elderly* mice were running ultramarathons.

Why? One of our key findings, in a study we published in 2018,[42] was that when treated with an NAD-boosting molecule that activated the SIRT1 enzyme, the elderly mice's endothelial cells, which line the blood vessels, were pushing their way into areas of the muscle that weren't getting very much blood flow. New tiny blood vessels, capillaries, were formed, supplying badly needed oxygen, removing lactic acid and toxic metabolites from muscles, and reversing one of the most significant causes of frailty in mice and in humans. That was how these old mice suddenly became such mighty marathoners.

Because the sirtuins had been activated, the mice's epigenomes were becoming more stable. The valley walls were growing higher. Gravity was growing stronger. And Waddington's marbles were being pushed back to where they belonged. The lining of the capillaries was responding as if the mice were exercised. It was an exercise mimetic, the first of its kind, and a sure sign that some aspects of age *reversal* are possible.

We still don't know everything about why this happens. We don't know what sorts of molecules will work best for activating sirtuins or in what doses. Hundreds of different NAD precursors have been synthesized, and there are clinical trials in progress to answer that question and more.

But that doesn't mean we need to wait to take advantage of all that we've learned about engaging the epigenetic survival circuit and living longer and healthier lives. We don't need to wait to take advantage of the Information Theory of Aging.

There are steps we can take right now to live much longer and much healthier lives. There are things we can do to slow, stop, and even reverse aspects of aging.

But before we talk about what steps we might take to combat aging, before I can explain the science-backed interventions that have the greatest promise for fundamentally changing the way we think about growing old, before we even begin to talk about the treatments and therapies that will be game changers for our species, we need to answer one very important question:

Should we?

THREE

THE BLIND EPIDEMIC

IT WAS MAY 10, 2010, AND LONDON WAS ABUZZ. CHELSEA FOOTBALL CLUB HAD just won its fourth national championship by devastating Wigan Athletic, 8–0, on the final day of Premier League play. Meanwhile, Gordon Brown announced that he would be stepping down as prime minister in response to a disastrous parliamentary result for his Labour Party, which had lost more than ninety seats in the previous week's general election.

With the eyes of the English sports world on one part of London and the attention of the British political universe on another, the goings-on at Carlton House Terrace were missed by all but the most attentive observers of the president, council, and fellows of the Royal Society of London for Improving Natural Knowledge.

More simply known as the Royal Society, the world's oldest national scientific organization was established in 1660 to promote and disseminate "new science" by big thinkers of the day such as Sir Francis Bacon, the Enlightenment's promulgator of "the prolongation of life."[1] Befitting its rich scientific history, the society has held annual scientific events ever

since. Highlights have included lectures by Sir Isaac Newton on gravity, Charles Babbage on his mechanical computer, and Sir Joseph Banks,
who had just arrived back from Australia with a bounty of more than a
thousand preserved plants that were all new to science.

Even today, in a post-Enlightenment world, most of the events
at the society are fascinating if not world changing. But the two-day
meeting that commenced in the spring of 2010 was nothing short of
that, for gathered together on that Monday and Tuesday was a motley
group of researchers who were meeting to discuss an important "new
science."

The gathering had been called by geneticist Dame Linda Partridge,
bioanalytics pioneer Janet Thornton, and molecular neuroscientist Gil-
lian Bates, all luminaries in their respective fields. The attendee list was
no less impressive. Cynthia Kenyon spoke about her landmark work on a
single mutation in the IGF-1 receptor gene that had doubled the lifespan
of roundworms by activating DAF-16[2]—work that was first suggested
by Partridge to be a worm-specific aberration[3] but soon forced her and
other leading researchers to confront long-held beliefs that aging could
be controlled by a single gene. Thomas Nyström, from the University
of Gothenburg, reported his discovery that Sir2 not only is important
for genomic and epigenomic stability in yeast, it also prevents oxidized
proteins from being passed on to young daughter cells.

Brian Kennedy, a former Guarente student who was about to assume
the presidency of the Buck Institute for Research on Aging, explained
the ways in which genetic pathways that had been similarly conserved in
a diverse array of species were likely to play similar roles in mammalian
aging. Andrzej Bartke from Southern Illinois University, former PhD adviser to Michael "Marathon Mouse" Bonkowski, talked about how dwarf
mice can live up to twice as long as normal mice, a record. Molecular biologist María Blasco explained how old mammalian cells are more
likely than young cells to lose their identity and become cancerous. And
geneticist Nir Barzilai spoke of genetic variants in long-lived humans and
his belief that all aging-related diseases can be substantially prevented

and human lives can be considerably extended with one relatively easy pharmaceutical intervention.

Over the course of those two days, nineteen presenting scientists from some of the best research institutions in the world moved toward a provocative consensus and began to build a compelling case that would challenge conventional wisdom about human health and disease. Summarizing the meeting for the society later that fall, the biogerontologist David Gems would write that advances in our understanding of organismal senescence are all leading to a momentous singular conclusion: that aging is not an inevitable part of life but rather a "disease process with a broad spectrum of pathological consequences."[4] In this way of thinking, cancer, heart disease, Alzheimer's, and other conditions we commonly associate with getting old are not necessarily diseases themselves but symptoms of something greater.

Or, put more simply and perhaps even more seditiously: aging itself is a disease.

THE LAW OF HUMAN MORTALITY

If the idea that aging is a disease sounds strange to you, you're not alone. Physicians and researchers have been avoiding saying that for a long time. Aging, we've long been told, is simply the process of growing old. And growing old has long been seen as an inevitable part of life.

We see aging, after all, in nearly everything around us and, in particular, the things around us that look anything like us. The cows and pigs in our farms age. The dogs and cats in our homes do, too. The birds in the sky. The fish in the sea. The trees in the forest. The cells in our petri dishes. It always ends the same way: dust to dust.

The connection between death and aging is so strong that the inevitability of the former governed the way we came to define the latter. When European societies first began keeping public death certificates in the 1600s, aging was a respected cause of death. Descriptions such as "decrepitude" or "feebleness due to old age" were commonly accepted explanations for death. But according to the seventeenth-century English demographer John Graunt, who wrote *Natural and Political Observations Mentioned in a Following Index, and Made upon the Bills of Mortality*, so were "fright," "grief," and "vomiting."

As we've moved forward in time, we've moved away from blaming death on old age. No one dies anymore from "getting old." Over the past century, the Western medical community has come to believe not only that there is *always* a more immediate cause of death than aging but that it is imperative to identify that cause. In the past few decades, in fact, we've become rather fussy about this.

The World Health Organization's *International Classification of Diseases*, a list of illnesses, symptoms, and external causes of injury, was launched in 1893 with 161 headings. Today there are more than 14,000, and in most places where records of death are kept, doctors and public health officials use these codes to record both immediate and underlying causes of disability and death.[5] That, in turn, helps medical leaders and policy makers around the globe make public health decisions. Broadly speaking, the more often a cause shows up on a death certificate, the more attention society gives to fighting it. This is why heart disease, type 2 diabetes, and dementia are major focuses of research and interventionary medical care, while aging is not, even though aging is the greatest cause of all those diseases.

Age is *sometimes* considered an underlying factor at the end of someone's life, but doctors never cite it as an immediate reason for death. Those who do run the risk of raising the ire of bureaucrats, who are prone to send the certificate back to the doctor for further information. Even worse, they are likely to endure the ridicule of their peers. David Gems, the deputy director of the Institute of Healthy Ageing at University College London and the same man who wrote the report from the Royal Society meeting on "the new science of aging," told Medical Daily in 2015 that "the idea that people die of pure aging, without pathology, is nuts."[6]

But this misses the point. Separating aging from disease obfuscates a truth about how we reach the ends of our lives: though it's certainly important to know why someone fell from a cliff, it's equally important to know what brought that person to the precipice in the first place.

Aging brings us to the precipice. Give any of us 100 years or so, and it brings us all there.

In 1825, the British actuary Benjamin Gompertz, a learned member of the Royal Society, tried to explain this upward limit with a "Law of Human Mortality," essentially a mathematical description of aging. He wrote, "It is possible that death may be the consequence of two generally co-existing causes; the one, chance, without previous disposition to death or deterioration; the other, a deterioration, or an increased inability to withstand destruction."[7]

The first part of the law says that there is an internal clock that ticks away at random, like the chance a glass at a restaurant will break; essentially a first-order rate reaction, similar to radioactive decay, with some glasses lasting far longer than most. The second part says that, as time passes, due to an unknown runaway process, humans experience an exponential increase in their probability of death. By adding these two components together, Gompertz could accurately predict deaths due to aging: the number of people alive after 50 drops precipitously, but there is a tail at the end where some "lucky" people remain alive beyond what you'd expect. His equations made his relatives, Sir Moses Montefiore and Nathan Mayer Rothschild, owners of the Alliance Insurance Company, a lot of money.

What Gompertz could not have known, but would have appreciated, is that most organisms obey his law: flies, roundworms, mice, even yeast cells. For larger organisms, we don't know exactly what the two clocks are, but we do know in yeast cells: the *chance* clock is the formation of an rDNA circle, and the *exponential* clock is the replication and exponential increase in the numbers of rDNA circles, with the resulting movement of Sir2 away from the silent mating-type genes that causes sterility.[8]

Humans are more complicated, but in the nineteenth century, British mortality rates were becoming amenable to simple mathematical modeling because they were increasingly avoiding not-from-aging deaths: childbirth, accidents, and infections. This increasingly revealed the underlying and exponential incidence of death due to internal clocks as being the same as it ever was. During those times, the probability of dying doubled every eight years, an equation that left very little room for survivors after the age of 100.

That cap has generally held true ever since, even as the global average life expectancy jumped twenty years between 1960 and today.[9] That's because all that doubling adds up quickly. So even though most people who live in developed nations can now feel confident that they will make it to 80, these days the chances that any of us will reach a century is just 3 in 100. Getting to 115 is a 1-in-100-million proposition. And reaching 130 is a mathematical improbability of the highest order.

At least it is right now.

THE MORTAL BREEZE

Back in the mid-1990s, when I was pursuing my PhD at Australia's University of New South Wales, my mother, Diana, was found to have a tumor the size of an orange in her left lung.

As she was a lifelong smoker, I'd suspected it was coming. It was the one thing we had argued about more than anything else. When I was a young boy, I used to steal her cigarettes and hide them. It infuriated her. The fact that she didn't respond to my pleas to stop smoking infuriated me, too.

"I have lived a good life. The rest is a bonus," she would say to me in her early 40s.

"Do you know how lucky you are to have been born? You're throwing your life away! I won't come visit you in hospital when you get cancer," I would say.

When the cancer finally arrived about a decade later, I wasn't angry. Tragedy has a way of vanquishing anger. I drove to the hospital, determined to solve any problem.

My mother was responsible for her own actions, but she was also a victim of an unscrupulous industry. Tobacco alone doesn't kill people; it's the combination of tobacco, genetics, and time that most often leads to death. She was diagnosed with cancer at the age of 50. That's twenty-one years earlier than the first diagnosis in the average lung cancer patient. It's also how old I am now.

In one way of thinking, my mother was unfortunate to develop cancer at such a young age. After her back was opened up, rows of ribs were cut from her spine, and major arteries were rerouted, she lived the rest of her life with just one lung, which certainly impacted her quality of life and ensured that she had only a few years of *good* life left.

On the genetics front, my mother was also unfortunate. Everyone in my family, from my grandmother to my youngest son, has had their genes analyzed by one of the companies that offer these services. When my mother had hers done, she learned, albeit after she had cancer, that she had inherited a mutation in the SERPINA1 gene, which is implicated in chronic obstructive pulmonary disease or emphysema. That meant her clock was ticking even faster. After her left lung was removed, her right lung was the sole provider of oxygen, but the deficiency in SERPINA1 meant that white blood cells attacked her remaining lung, destroying the tissue as if it were an invader. Eventually the lung gave out.[10]

In another way of thinking, though, my mother was *very lucky*—she had the come-to-God moment that many smokers need to go to battle with the tremendously powerful forces of addiction in time to save herself, and she spent another two decades on this planet. She traveled the world, visiting eighteen different countries. She met her grandchildren. She saw me give a TED Talk at the Sydney Opera House. For this we must certainly credit the doctors who removed her cancerous lung, but we should also acknowledge the positive impact of her age. One of the best ways to predict whether someone will survive a disease, after all, is to take a look at how old he or she is when diagnosed—and my mother was, relatively speaking, very young.

This is key. We know that smoking accelerates the aging clock and makes you more likely to die than a nonsmoker—15 years earlier, on average. So, we have fought it with public health campaigns, class action lawsuits, taxes on tobacco products, and legislation. We know that cancer makes you more likely to die, and we've fought it with billions of dollars' worth of research aimed at ending it once and for all.

We know that aging makes you more likely to die, too, but we've accepted it as part of life.

It's also worth noting that even before my mother was diagnosed with lung cancer—indeed, even before the cancerous cells in her lungs began growing out of control—she was aging. And in that way, of course, she was hardly unique. We know that the process of aging begins long before we notice it. And with the unfortunate exceptions of those whose lives are taken by the early onset of a hereditary ailment or a deadly pathogen, most people begin to experience at least some of the effects of aging long before they are impacted by the accumulation of diseases we commonly associate with growing old. At the molecular level, this starts to happen at a time in our lives that many of us still look and feel young. Girls who go through puberty earlier than normal, for example, have an accelerated epigenetic clock. At that age, we can't hear the mistakes of the concert pianist.[11] But they are there, even as a teenager.

In our 40s and 50s, we don't often think about what it feels like to grow old. When I give talks about my research, sometimes I bring an "age suit" and ask a young volunteer to wear it. A neck brace reduces mobility in the neck, lead-lined jackets and wraps all over the body simulate weak muscles, earplugs reduce hearing, and ski goggles simulate cataracts. After a few minutes of walking around in the suit, the test subject is very relieved to take it off—and fortunately can do so.

"Imagine wearing it for a decade," I say.

To put yourself into an aged mind-set, try this little experiment. Using your nondominant hand, write your name, address, and phone number while circling your opposite foot counterclockwise. That's a rough approximation of what it feels like.

Different functions peak at different times for different people, but physical fitness, in general, begins to decline in our 20s and 30s. Men who run middle-distance races, for instance, are fastest around the age of 25, no matter how hard they train after that. The best female marathoners can stay competitive well into their late 20s and early 30s, but their times begin to rise quickly after 40. Occasionally, exceptionally fit outliers—such as National Football League quarterback Tom Brady, National Women's Soccer League defender Christie Pearce, Major League Baseball outfielder Ichiro Suzuki, and tennis legend Martina Navratilova—demonstrate that professional athletes can stay competitive into their 40s, but almost no one remains at the highest levels of these or most other professional sports much past their mid-40s. Even someone as resilient as Navratilova peaked when she was in her early 20s through her early 30s.

There are some simple tests to determine how biologically old you probably are. The number of push-ups you can do is a good indicator. If you are over 45 and can do more than twenty, you are doing well. The other test of age is the sitting-rising test (SRT). Sit on the floor, barefooted, with legs crossed. Lean forward quickly and see if you can get up in one move. A young person can. A middle-aged person typically needs to push off with one of their hands. An elderly person often needs to get onto one knee. A study of people 51 to 80 years found that 157 out of 159 people who passed away in 75 months had received less than perfect SRT scores.

Physical changes happen to everyone. Our skin wrinkles. Our hair grays. Our joints ache. We start groaning when we get up. We begin to lose resilience, not just to diseases but to all of life's bumps and bruises.

Fortunately, a hip fracture for a teenager is a very rare event that nearly everyone is expected to bounce back from. At 50, such an injury could be a life-altering event but generally not a fatal one. It's not long after that, though, that the risk factor for people who suffer a broken hip becomes terrifyingly high. Some reports show that up to half of those over the age of 65 who suffer a hip fracture will die within six months.[12] And those who survive often live the rest of their lives in pain and with limited mobility. At 88, my grandmother Vera tripped on a rumpled carpet and broke her upper femur. During surgery to repair the damage, her heart stopped on the operating table. Though she survived, her brain had been starved for oxygen. She never walked again and died a few years later.

Wounds also heal much more slowly with age—a phenomenon first scientifically studied during World War I by the French biophysicist Pierre Lecomte du Noüy, who noted a difference in the rate of healing between younger and older wounded soldiers. We can see this in even starker relief when we look at the differences in the ways children and the elderly heal from wounds. When a child gets a cut on her foot, a noninfected wound will heal quite quickly. The only medicine most kids need when they get hurt like this is a kiss, a Band-Aid, and some assurance that everything will be okay. For an elderly person, a foot injury is not just painful but dangerous. For older diabetics, in particular, a small wound can be deadly: The five-year mortality rate for a foot ulcer in a diabetic is greater than 50 percent. That's higher than the death rates for many kinds of cancer.[13]

Chronic foot wounds, by the way, are not rare; we just don't hear much about them. They almost always begin with seemingly benign rubbing on increasingly numb and fragile soles—but not always. My friend David Armstrong, at the University of Southern California, a passionate advocate for increasing our focus on preventing diabetic foot injuries, often tells the story of one of his patients, who had a nail stuck in his foot for four days. The patient noticed it only because he wondered where the tapping sound on the floor was coming from.

Small and large diabetic foot wounds rarely heal. They can look as though someone has taken an apple corer to the balls of both feet. The body doesn't have enough blood flow and cell regeneration capacity, and bacteria thrive in this meaty, moist environment. Right now, 40 million people, bedridden and waiting for death, are living this nightmare. There's almost nothing that can be done for them except to cut back the dead and dying tissue, then cut some more, and then some more. From there, robbed of upright mobility, misery is your bedfellow and thankfully death is nigh. In the United States alone, each year, 82,000 elderly people have a limb amputated. That's ten every hour. All this pain, all this cost, comes from relatively minor initial injuries: foot wounds.

The older we get, the less it takes for an injury or illness to drive us to our deaths. We are pushed closer and closer to the precipice until it takes nothing more than a gentle wind to send us over. This is the very definition of frailty.

If hepatitis, kidney disease, or melanoma did the sorts of things to us that aging does, we would put those diseases on a list of the deadliest illnesses in the world. Instead, scientists call what happens to us a "loss of resilience," and we generally have accepted it as part of the human condition.

There is nothing more dangerous to us than age. Yet we have conceded its power over us. And we have turned our fight for better health in other directions.

WHACK-A-MOLE MEDICINE

There are three large hospitals within a few minutes walk of my office. Brigham and Women's Hospital, Beth Israel Deaconess Medical Center, and Boston Children's Hospital are focused on different patient populations and medical specialties, but they're all set up the same way.

If we were to take a walk into the lobby of Brigham and Women's and head over to the sign by the elevator, we'd get a lay of this nearly universal medical landscape. On the first floor is wound care. Second floor: orthopedics. Third floor: gynecology and obstetrics. Fourth floor: pulmonary care.

At Boston Children's, the different medical specialties are similarly separated, though they are labeled in a way more befitting the young patients at this amazing hospital. Follow the signs with the boats for psychiatry. The flowers will take you to the cystic fibrosis center. The fish will get you to immunology.

And now over to Beth Israel. This way to the cancer center. That way to dermatology. Over here for infectious diseases.

The research centers that surround these three hospitals are set up in much the same way. In one lab you'll find researchers working to cure cancer. In another they're fighting diabetes. In yet another they're working on heart disease. Sure, there are geriatricians, but they almost always take care of the already sick, thirty years too late. They treat the aged— not the aging. No wonder so few doctors today are choosing to specialize in this area of medicine.

There's a reason why hospitals and research institutions are organized in this way. Most of our modern medical culture has been built to address medical problems one by one—a segregation that owes itself in no small part to our obsession with classifying the specific pathologies leading to death.

There was nothing wrong with this setup when it was established hundreds of years ago. And by and large, it still works today. But what this approach ignores is that stopping the progression of one disease doesn't make it any less likely that a person will die of another. Sometimes, in fact, the treatment for one disease can be an aggravating factor for another. Chemotherapy can cure some forms of cancer, for instance, but it also makes people's bodies more susceptible to other forms of cancer. And as we learned in the case of my grandmother Vera, something as seemingly routine as orthopedic surgery can make patients more susceptible to heart failure.

Because the stakes are so exceptionally high for the individual patients being treated in these places, a lot of people don't recognize that a battle won on any of these individual fronts won't make much of a difference against the Law of Human Mortality. Surviving cancer or heart disease doesn't substantially increase the average human lifespan, it just decreases the odds of dying of cancer or heart disease.

The way doctors treat illness today "is simple," wrote S. Jay Olshansky, a demographer at the University of Illinois. "As soon as a disease appears, attack that disease as if nothing else is present; beat the disease down, and once you succeed, push the patient out the door until he or she faces the next challenge; then beat that one down. Repeat until failure."[14]

The United States spends hundreds of billions of dollars each year fighting cardiovascular disease.[15] But if we could stop all cardiovascular disease—every single case, all at once—we wouldn't add many years to the average lifespan; the gain would be just 1.5 years. The same is true for cancer; stopping all forms of that scourge would give us just 2.1 more years of life on average, because all other causes of death still increase exponentially. We're still aging, after all.

Aging in its final stages is nothing like a bushwalk, where a bit of rest, a drink of water, a nutritional bar, and some fresh socks can get you another dozen miles before sunset. It's more like a fast sprint over

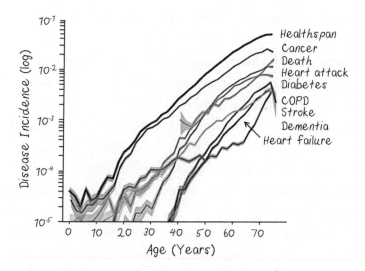

WHY TREATING ONE DISEASE AT A TIME HAS LITTLE IMPACT ON LIFESPAN. The graph shows an exponential increase in disease as each year passes after the age of 20. It's hard to appreciate exponential graphs. If I were to draw this graph with a linear Y-axis, it would be two stories tall. What this means is your chance of developing a lethal disease increases by a thousandfold between the ages of 20 and 70, so preventing one disease makes little difference to lifespan.

Source: Adapted from A. Zenin, Y. Tsepilov, S. Sharapov, et al., "Identification of 12 Genetic Loci Associated with Human Healthspan," *Communications Biology* 2 (January 2019).

an ever-higher and ever-closer set of hurdles. One of those hurdles will eventually send you for a tumble. And once you've fallen one time, if you do get up, the odds of falling again just keep getting higher. Take away one hurdle, and the path forward is really no less precarious. That's why the current solutions, which are focused on curing individual diseases, are both very expensive and very ineffective when it comes to making big advances in prolonging our healthspans. What we need are medicines that knock down *all* the hurdles.

Thanks to statins, triple-bypass surgeries, defibrillators, transplants, and other medical interventions, our hearts are staying alive longer than ever. But we haven't been nearly so attentive to our other organs, including the most important one of all: our brains. The result is that more of us are spending more years suffering from brain-related maladies, such as dementia.

Eileen Crimmins, who studies health, mortality, and global aging at the University of Southern California, has observed that even though average lifespans in the United States have increased in recent decades, our healthspans have not kept up. "We have reduced mortality more than we prevented morbidity," she wrote in 2015.[16]

So prevalent is the combined problem of early mortality and morbidity that there is a statistic for it: the disability-adjusted life year, or DALY, which measures the years of life lost from both premature death and poor state of health. The Russian DALY is the highest in Europe, with twenty-five lost years of healthy life per person. In Israel, it is an impressive ten years. In the United States, the number is a dismal twenty-three.[17]

The average age of death can vary rather significantly over time, and is affected by many factors, including the prevalence of obesity, sedentary lifestyles, and drug overdoses. Similarly, the very idea of poor health is both subjective and measured differently from place to place, and so researchers are divided on whether the DALY is rising or declining in the United States. But even the more optimistic assessments suggest that the numbers have largely been static in recent years. To me, that in itself is an indictment of the US system; like other advanced countries, we should be making tremendous progress toward reducing the DALY and other measures of morbidity, yet, at best, it seems we're treading water. We need a new approach.

It doesn't take studies and statistics to know what's happening, though. It's all around us, and the older we get, the more obvious it becomes. We get to 50 and begin to notice we look like our parents, with graying hair and an increasing number of wrinkles. We get to 65, and if we haven't faced some form of disease or disability yet, we consider ourselves fortunate. If we're still around at 80, we are almost guaranteed to be combating an ailment that has made life harder, less comfortable, and less joyful. One study found that 85-year-old men are diagnosed with an average of four different diseases, with women of that age suffering from five. Heart disease and cancer. Arthritis and Alzheimer's. Kidney disease and diabetes. Most patients have several additional undiagnosed diseases, including hypertension, ischemic heart disease, atrial fibrillation, and dementia.[18] Yes, these are different ailments with different pathologies, studied in different buildings at the National Institutes of Health and in different departments within universities.

But aging is a risk factor for all of them.

In fact, it's *the* risk factor. Truly, by comparison, little else matters.

The final years of my mother's life serve as a good example. Like almost everyone else, I recognized that smoking would increase my mother's chances of getting lung cancer. I also knew why: cigarette smoke contains a chemical called benzo(a)pyrene, which binds to guanine in DNA, induces double-strand breaks, and causes mutations. The repair process also causes epigenetic drift and metabolic changes that cancer cells thrive on, in a process we've called *geroncogenesis*.[19]

The combination of genetic and epigenetic changes induced by years of exposure to cigarette smoke increases the chances of developing lung cancer about fivefold.

That's a big increase. And because of it—and the devastatingly high health costs associated with treating cancer—the majority of the world's nations sponsor smoking cessation programs. Most countries also put health warnings on cigarette packaging, some with horrific color pictures of tumors and blackened extremities. Most countries have passed laws against certain kinds of tobacco advertising. And most have sought to decrease consumption through punitive taxes.[20]

All of that to prevent a fivefold increase in a few kinds of cancer. And having watched my mother suffer from that kind of cancer, I'll be the first to say it's totally worth it. From both an economic and emotional point of view, these are good investments.

But consider this: though smoking increases the risk of getting cancer fivefold, being 50 years old increases your cancer risk a hundredfold. By the age of 70, it is a thousandfold.[21]

Such exponentially increasing odds also apply to heart disease. And diabetes. And dementia. The list goes on and on. Yet there is not a country in the world that has committed any significant resources to help its citizens combat aging. In a world in which we seem to agree on very little, the feeling that "it's just the way it goes" is almost universal.

A GLORIOUS FIGHT

Aging results in physical decline.

It limits the quality of life.

And it has a specific pathology.

Aging does all this, and in doing so it fulfills every category of what we call a disease except one: it impacts more than half the population.

According to *The Merck Manual of Geriatrics*, a malady that impacts less than half the population is a disease. But aging, of course, impacts everyone. The manual therefore calls aging an "inevitable, irreversible decline in organ function that occurs over time even in the absence of injury, illness, environmental risks, or poor lifestyle choices."

Can you imagine saying that cancer is inevitable and irreversible? Or diabetes? Or gangrene?

I can. Because we used to say that.

All of these may be natural problems, but that doesn't make them inevitable and irreversible—and it sure doesn't make them acceptable. The manual is wrong about aging.

But being wrong has never stopped conventional wisdom from negatively impacting public policy. And because aging isn't a disease by the commonly accepted definition, it doesn't fit nicely into the system we've built for funding medical research, drug development, and the reimbursement of medical costs by insurance companies. Words matter. Definitions matter. Framing matters. And the words, definitions, and framing we use to describe aging are all about inevitability. We didn't just throw in the towel before the fight began, we threw it in before we even knew there was a fight to be had.

But there *is* a fight. A glorious and global one. And, I think, a winnable one.

There's no good reason why we have to say that something that happens to 49.9 percent of the population is a disease while something that happens to 50.1 percent of the population is not. In fact, that's a backward way of approaching problems that lends itself to the whack-a-mole system of medicine we've set up in hospitals and research centers around the world.

Why would we choose to focus on problems that impact small groups of people if we could address the problem that impacts everyone—especially if, in doing so, we could significantly impact all those other, smaller problems?

We can.

I believe that aging is a disease. I believe it is treatable. I believe we can treat it within our lifetimes. And in doing so, I believe, everything we know about human health will be fundamentally changed.

If you are not yet convinced that aging is a disease, I want to let you in on a secret. I have a window into the future. In 2028, a scientist will discover a new virus, called LINE-1. It will turn out that we are all infected with it. We get it from our parents. It will turn out that the LINE-1 virus is responsible for most other major diseases: diabetes, heart disease, cancer, dementia. It causes a slow, horrible chronic disorder, and all humans eventually succumb to it, even if they have a low-grade infection. Fortunately, the world pours billions of dollars into finding a cure. In 2033, a company will succeed in making a vaccine that prevents LINE-1 infections. New generations who are vaccinated at birth will live fifty years longer than their parents did—it will turn out that that's our *natural* lifespan and we had no idea. The new generation of healthy humans will pity previous generations, who blindly accepted that physical decline at 50 was natural and an 80-year life was a life well lived.

Of course, this is a science fiction story I just invented. But it might be truer than you think.

A few recent studies have suggested that the so-called selfish genes we all carry in our genome, actually called LINE-1 elements, replicate and cause cellular havoc as we get older, accelerating our physical demise. We'll discuss them in more detail later, but for now, it's the idea I want to focus on because it raises important questions: Does it matter whether LINE-1 comes from your parents directly or via a virus? Would you want to eradicate LINE-1 from humanity or let it grow in your kids and inflict horrible diseases on them? Would you say that LINE-1 causes a disease or not?

If not, is it simply because more than half of all people carry it?

Whether it's a virus, a selfish DNA element, or simply the makeup of our cells that causes these health problems, what's the difference? The end result is the same.

The belief that aging is a *natural* process is deep-rooted. So even if I've somewhat convinced you that aging should be considered a disease, let's do another thought experiment.

Imagine that everyone on our planet typically lives to 150 years in good health. Your family, though, doesn't. You become wrinkled, gray-haired, diabetic, and frail at 80. Upon seeing these poor, unfortunate souls in this poor, unfortunate state of existence, what doctor would not diagnose your family with a disease, name it after him- or herself, and publish horrid photos of you with your eyes blacked out in medical journals? Communities would raise money to understand and find a cure for your family's wretched inheritance.

That was exactly what happened when the German physician Otto Werner first described a condition that causes people to look and feel as though they are 80 when they are in their 40s. That's Werner syndrome, the disease I was studying when I first arrived at MIT in the 1990s. Nobody said I was studying something that is inevitable or irreversible. Nobody said it was crazy to call Werner syndrome a disease or to work to find a breakthrough therapy. Nobody told me or the Werner patients that "that's just the way it goes."

In front of us is the deadliest and costliest disease on the planet, a disease that almost no one is working on. It is as if the planet is in a stupor. If your first thought is "But I don't want to live past 90," let me assure you: I don't want you to live a year longer than you wish.

But before you make your decision, let's do one final thought experiment.

Imagine that a clerk at City Hall has found a mistake on your birth certificate. It turns out that you are actually 92 years old.

"You'll get a new one in the mail," the clerk says. "Have a nice day."

Do you feel any different now that you are 92? Nothing else has changed in your life—just a few numbers on your identification. Do you suddenly want to kill yourself?

Of course not. When we stay healthy and vibrant, as long as we feel young physically and mentally, our age doesn't matter. That's true whether you are 32, 52, or 92. Most middle-aged and older adults in the United States report feeling ten to twenty years younger than their age, because they still feel healthy. And feeling younger than your age predicts lower mortality and better cognitive abilities later in life.[22] It's a virtuous cycle, as long as you keep pedaling.

But no matter how you feel at this moment in your life, even with a positive outlook and a healthy lifestyle, you have a disease. And it's going to catch up to you, sooner rather than later, unless something is done.

I acknowledge that calling aging a *disease* is a radical departure from the mainstream view of health and well-being, which has established an array of medical interventions addressing the various causes of death. That framework evolved, however, largely because we didn't understand why aging occurs. Up until very recently, the best thing we had was a list of aging hallmarks. The Information Theory of Aging could change that.

There is nothing wrong with using the hallmarks to guide interventions. We can probably have a positive impact on people's lives by addressing each of them. It's possible that interventions aimed at slowing telomere deterioration will improve people's long-term well-being. Maintaining proteostasis, preventing deregulation of nutrient sensing, thwarting mitochondrial dysfunction, stopping senescence, rejuvenating stem cells, and decreasing inflammation might all be ways to delay the inevitable. Indeed, I work with students, postdocs, and companies around the globe that are developing solutions to each one of these hallmarks and hope to continue.[23] Anything we can do to alleviate suffering we should do.

But we're still building nine dams on nine tributaries.

In coming together to tackle the "new science of aging," as the attendees of the Royal Society meeting termed this fight in their 2010 meeting, increasing numbers of scientists are starting to acknowledge the possibility and potential inherent in heading upstream.

Together we can build a single dam—at the source. Not just intervene when things go wrong. Not just slow things down. We can eliminate the symptoms of aging altogether.

This disease is treatable.

PART II

WHAT WE'RE LEARNING

(THE PRESENT)

FOUR

LONGEVITY NOW

EVERY DAY I WAKE UP TO AN INBOX FULL OF MESSAGES FROM PEOPLE FROM ALL over the world. The tide ebbs and flows but always takes the form of a flash flood in the wake of newly announced research from my team or others.

"What should I be taking?" they ask.

"Can you tell me what I need to do to get admitted into one of the human trials?" they implore.

"Can you extend the lifespan of my daughter's hamster?" I kid you not.

Some of the letters are much sadder than others. One man recently wrote to offer to contribute a donation to my lab in honor of his mother, who had passed away after suffering terribly through many years of age-related illness. "I feel compelled to help, even in some small way, to prevent this from happening to someone else," he wrote. The next day, a woman whose father had been diagnosed with Alzheimer's wrote to ask if there were any way to get him admitted into a study. "I would do anything, take him anywhere, spend every last cent I have," she pleaded. "He is the only family I have and I cannot bear the thought of what is about to happen to him."

There is great reason for hope on the not-so-distant horizon, but those battling against the ravages of aging *right now* must do so in a world in which most doctors have never even thought about *why* we age, let alone how to *treat* aging.

Some of the medical therapies and life-extending technologies discussed in this book are already here. Others are a few years away. And there are more to discuss that are a decade or so down the road; we'll get to those as well.

But even without access to this developing technology, no matter who you are, where you live, how old you are, and how much you earn, you can engage your longevity genes, starting *right now.*

That's what people have been doing for centuries—without even knowing it—in centenarian-heavy places such as Okinawa, Japan; Nicoya, Costa Rica; and Sardinia, Italy. These are, you might recognize, some of the places the writer Dan Buettner introduced to the world as so-called Blue Zones starting in the mid-2000s. Since that time, the primary focus for those seeking to apply lessons from these and other longevity hot spots has been on what Blue Zone residents eat. Ultimately this resulted in the distillation of "longevity diets" that are based on the commonalities in the foods eaten in places where there are lots of centenarians. And overwhelmingly that advice comes down to eating more vegetables, legumes, and whole grains, while consuming less meat, dairy products, and sugar.

And that's not a bad place to start—in fact, it's a *great* place to start. There is widespread disagreement, even among the best nutritionists in the world, as to what constitutes the "best" diet for *H. sapiens.* That's likely because there is no best diet; we're all different enough that our diets need to be subtly and sometimes substantially different, too. But we're also all similar enough that there are some very broad commonalities: more veggies and less meat; fresh food versus processed food. We all know this stuff, though applying it can be a challenge.

A big part of the reason so many people aren't willing to face up to that challenge is because we've always thought of aging as an inevitable part of life. It might come a little earlier for some and a little later for others, but we've always been told that it's coming for us all.

That's what we used to say about pneumonia, influenza, tuberculosis, and gastrointestinal conditions, too. In 1900, those four illnesses accounted for about *half* of the deaths in the United States and—if you managed to live long enough—you could be virtually assured that one of them would get you eventually.

Today, deaths among people suffering from tuberculosis and gastrointestinal conditions are exceedingly rare. And pneumonia and influenza claim less than 10 percent of the lives taken by those conditions a little more than a century ago—with most of those deaths now among individuals weakened by aging.

What changed? In no small part it was framing. Advances in medicine, innovations in technology, and better information to guide our lifestyle decisions resulted in a world in which we didn't have to accept the idea that these diseases were "just the way it goes."

We don't have to accept aging like that, either.

But even among those who will have the most immediate access to pharmaceuticals and technologies that will be emerging to offer longer and healthier lives in the next few decades, reaching an optimal lifespan and healthspan won't be as easy as flipping a switch.

There will *always* be good and bad choices. And that starts with what we put into our bodies.

And what we don't.

GO, FAST

After twenty-five years of researching aging and having read thousands of scientific papers, if there is one piece of advice I can offer, one surefire way to stay healthy longer, one thing you can do to maximize your lifespan right now, it's this: eat less.

This is nothing revolutionary, of course. As far back as Hippocrates, the ancient Greek physician, doctors have been espousing the benefits of limiting what we eat, not just by rejecting the deadly sin of gluttony, as the Christian monk Evagrius Ponticus counseled in the fourth century, but through "intentional asceticism."

Not malnutrition. Not starvation. These are not pathways to more years, let alone better years. But fasting—allowing our bodies to exist in a state of want, more often than most of us allow in our privileged world of plenty—is unquestionably good for our health and longevity.

Hippocrates knew this. Ponticus knew this. So, too, did Luigi Cornaro, a fifteenth-century Venetian nobleman who could, and probably should, be considered the father of the self-help book.

The son of an innkeeper, Cornaro made a fortune as an entrepreneur and lavishly spent his money on wine and women. By his mid-30s, he was exhausted by food, drink, and sex—the poor guy—and resolved to limit himself in each regard. The historical record is a bit vague on the details of his sex life after that fateful decision,[1] but his diet and drinking habits have been well documented: he ate no more than twelve ounces of food and drank two glasses of wine each day.

"I accustomed myself to the habit of never fully satisfying my appetite, either with eating or drinking," Cornaro wrote in his *First Discourse on the Temperate Life*, "always leaving the table well able to take more."[2]

Cornaro's discourses on the benefits of *la vita sobria* might have fallen into obscurity had he not provided such compelling personal proof that his advice had merit: he published his guidance when he was in his 80s, and in exceptional health, no less, and he died in 1566 at nearly (and some sources say more than) 100 years old.

In more recent times, Professor Alexandre Guéniot, the president of the Paris Medical Academy just after the turn of the twentieth century, was famed for living on a restricted diet. It is said that his contemporaries mocked him—for there was no science at that time to back his suspicion that hunger would lead to good health, just his gut hunch—but he outlived them, one and all. He finally succumbed at the age of 102.

The first modern scientific explorations of the lifelong effects of a severely restricted diet began during the last days of World War I. That's when the longtime biochemical collaborators Lafayette Mendel and Thomas Osborne—the duo who had discovered vitamin A—discovered, along with researcher Edna Ferry, that female rats whose growth was stunted due to lack of food early in life lived much longer than those that ate plenty.[3]

Picking up on that evidence in 1935, a now-famous Cornell University professor named Clive McCay demonstrated that rats fed a diet containing 20 percent indigestible cellulose—cardboard, essentially—lived significantly longer lives than those that were fed a typical lab diet. Studies conducted over the next eighty years demonstrated again and again that calorie restriction without malnutrition, or CR, leads to longevity for all sorts of life-forms. Hundreds of mouse studies have been done since to test the effects of calories on health and lifespan, mostly on male mice.

Reducing calories works even in yeast. I first noticed this in the late 1990s. Cells fed with lower doses of glucose were living longer, and their DNA was exceptionally compact—significantly delaying the inevitable ERC accumulation, nucleolar explosion, and sterility.

If this happened only in yeast, it would merely be interesting. But because we knew that rodents also lived longer when their food was restricted—and later learned that this was the case for fruit flies, as well[4]—it was apparent that this genetic program was very old, perhaps nearly as old as life itself.

In animal studies, the key to engaging the sirtuin program appears to be keeping things on the razor's edge through calorie restriction—just enough food to function in healthy ways and no more. This makes sense. It engages the survival circuit, telling longevity genes to do what they have been doing since primordial times: boost cellular defenses, keep organisms alive during times of adversity, ward off disease and deterioration, minimize epigenetic change, and slow down aging.

But this has, for obvious reasons, proven a challenge to test on humans in a controlled scientific setting. Sadly, it's not hard to find instances in which humans have had to go without food, but those periods are generally times in which food insecurity results in malnutrition, and it would be a challenge to keep a test group of humans on the razor's edge for the long periods of time that would be required for comprehensive controlled studies.

As far back as the 1970s, though, there have been observational studies that strongly suggested long-term calorie restriction could help humans live longer and healthier lives, too.

In 1978 on the island of Okinawa, famed for its large number of centenarians, bioenergetics researcher Yasuo Kagawa learned that the total number of calories consumed by schoolchildren was less than two-thirds of what children were getting in mainland Japan. Adult Okinawans were also leaner, taking in about 20 percent fewer calories than their mainland counterparts. Kagawa noted that not only were the lifespans of Okinawans longer, but their healthspans were, too—with significantly less cerebral vascular disease, malignancy, and heart disease.[5]

In the early 1990s, the Biosphere 2 research experiment provided another piece of evidence. For two years, from 1991 to 1993, eight people lived inside a three-acre, closed ecological dome in southern Arizona, where they were expected to be reliant on the food they were growing inside. Green thumbs they weren't, though, and the food they farmed turned out to be insufficient to keep the participants on a typical diet. The lack of food wasn't bad enough to result in malnutrition, but it did mean that the team members were frequently hungry.

One of the prisoners (and by "prisoners" I mean "experimental subjects") happened to be Roy Walford, a researcher from California whose studies on extending life in mice are still required reading for scientists entering the aging field. I have no reason to suspect that Walford sabotaged the crops, but the coincidence was rather fortuitous for his research; it gave him an opportunity to test his mouse-based findings on human subjects. Because they were thoroughly medically monitored before, during, and after their two-year stint inside the dome, the participants gave Walford and other researchers a unique opportunity to observe the numerous biological effects of calorie restriction. Tellingly, the biochemical changes they saw in their bodies closely mirrored those Walford had seen in his long-lived calorie-restricted mice, such as decreased body mass (15 to 20 percent), blood pressure (25 percent), blood sugar level (21 percent), and cholesterol levels (30 percent), among others.[6]

In recent years, formal human studies have begun, but it has turned out to be quite difficult to get volunteer human subjects to reduce their food intake and maintain that level of consumption over long periods. As my colleagues Leonie Heilbronn and Eric Ravussin wrote in *The American Journal of Clinical Nutrition* in 2003, "the absence of adequate information on the effects of good-quality, calorie-restricted diets in non-obese humans reflects the difficulties involved in conducting long-term studies in an environment so conducive to overfeeding. Such studies in free-living persons also raise ethical and methodologic issues."[7] In a report published in *The Journals of Gerontology* in 2017, a Duke University research team described how it sought to limit 145 adults to a diet of 25 percent fewer calories than is typically recommended for a healthy lifestyle. People being people, the actual calorie restriction achieved was, on average, about 12 percent over two years. Even that was enough, however, for the scientists to see a significant improvement in health and a slowdown in biological aging based on changes in blood biomarkers.[8]

These days, there are many people who have embraced a lifestyle that permits significantly reduced caloric intake; about a decade ago, before fasting's most recent revival, some of them visited my lab at Harvard.

"Isn't it hard to do what you do?" I asked Meredith Averill and her husband, Paul McGlothin, at the time members of CR Society International and still very much advocates for calorie restriction, who limit themselves to about 75 percent of the calories typically recommended by doctors and sometimes quite a bit less than that. "Don't you just feel hungry all the time?"

"Sure, at first," McGlothin told me. "But you get used to it. We feel great!"

At lunch that day, McGlothin expounded upon the merits of eating organic baby food and slurped down something that looked to me like orange mush. I also noticed that both he and Averill were wearing turtlenecks. It wasn't winter. And most folks in my lab are perfectly comfortable in T-shirts. But with so little fat on their bodies, they needed the extra warmth. Then in his late 60s, McGlothin showed no signs that his diet might slow him down. He was the CEO of a successful marketing company and a former New York State chess champion. He didn't look much younger than his age, though; in large part, I suspect this was because a lack of fat exposes wrinkles, but his blood biochemistry suggested otherwise. On his 70th birthday, his health indicators, from blood pressure and LDL cholesterol to resting heart rate and visual acuity, were typical of those of a much younger person.[9] Indeed, they resembled those seen in the long-lived rats on calorie restriction.

It's true that what we know about the impact of lifelong calorie restriction in humans comes down to short-term studies and anecdotal experiences. But one of our close relatives has offered us insights into the longitudinal benefits of this lifestyle.

Since the 1980s, a long-term study of calorie restriction in rhesus monkeys—our close genetic cousins—has produced stunningly compelling results. Before the study, the maximum known lifespan for *any* rhesus monkey was 40 years. But of twenty monkeys in the study that lived on calorie-restricted diets, six reached that age, which is roughly equivalent to their reaching 120 in human terms.

To hit that mark, the monkeys didn't need to live on a calorie-restricted diet for their entire lives. Some of the test subjects were started on a 30 percent reduction regimen when they were middle-aged monkeys.[10]

CR works to extend the lifespan of mice, even when initiated at 19 months of age, the equivalent of a 60- to 65-year-old human, but the earlier the mice start on CR, the greater the lifespan extension.[11] What these and other animal studies tell us is that it's hard to "age out" of the longevity benefits of calorie restriction, but it's probably better to start earlier than later, perhaps after age 40, when things really start to go downhill, molecularly speaking.

That doesn't make a CR diet a good plan for everyone. Indeed, even Rozalyn Anderson, a former trainee of mine who's now a famous professor at the University of Wisconsin and a lead researcher in the rhesus study, says a 30 percent calorie-reduced diet for humans, long term, amounted in her mind to a "bonkers diet."[12]

It's certainly not bonkers for everyone, though, especially considering that calorie restriction hasn't been demonstrated only to lengthen life but also to forestall cardiac disease, diabetes, stroke, and cancer. It's not just a longevity plan; it's a vitality plan.

It's nonetheless a hard sell for many people. It takes strong willpower to avoid the fridge at home or snacks at work. There's an adage in my field: if calorie restriction doesn't make you live longer, it will certainly make you feel that way.

But it turns out that's okay, because research is increasingly demonstrating that many of the benefits of a life of strict and uncompromising calorie restriction can be obtained in another way. In fact, that way might be even *better*.

THE PERIODIC TABLE

To ensure a genetic response to a lack of food, hunger doesn't need to be the status quo. Once we've grown accustomed to stress, after all, it's no longer as stressful.

Intermittent fasting, or IF—eating normal portions of food but with periodic episodes without meals—is often portrayed as a new innovation in health. But long before my friend Valter Longo at the University of California, Los Angeles, began touting the benefits of IF, scientists had been studying the effects of periodic calorie restriction for the better part of a century.

In 1946, University of Chicago researchers Anton Carlson and

Frederick Hoelzel subjected rats to periodic food restriction and found, when they did, that those that went hungry every third day lived 15 to 20 percent longer than their cousins on a regular diet.[13]

At the time it was believed that fasting provided the body with a "rest."[14] That's very much the opposite of what we now know about what happens at a cellular level when we subject our bodies to the stress of going without food. Either way, Carlson and Hoelzel's work provided valuable information on the long-term results of irregular calorie restriction.

It's not clear whether the pair applied what they'd learned to their own lives, but both lived relatively long lives for their time. Carlson died at the age of 81. Hoelzel made it to 74, despite having subjected himself over the years to experiments that included swallowing gravel, glass beads, and ball bearings to study how long it would take for such objects to pass through his system. And people say I'm crazy.

Today, human studies are confirming that once-in-a-while calorie restriction can have tremendous health results, even if the times of fasting are quite transient.

In one such study, participants ate a normal diet most of the time but turned to a significantly restricted diet consisting primarily of vegetable soup, energy bars, and supplements for five days each month. Over the course of just three months, those who maintained the "fasting mimicking" diet lost weight, reduced their body fat, and lowered their blood pressure, too. Perhaps most important, though, the participants had lower levels of a hormone made primarily in the liver called insulin-like growth factor 1, or IGF-1. Mutations in IGF-1 and the IGF-1 receptor gene are associated with lower rates of death and disease and found in abundance in females whose families tend to live past 100.[15]

Levels of IGF-1 have been closely linked to longevity. The impact is so strong, in fact, that in some cases it can be used to predict—with great accuracy—how long someone will live, according to Nir Barzilai and Yousin Suh, who research aging at the Albert Einstein College of Medicine at Yeshiva University in New York.

Barzilai and Suh are geneticists whose research focuses on centenarians who have made it to 100—and beyond—without suffering from any age-related diseases. That unique population is a vital study group, because its members provide a model for aging the way most people say they want to age—not accepting that additional years of life need to come with additional years of misery.

When we find clusters of these people, we see that in some cases it doesn't actually matter what they put into their bodies. They carry gene variants that seem to put them into a state of fasting no matter what they eat. As anyone who has ever known a centenarian can attest, it doesn't take a lifetime of making 100 percent healthy decisions to reach 100. When Barzilai's team studied nearly 500 Ashkenazi Jews over the age of 95, they saw that many engaged in the same sorts of behaviors doctors have long been telling us to shun: eating fried foods, smoking, and just sitting around and drinking a little too much. Barzilai once asked one of his centenarian study subjects why she hadn't listened to her doctors over the years when they had strongly advised her to end her lifelong smoking habit. "I've had four doctors tell me smoking would kill me," she said with a wry smile, "and well, all four are dead now, aren't they?"

Some people are simply winners in the genetic lottery. The rest of us have some extra work to do. But the good news is that the epigenome is malleable. Since it's not digital, it's easier to impact. We can control the behavior of this analog element of our biology by how we live our lives.

The important thing is not just *what* we eat but the *way* we eat. As it turns out, there is a strong correlation between fasting behavior and longevity in Blue Zones such as Ikaria, Greece, "the island where people forget to die," where one-third of the population lives past the age of 90 and almost every older resident is a staunch disciple of the Greek Orthodox church and adheres to a religious calendar that calls for some manner of fasting more than half the year.[16] On many days, that means no meat, dairy products, or eggs and sometimes no wine or olive oil, either—for some Greeks, that's just about everything. Additionally, many Greeks observe periods of total fasting before taking Holy Communion.[17]

Other longevity hot spots, such as Bama County in southern China, are places where people have access to good, healthy food but choose to forgo it for long periods each day.[18] Many of the centenarians in this region have spent their lives eschewing a morning meal. They generally eat their first small meal of the day around noon, then share a larger meal with their families at twilight. In this way, they typically spend sixteen hours or more of each day without eating.

When we investigate places like this, and as we seek to apply research about fasting to our modern lives, we find that there are scores of ways to calorie restrict that are sustainable, and many take the form of what has come to be known as periodic fasting—not being hungry all the time but using hunger some of the time to engage our survival circuit.

Over time, some of these ways of limiting food will prove to be more effective than others. A popular method is to skip breakfast and have a late lunch (the 16:8 diet). Another is to eat 75 percent fewer calories for two days a week (the 5:2 diet). If you're a bit more adventurous, you can try skipping food a couple of days a week (Eat Stop Eat), or as the health pundit Peter Attia does, go hungry for an entire week every quarter. The permutations of these various models for extending life and health are being worked out in animals and will be worked out in people, too. The short-term studies are promising. I suspect the long-term research will be, too. In the meantime, however, almost any periodic fasting diet that does not result in malnutrition is likely to put your longevity genes to work in ways that will result in a longer, healthier life.

It doesn't take any money to eat this way. In fact, it saves money. Moreover, people who are not accustomed to being able to gorge themselves whenever they want might be in a better position to be successful at going a few days each month with a lot less food.

At least at this juncture in the evolution of our customs around food, though, for many people *any* form of fasting is a nonstarter.

I've tried calorie restriction. I can't do it. Feeling hungry isn't fun, and food is just too pleasurable. Lately, I have taken to periodic fasting—skipping a meal or two each day—but I admit that it's mostly unintentional. I simply forget to eat.

So far, though, we've talked only about engaging the survival circuit by limiting how much we eat, but *what* we eat is also important.

AMINO RIGHT

We'd die quite quickly without amino acids, the organic compounds that serve as the building blocks for every protein in the human body. Without them—and in particular the nine essential amino acids that our bodies cannot make on their own—our cells can't assemble the life-giving enzymes needed for life.

Meat contains all nine of the essential amino acids. That's easy energy, but it doesn't come without a cost. Actually, a lot of costs. Because no matter how you feel about the morals of the matter, meat is murder—on our bodies. So can we just avoid protein? Ironically, protein is what satiates us. Same for mice. Same for swarming locusts in need of nutrients, which is why they eat each other.[19] It would appear that animal life can't easily limit protein in the diet without some hunger pains.

There isn't much debate on the downsides of consumption of animal protein. Study after study has demonstrated that heavily animal-based diets are associated with high cardiovascular mortality and cancer risk. Processed red meats are especially bad. Hot dogs, sausage, ham, and bacon might be gloriously delicious, but they're ingloriously carcinogenic, according to hundreds of studies that have demonstrated a link between these foods and colorectal, pancreatic, and prostate cancer.[20] Red meat also contains carnitine, which gut bacteria convert to trimethylamine N-oxide, or TMAO, a chemical that is suspected of causing heart disease.

That doesn't mean a little red meat will kill you—the diet of hunter-gatherers is a mix of plants packed with fiber and nutrients, mixed with some red meat and fish in moderation[21]—but if you're interested in a long and healthy life, your diet probably needs to look a lot more like a rabbit's lunch than a lion's dinner. When we substitute animal protein with more plant protein, studies have shown, all-cause mortality falls significantly.[22]

From an energy perspective, the good news is that there isn't a single amino acid that can't be obtained by consuming plant-based protein sources. The bad news is that, unlike most meats, weight for weight, any given plant usually delivers limited amounts of amino acids.

From a vitality perspective, though, that's great news. Because a body that is in short supply of amino acids overall, or any single amino acid for a spell, is a body under the very sort of stress that engages our survival circuits.

You'll recall that when the enzyme known as mTOR is inhibited, it forces cells to spend less energy dividing and more energy in the process of autophagy, which recycles damaged and misfolded proteins. That act of hunkering down ends up being good for prolonged vitality in every organism we've studied. What we're coming to learn is that mTOR isn't impacted only by caloric restriction.[23] If you want to keep mTOR from being activated too much or too often, limiting your intake of amino acids is a good way to start, so inhibiting this particular longevity gene is really as simple as limiting your intake of meat and dairy.

It's also increasingly clear that all essential amino acids aren't equal. Rafael de Cabo at the National Institutes of Health, Richard Miller at the University of Michigan, and Jay Mitchell at Harvard Medical School have found over the years that feeding mice a diet with low levels of the amino acid methionine works particularly well to turn on their bodily defenses, to protect organs from hypoxia during surgery, and to increase healthy lifespan by 20 percent.[24] One of my former students, Dudley Lamming, who now runs a lab at the University of Wisconsin, demonstrated that methionine restriction causes obese mice to shed most of their fat—and fast. Even as the mice, which Lamming called "couch potatoes," continued to eat as much as they wanted and shun exercise, they still lost about 70 percent of their fat in a month, while also lowering their blood glucose levels.[25]

We can't live without methionine. But we can do a better job of restricting the amount of it we put into our bodies. There's a lot of methionine in beef, lamb, poultry, pork, and eggs, whereas plant proteins, in general, tend to contain low levels of that amino acid—enough to keep the light on, as it were, but not enough to let biological complacency set in.

The same is true for arginine and the three branched-chain amino acids, leucine, isoleucine, and valine, all of which can activate mTOR. Low levels of these amino acids correlate with increased lifespan[26] and in human studies, a decreased consumption of branched-chain amino acids has been shown to improve markers of metabolic health significantly.[27]

We can't live without them, but most of us can definitely stand to get less of them, and we can do that by lowering our consumption of foods that many people consider to be the "good animal proteins," chicken, fish, and eggs—particularly when those foods aren't being used to recover from physical stress or injury.

All of this might seem counterintuitive; amino acids, after all, are often considered helpful. And they can be. Leucine, for instance, is well known to boost muscle, which is why it's found in large quantities in the protein drinks that bodybuilders often chug before, during, and after workouts. But that muscle building is coming in part because leucine is activating mTOR, which essentially calls out to your body, "Times are good right now, let's disengage the survival circuit."[28] In the long run, however, protein drinks may be preventing the mTOR pathway from providing its longevity benefits. Studies in which leucine is completely eliminated from a mouse's diet have demonstrated that just one week without this particular amino acid significantly reduces blood glucose levels, a key marker of improved health.[29] So a little leucine is necessary, of course, but a little goes a long way.

All of these findings may explain why vegetarians suffer significantly lower rates of cardiovascular disease and cancer than meat eaters.[30] The reduction of amino acids—and thus the inhibition of mTOR—isn't the only thing at play in that equation. The lower calorie content, increased polyphenols, and feeling of superiority over your fellow human beings are also helpful. All of these, except the last, are valid explanations for why vegetarians live longer and stay healthier.

Even if we eat a low-protein, vegetable-rich diet, we may live longer, but we won't maximize our lifespans—because putting our bodies into nutritional adversity isn't going to maximally trigger our longevity genes. We need to induce some physical adversity, too. If that doesn't happen, we miss a key opportunity to trigger our survival circuits further. Like a beautiful sports car driven only a block and back on Sunday mornings, our longevity genes will go tragically underutilized.

With so much horsepower under the hood, we just have to fire up the engine and take it out for a spin.

DO SWEAT IT

There's a reason why for centuries exercise has been the go-to prescription for vitality. But that reason isn't what most people—or even many doctors—think.

In the nearly four hundred years since the English physician William Harvey discovered that blood flows around the body in an intricate network of tubes, doctors thought that exercise improves health by moving blood through the circulatory system faster, flushing out the buildup of plaque.

That's not how it works.

Yes, exercise improves blood flow. Yes, it improves lung and heart health. Yes, it gives us bigger, stronger muscles. But more than any of that—and indeed, what is responsible for much of that—is a simple thing that happens at a much smaller scale: the cellular scale.

When researchers studied the telomeres in the blood cells of thousands of adults with all sorts of different exercise habits, they saw a striking correlation: those who exercised more had longer telomeres. And according to one study funded by the Centers for Disease Control and Prevention and published in 2017, individuals who exercise more—the equivalent of at least a half hour of jogging five days a week—have telomeres that appear to be nearly a decade younger than those who live a more sedentary life.[31] But why would exercising delay the erosion of telomeres?

If you think about how our longevity genes work—employing those ancient survival circuits—this all makes sense. Limiting food intake and reducing the heavy load of amino acids in most diets aren't the only ways to activate longevity genes that order our cells to shift into survival mode. Exercise, by definition, is the application of stress to our bodies. It raises NAD levels, which in turn activates the survival network, which turns up energy production and forces muscles to grow extra oxygen-carrying capillaries. The longevity regulators AMPK, mTOR, and sirtuins are all modulated in the right direction by exercise, irrespective of caloric intake, building new blood vessels, improving heart and lung health, making people stronger, and, yes, extending telomeres. SIRT1 and SIRT6, for example, help extend telomeres, then package them up so they are protected from degradation. Because it's not the absence of food or any particular nutrient that puts these genes into action; instead it is the hormesis program governed by the survival circuit, the mild kind of adversity that wakes up and mobilizes cellular defenses without causing too much havoc.

There's really no way around this. We all need to be pushing ourselves, especially as we get older, yet only 10 percent of people over the age of 65 do.[32] The good news is that we don't have to exercise for hours on end. One recent study found that those who ran four to five miles a week—for most people, that's an amount of exercise that can be done in less than 15 minutes per day—reduce their chance of death from a heart attack by 40 percent and all-cause mortality by 45 percent.[33] That's a massive effect.

In another study, researchers reviewed the medical records of more than 55,000 people and cross-referenced those documents with death certificates issued over fifteen years.[34] Among 3,500 deaths, they weren't particularly surprised to see that those who had told their doctors they were runners were far less likely to die of heart disease. Even when the researchers adjusted for obesity and smoking, the runners were less likely to have died during the years of the study. The big shock was that the health benefits were remarkably similar no matter how much running the people had done. Even about ten minutes of moderate exercise a day added years to their lives.[35]

There is a difference between a leisurely walk and a brisk run, however. To engage our longevity genes fully, intensity does matter. Mayo Clinic researchers studying the effects of different types of exercise on different age groups found that although many forms of exercise have positive health effects, it's high-intensity interval training (HIIT)—the sort that significantly raises your heart and respiration rates—that engages the greatest number of health-promoting genes, and more of them in older exercisers.[36]

You'll know you are doing vigorous activity when it feels challenging. Your breathing should be deep and rapid at 70 to 85 percent of your maximum heart rate. You should sweat and be unable to say more than a few words without pausing for breath. This is the hypoxic response, and it's great for inducing just enough stress to activate your body's defenses against aging without doing permanent harm.[37]

We're still working to understand what all of the longevity genes do, but one thing is already clear: many of the longevity genes that are turned on by exercise are responsible for the health benefits of exercise, such as extending telomeres, growing new microvessels that deliver oxygen to cells, and boosting the activity of mitochondria, which burn oxygen to make chemical energy. We've known for a long time that these bodily activities fall as we age. What we also know now is that the genes most impacted by exercise-induced stress can bring them back to the levels associated with youth. In other words: exercise turns on the genes to make us young again at a cellular level.

Often I'm asked, "Can I just eat what I want and run off the extra calories?" My answer is "Unlikely." When you give rats a high-calorie diet and allow them to burn off the energy, lifespan extension is minimal.

Same for a CR diet. If you make food filling but not as calorific, some of the health benefits are lost. Being hungry is necessary for CR to work because hunger helps turn on genes in the brain that release longevity hormones, at least according to a recent study by Dongsheng Cai at the Albert Einstein College of Medicine.[38]

Would a *combination* of fasting and exercise lengthen your lifespan? Absolutely. If you manage to do both these things: congratulations, you are well on your way.

But there is plenty more you can do.

THE COLD FRONT

Before arriving in Boston in my early 20s, I'd spent my whole life in Australia. Culturally, everything worked out just fine. Within a week, I'd figured out which markets carried Vegemite, the black yeast spread that some might say requires some pretty significant epigenetic programming as a child to enjoy as an adult. It took a bit longer to track down the best places for meat pies, Violet Crumble, Tim Tams, and musk sticks, but eventually I figured out how to get all of those tastes of home, too. And it didn't take long before I stopped caring that folks in the United States seem to have a hard time differentiating between Australian and British accents. (It's not that hard; Aussie accents are sexier.)

The toughest part was the cold.

As a boy, I thought I knew what cold was. When the temperature at Observatory Hill, Sydney's official weather station for more than a century, *approached* freezing (it hasn't actually fallen *below* freezing in modern history), *that* was cold.

Boston was a whole different world. A really frigid one.

I invested in coats, sweaters, and long underwear and spent a lot of time indoors. Like a lot of postdoctoral fellows, I often worked through the night. I truly was committed to my work, but the truth is that part of the calculus for not going home, on many nights, was that I didn't want to go outside.

These days I wish I'd taken a different approach. I wish I'd just told myself to tough it out. To take a walk in the bitter cold. To dip my toes into the Charles River in the middle of January. Because as it turns out, exposing your body to less-than-comfortable temperatures is another very effective way to turn on your longevity genes.

When the world takes us out of the thermoneutral zone—the small range of temperatures that don't require our bodies to do any extra work to stay warm or cool off—all sorts of things happen. Our breathing patterns shift. The blood flow to and through our skin—the largest organ in our body—changes. Our heart rates speed up or slow down. These reactions aren't happening just "because." All of these reactions have genetic roots dating back to *M. superstes*'s fight for survival all those billions of years ago.

Homeostasis, the tendency for living things to seek a stable equilibrium, is a universal biological principle. Indeed, it is the guiding force of the survival circuit. And thus we see it everywhere we look—especially on the low end of the thermometer.

As scientists have increasingly turned their attention to the impacts of reduced food intake on the human body, it has quickly become clear that calorie restriction has the effect of reducing core body temperature. It wasn't at first clear whether this contributed to prolonged vitality or was simply a by-product of all of the changes happening in the bodies of organisms exposed to this particular sort of stress.

Back in 2006, though, a team from the Scripps Research Institute genetically engineered some lab mice to live their lives a half degree cooler than normal—a feat they accomplished by playing a trick on the mice's biological thermostat. The team inserted copies of the mouse UCP2 gene into the mice's hypothalamus, which regulates the skin, sweat glands, and blood vessels. UCP2 short-circuited mitochondria in the hypothalamus so they produced less energy but more heat. That, in turn, caused the mice to cool down about half a degree Celsius. The result was a 20 percent longer life for female mice, the equivalent of about seven additional healthy human years, while male mice got an extension of 12 percent.[39]

COLD ACTIVATES LONGEVITY GENES. Sirtuins are switched on by cold, which in turn activates protective brown fat in our back and shoulders. Image: The author enduring "cold therapy" at the Massachusetts Institute of Technology in 1999.

The gene involved—which has a human analog—wasn't just a piece of the complex machinery that tricked the hypothalamus into thinking the mice's bodies were warmer than they were. It was also a gene that has been connected time and time again to longevity. Five years earlier, a joint team of researchers from Beth Israel Deaconess Medical Center and Harvard Medical School showed that mice age faster when their UCP2 gene is nullified.[40] And in 2005, Stephen Helfand and his team, then at the University of Connecticut Health Center, had demonstrated that targeted upregulation of an analogous gene could extend the lifespans of fruit flies by 28 percent in females and 11 percent in males.[41] Then, in 2017, the connection between the UCP2 gene and aging came full circle, thanks to researchers from Université Laval in Quebec: not only could UCP2 make mice "run cold," the Canadian team demonstrated, but colder temperatures could change the way the gene operated, too—through its ability to rev up brown adipose tissue.[42]

Also known as "brown fat," this mitochondria-rich substance was, until recently, thought to exist only in infants. Now we know that it is found in adults, too, although the amount of it decreases as we age. Over time, it becomes harder and harder to find; it mingles with white fat and is spread out even more unevenly across the body. It "hangs out" in different areas in different people, sometimes in the abdomen, sometimes across the upper back. That makes researching it in humans a bit of a challenge: it generally takes a PET scan—which requires the injection of radioactive glucose—to locate it. Rodent studies, however, have provided significant insights into the correlation between brown fat and longevity.

One study of genetically engineered Ames dwarf mice, for instance, demonstrated that the function of brown fat is enhanced in these remarkably long-lived animals.[43] Other studies have shown that animals with abundant brown fat or subjected to shivering cold for three hours a day have much more of the mitochondrial, UCP-boosting sirtuin, SIRT3, and experience significantly reduced rates of diabetes, obesity, and Alzheimer's disease.[44]

That is why we need to learn more about how to chemically substitute

for brown adipose tissue thermogenesis.[45] Chemicals called mitochondrial uncouplers can mimic the effects of UCP2, allowing protons to
leak through mitochondrial membranes, like drilling holes in a dam at a
hydroelectric plant. The result is not cold but heat as a by-product of the
mitochondrial short circuit.

The sweet-smelling mitochondrial uncoupler called 2,4-dinitrophenol
(DNP) was used for making explosives in the First World War, and it
soon became apparent that employees exposed to the chemical were rapidly losing weight, with one employee even dying from overexposure.[46]
In 1933, doctors Windsor Cutting and Maurice Tainter, from the Stanford University School of Medicine, summarized a series of their papers
showing that DNP markedly increases metabolic rate.[47] That same year,
despite Tainter and Cutting's warnings about "certain potential dangers,"
twenty companies started selling it in the United States, as did others in
Great Britain, France, Sweden, Italy, and Australia.

It worked well—too well, in fact.

Just one year later, speaking before the American Public Health Association, Tainter said, "The interest in and enthusiasm for this product were so great that its wide-spread use has become a matter of some
concern in public health. The total amount of the drug being used is
astonishing."

Moments later he dropped a bombshell: "during the past year, the
Stanford Clinics have supplied . . . over 1,200,000 capsules of dinitrophenol of 0.1 gm. each."[48]

Over 1 million capsules? From one university? In one year? That *is*
astonishing. And that was in 1933, when California had an eighth of its
present population. Three pounds of weight per person per week were reportedly being shed. The public was relieved—something *finally* worked.
Obesity was going to be a thing of the past.

But the metabolic party didn't last long. People began to die from
overdoses, and other long-term side effects showed up. DNP was declared "extremely dangerous and not fit for human consumption" in
the United States Federal Food, Drug, and Cosmetic Act of 1938. As a

curious aside, the legislation was written by Senator Royal Copeland, a homeopathic physician who, only days before he died, entrenched protections for natural supplements that today fuel a largely unregulated industry with revenues of $122 billion.

The act rightly banned a dangerous substance but dashed hopes that obesity would be a thing of the past.[49] Anecdotally, DNP continued to be prescribed to Russian soldiers during World War II to keep them warm,[50] and today some unscrupulous people sell it on the internet. But they do so at their peril. In 2018, Bernard Rebelo was sentenced to seven years in prison for the death of a woman to whom he sold DNP. In the United States, there have been sixty-two documented deaths since 1918, though there were likely many more than that.[51]

One thing is clear: DNP is extremely dangerous. Eating less at each meal, moving more, and focusing on plant-based foods are much safer options.

Another thing you can try is activating the mitochondria in your brown fat by being a bit cold. The best way to do this might be the simplest—a brisk walk in a T-shirt on a winter day in a city such as Boston will do the trick. Exercising in the cold, in particular, appears to turbocharge the creation of brown adipose tissue.[52] Leaving a window open overnight or not using a heavy blanket while you sleep could help, too.

This hasn't gone unnoticed by the health and wellness industry. Being cold is hot right now. Cryotherapy—a few minutes in a box superchilled to –110°C or –166°F—is an increasingly popular method of inducing a helping of this sort of stress to our bodies, although the research is still a ways away from being conclusive as to how, why, and even whether it truly works.[53] That didn't stop me from accepting an invitation from Joe Rogan, the media mogul and comedian, to go with him to a cryotherapy spa. Three minutes standing in my underwear at Mars temperatures may have activated my brown fat and all the great health benefits that go with that. At the very least, it left me invigorated and grateful to be alive.

As with most things in life, it's probably best to change your lifestyle when you are young, because making brown fat becomes harder as you get older. If you choose to expose yourself to the cold, moderation will be key. Similar to fasting, the greatest benefits are likely to come for those who get close to, but not beyond, the edge. Hypothermia is not good for our health. Neither is frostbite. But goose bumps, chattering teeth, and shivering arms aren't dangerous conditions—they're simply signs that you're not in Sydney. And when we experience these conditions often enough, our longevity genes get the stress they need to order up some additional healthy fat.

What happens on the other side of the thermostat? The picture is a bit less clear, but we have some promising leads from our friend *S. cerevisiae*. We know from work in my lab that raising the temperature of yeast—from 30°C to 37°C, just below the limits of what those single-celled organisms can sustain—turns on the *PNC1* gene and boosts their NAD production, so their Sir2 proteins can work that much harder. What's fascinating is not so much that these temperature-stressed cells lived 30 percent longer but that the mechanism was the same as that evoked by calorie restriction.

Is heat good for human bodies, too? Possibly, but not exactly in the same way. Because we are warm-blooded animals, our enzymes haven't evolved a tolerance for large changes in temperature. You can't just raise your core body temperature and expect to live longer. But as my northern German wife, Sandra, likes to point out, there are a lot of benefits to exposing your skin and lungs to high temperatures, at least temporarily.

Continuing an ancient Roman tradition, many northern and eastern Europeans regularly partake in "sauna bathing" for relaxation and health reasons. The Finns are the most dedicated, with the majority of men reporting using a sauna once a week, year round. Sandra tells me it's pronounced "ZOW-na" not "saw-nah," and that no home should be without one. I'm sticking with saw-nah to avoid sounding like a snicklefritz, but when it comes to housing construction, Sandra may be on to something.

A 2018 study conducted in Helsinki found that "physical function, vitality, social functioning, and general health were significantly better among sauna users than non-users," although the researchers were correct to point out that part of the effect could be due to the fact that those who are sick or disabled don't go to the sauna.[54]

A more convincing study followed a group of more than 2,300 middle-aged men from eastern Finland for more than twenty years.[55] Those who used a sauna with great frequency—up to seven times a week—enjoyed a twofold drop in heart disease, fatal hearts attacks, and all-cause mortality events over those who heat bathed once per week.

None of the sauna studies dug deep enough to tell us why temporary heat exposure may be so good for us. If yeast is any guide, NAMPT, the gene in our bodies that recycles NAD, may be in on the act. NAMPT is turned on by a variety of adversity triggers, including fasting and exercise, which makes more NAD so the sirtuins can work hard at making us healthier.[56] We have never tested if NAMPT is turned on by heat, but that would be something to do. Either way, one thing is clear: it does us little good to spend our entire lives in the thermoneutral zone. Our genes didn't evolve for a life of pampered comfort. A little stress to induce hormesis once in a while likely goes a long way.

But dealing with biological adversity is one thing. Overwhelming genetic damage is another.

DON'T ROCK THE LANDSCAPE

A bit of adversity or cellular stress is good for our epigenome because it stimulates our longevity genes. It activates AMPK, turns down mTOR, boosts NAD levels, and activates the sirtuins—the disaster response teams—to keep up with the normal wear and tear that comes from living on planet Earth.

But "normal" is the operative word, because when it comes to aging, "normal" is bad enough. When our sirtuins have to respond to many disasters—especially those that cause double-strand DNA breaks—these epigenetic signalers are forced to leave their posts and head to other places on the genome where DNA breaks have occurred. Sometimes they make their way back home. Sometimes they don't.

We can't prevent all DNA damage—and we wouldn't want to because it's essential for the function of the immune system and even for consolidating our memories[57]—but we do want to prevent extra damage.

And there's a lot of extra damage to be had out there.

Cigarettes, for starters. There aren't many legal vices out there that are worse for your epigenome than the deadly concoction of thousands of chemicals smokers put into their bodies every day. There's a reason why smokers seem to age faster: they *do* age faster. The DNA damage that results from smoking keeps the DNA repair crews working overtime, and likely the result is the epigenetic instability that causes aging. And although I'm not likely to be the first person you'll hear this from, it nonetheless bears repeating: smoking is not a private, victimless activity. The levels of DNA-damaging aromatic amines in cigarette smoke are about fifty to sixty times as high in secondhand as in firsthand smoke.[58] If you do smoke, it is worth trying to quit.

Don't smoke? That's great, but even without smoke there's fire. In much of the developed world—and increasingly in the developing world as well—we're practically bathing in DNA-damaging chemicals. In some places—cities with lots of people and lots of cars, especially—the simple act of breathing is enough to do extra damage to your DNA. But it would also be wise to be wary of the PCBs and other chemicals found in plastics, including many plastic bottles and take-out containers.[59] (Avoid microwaving these; it releases even more PCBs.) Exposure to azo dyes, such as aniline yellow, which is used in everything from fireworks to the yellow ink in home printers, can also damage our DNA.[60] And organohalides—compounds that contain substituted halogen atoms and are used in solvents, degreasers, pesticides, and hydraulic fluid—can also wreak havoc on our genomes.

Nobody in his right mind would purposefully ingest solvents, degreasers, pesticides, and hydraulic fluid, of course, but there's plenty of damage to be had in some of the things we do intentionally eat and

drink. We've known for more than half a century that N-nitroso compounds are present in food treated with sodium nitrite, including some
beers, most cured meats, and especially cooked bacon. In the decades
since, we've learned that these compounds are potent carcinogens.[61]
What we've also come to understand is that cancer is just the start of
our nitrate-treated woes, because nitroso compounds can inflict DNA
breakage as well[62]—sending those overworked sirtuins back to work
some more.

Then there's radiation. Any source of natural or human-inflicted
radiation, such as UV light, X-rays, gamma rays, and radon in homes
(which is the second most frequent cause of lung cancer besides smoking[63]) can cause additional DNA damage, necessitating the call-up of an
epigenetic fix-it team. As someone who flies a lot for work, I think about
this quite a bit—every time I go through security, in fact. Most of the
research on the current versions of airport scanners suggests that they
probably don't do tremendous damage to our DNA, but there's been
little attention given to their long-term impact on our epigenome and
the aging process. No one has ever tested what a mouse looks like two
years after being repeatedly exposed to these devices. The ICE mice tell
us that chromosome tickling is all that's needed to accelerate aging. I'm
aware the radiation exposure from millimeter-wave scanners is lower
than that from previous scanners. The security attendants at the machine
tell travelers the exposure is about the "same as the flight." But with millions of flight miles under my belt, why would I want to double the
damage? Whenever possible, I take the pre-check line or ask for a pat
down instead.

If all of this makes you feel as if it's impossible to completely avoid
DNA breaks and the epigenetic consequences of those breaks, well, that's
true. The natural and necessary act of replicating DNA causes DNA
breaks, trillions of them throughout your body every day. You can't avoid
radon particles or cosmic rays unless you live in a lead box at the bottom
of the ocean. And even if you were to move to a desert island, the fish
you'd have to eat would likely contain mercury, PCBs, PBDEs, dioxins,

and chlorinated pesticides, all of which can damage your DNA.[64] In our modern world, even with the most "natural" lifestyle you can follow, this sort of DNA damage is inevitable.

No matter how old you are, even if you are a teenager, it is already happening to you.[65] DNA damage has accelerated your clock, with implications at all stages of life. Embryos and babies experience aging. What, then, of people in their 60s, 70s, and 80s? What of those individuals who are already frail and cannot restrict their calories, go for a run, or make snow angels in the dead of winter? Is it too late for them?

Not at all.

But if we're *all* going to live longer and healthier lives—regardless of how much epigenetic drift and aging we have experienced at this moment in time—we might need some additional help.

FIVE

A BETTER PILL TO SWALLOW

THE DREAM OF EXTENDING HUMAN LIVES DID NOT BEGIN IN THE EARLY TWENTY-first century any more than the dream of human flight began in the early twentieth. Nothing begins with science; it all begins with stories.

From Gilgamesh the Sumerian king, who is said to have reigned over Uruk for 126 years, to Methuselah the patriarch in Hebrew scriptures, who is said to have lived to the age of 969, humanity's sacred stories testify to our deep-seated fascination with longevity. Outside myths and parables, though, we had little scientific evidence of anyone succeeding in extending their life far beyond the single century mark.

We had little hope of doing so without a deep understanding of how life works. That is knowledge, albeit still imperfect, that some of my colleagues and I believe we finally possess.

It wasn't until 1665 that "England's Leonardo," Robert Hooke, published *Micrographia*, in which he reported seeing cells in cork bark. That discovery launched us into the modern era of biology. But centuries would pass before we had any clue about how cells work at the molecular scale. That knowledge could come only from the combination of a series of great leaps in microscopy, chemistry, physics, genetics, nanoengineering, and computing power.

To understand how aging occurs, we must journey down into the subcellular nanoworld, heading down to the cell, piercing the outer membrane, and traveling into the nucleus. From there, we head down to the scale of amino acids and DNA. At this size, it is obvious why we don't live forever.

Until we understood life at the nanoscale, even why we live was a mystery. The brilliant Austrian theoretical physicist Erwin Schrödinger, the man who developed quantum physics (and yes, that famous thought experiment involving a both-dead-and-alive cat) was flummoxed when he tried to explain life. In 1944, he threw up his hands and declared that living matter "is likely to involve 'other laws of physics' hitherto unknown."[1] That was the best he could do at the time.

But things moved quickly in the decades to come. And today, the answer to Schrödinger's 1944 book, *What Is Life?*, if not fully answered, is certainly close to being so.

Turns out, there is no new law required to explain life. At the nanoscale, it is merely an ordered set of chemical reactions, concentrating and assembling atoms that would normally never assemble, or breaking apart molecules that would normally never disintegrate. Life does this using proteinaceous Pac-Men called enzymes made up of coils and layered mats of amino acid chains.

Enzymes make life possible by taking advantage of fortuitous molecular movements. Every second you are alive, thousands of glucose molecules are captured within each of your trillions of cells by an enzyme called glucokinase, which fuses glucose molecules to phosphorus atoms, tagging them for energy production. Most of the energy created is used by a multicomponent RNA and protein complex called a ribosome, whose primary job is to capture amino acids and fuse them with other amino acids to make fresh proteins.

Does this sort of talk make your eyes gloss over? You are not alone, and you are not to blame. We teachers have done society a great disservice by making cool science boring. Textbooks and scientific papers depict biology as a static, two-dimensional world. Chemicals are drawn as sticks, biochemical pathways are arrows, DNA is a line, a gene is a rectangle, and enzymes are ovals, drawn thousands of times larger relative to the cell than they actually are.

But once you understand how cells actually work, they are the most amazing things. The problem with conveying this wonder in a classroom is that cells exist in four dimensions and buzz around with speeds and on scales we humans cannot perceive or even conceive. To us, the second and the millimeter are short divisions of time and space, but to an enzyme about 10 nanometers across and vibrating every quadrillionth of a second, a millimeter is the size of a continent and a second is more than a year.[2]

Consider catalase, a ubiquitous, regular-sized enzyme that can break apart and detoxify 10,000 molecules of hydrogen peroxide per second. A million of them could fit inside an *E. coli* bacterium, a million of which could fit on the head of a pin.[3] These numbers aren't just hard to imagine; they are inconceivable.

In each cell are a total of 75,000 enzymes like catalase,[4] all thrown together, jostling around in a slightly salty sea. At the nanoscale, water is gelatinous, and molecular events are more violent than a category 5 hurricane, with molecules thrown together at speeds we would perceive as a thousand miles per hour. Enzymatic reactions are one-in-a-thousand events, but at the nanoscale one-in-a-thousand events can occur thousands of times a second, enough to sustain life.

If this sounds chaotic, it is, but we *need* this chaos for order to emerge. Without it, the molecules that must come together to sustain life would not find each other, and they would not fuse. The human sirtuin enzyme called SIRT1 serves as a good example. Precise vibrating sockets on SIRT1 simultaneously clasp onto an NAD molecule and the protein it wants to strip the acetyls from, such as a histone or FOXO3. The two captured molecules immediately lock together, just before SIRT1 rips them apart in a different way, producing vitamin B_3 and acetylated adenine ribose as waste products that are recycled back to NAD.

More important is the fact that the target protein has now been stripped of the acetyl chemical group that was holding it at bay. Now the histone can pack DNA more tightly to silence genes, and FOXO3 has had its shackles removed, allowing it to go turn on a defense program of protective genes.

If the chaos ended and our enzymes suddenly stopped doing what they do, we would all be dead within a few seconds. Without energy and cell defenses, there can be no life. *M. superstes* would never have emerged from the scum and its descendants would never have been capable of comprehending the words on this page.

And so, at the fundamental level, life is rather simple: we exist by the grace of an order created from chaos. When we toast to life, we really should be toasting to enzymes.

By studying life at this level, we've also learned something rather important—something the Nobel Prize–winning physicist Richard Feynman expressed succinctly: "There is nothing in biology yet found that indicates the inevitability of death. This suggests to me that it is not at all inevitable and that it is only a matter of time before biologists discover what it is that is causing us the trouble."[5]

It's true: there are no biological, chemical, or physical laws that say life must end. Yes, aging is an increase in *entropy*, a loss of information leading to disorder. But living things are *not* closed systems. Life can potentially last forever, as long as it can preserve critical biological information and absorb energy from somewhere in the universe. This doesn't mean we could be immortal tomorrow—no more than we could have flown to the moon on December 18, 1903. Science moves forward with small steps and big steps, but always one step at a time.

Here's the remarkable thing: the first steps have actually been available to us since the times of Gilgamesh and Methuselah, and indeed from the time of *M. superstes*. And, in the past few centuries, and by accident even earlier than that, we have discovered ways to chemically modulate enzymes with molecules we call medicines.

Now that we know how life works and have the tools to change it at a genetic and epigenetic level, we can build upon this very old wisdom. And when it comes to the goal of extending healthy lifespans, the easiest measures to use are the various drugs that we already know can impact human aging.

THE WORLD'S GREATEST EASTER EGG

Rapa Nui, a remote volcanic island 2,300 miles west of Chile, is commonly known as Easter Island and even better known for the nearly nine hundred giant stone heads that line the island's perimeter. What should be just as well known—and perhaps one day will be—is the story of how the island came to be the source of the world's most effective lifespan-extending molecule.

Back in the mid-1960s, a team of scientists traveled to the island. The researchers were not archaeologists seeking answers about the origins of the *moai* statues but rather biologists looking for endemic microorganisms.

In the dirt beneath one of the island's famed stone heads, they discovered a new actinobacterium. That single-celled organism was *Streptomyces hygroscopicus*, and when it was isolated by a pharmaceutical researcher, Suren Sehgal, it soon became clear that the actinobacterium secreted an antifungal compound. Sehgal named that compound rapamycin, in honor of the island where it was discovered, and began looking for ways to process it as a potential remedy for fungal conditions such as athlete's foot.[6] The compound looked promising for that purpose, but when the Montreal lab where Sehgal worked was shuttered in 1983, he was directed to destroy the compound.

He couldn't bring himself to do that, though. Instead he spirited a few vials of the bacterium out of the lab and kept them in his freezer at home until the late 1980s, when he convinced his bosses at a new lab in New Jersey to let him resume studying it.

It wasn't long before researchers discovered that the compound was an effective suppressor of the immune system. That would end its potential as an antifungal—there are plenty of remedies for athlete's foot that don't come at the cost of lowered immunity—but it gave scientists a new attribute to study.

Even in the 1960s, researchers knew that one of the most common reasons for an organ transplant to fail is that the recipient patient's body rejects it. Could rapamycin lower the immune response enough to ensure the organ would be accepted? Indeed it could.

It is for this reason that if you were to make a pilgrimage to Rapa Nui, you might come upon a small plaque at the site where *S. hygroscopicus* was discovered. "At this site," the plaque reads in Portuguese, "soil samples were obtained in January 1965 that allowed the production of rapamycin, a substance that inaugurated a new era for patients who need organ transplants."

I suspect that a larger plaque may soon be in order, because the discovery of *S. hygroscopicus* set into motion a tremendous amount of research, much of which is still ongoing and some of which has the potential to prolong vitality for countless other people. Because in recent years it has become clear that rapamycin isn't just an antifungal compound and it isn't just an immune system suppressor; it's also one of the most consistently successful compounds for extending life.

We know this from experiments on a diverse menagerie of model organisms in labs around the world. And much as my own research began with experiments with yeast, much of the initial work that has been done to understand rapamycin was completed on *S. cerevisiae*. If you put 2,000 normal yeast cells into a culture, a few will remain viable after six weeks. But if you feed those yeast cells rapamycin, in six weeks about half will still be healthy.[7] The drug will also increase the number of daughter cells mothers can produce by stimulating the production of NAD.

Fruit flies fed rapamycin live about 5 percent longer.[8] And small doses of rapamycin given to mice when they are already in the final months of their normal lives results in 9 to 14 percent longer lives, depending on whether they are male or female, which translates to about a decade of healthy human life.[9]

We've known for a long time that greater parental age is a risk factor for disease in the next generation. That's the power of epigenetics. But mice treated with rapamycin buck this trend. When researchers from the German Center for Neurodegenerative Diseases inhibited mTOR in mice born to older fathers, the negative impact of having an old parent went away.[10]

Want to know what the world's most prominent arbiters of great science think about the potential of TOR and the molecules that inhibit it to change the world? The three men who discovered TOR in yeast, Joseph Heitman, Michael Hall, and Rao Movva, are on a lot of people's shortlists for the Nobel Prize in Medicine or Physiology. My colleague across the river at MIT, David Sabatini, who identified mTOR, was named a Clarivate Citation Laureate for having his work cited most frequently in top-tier peer-reviewed journals; the Clarivate list has predicted more than forty Nobel Prize winners since 2002.[11]

Rapamycin isn't a panacea. Longer-lived animals might not fare as well on it as shorter-lived ones do; it's been shown to be toxic to kidneys at high doses over extended periods of time; and it might suppress the immune system over time. That doesn't mean TOR inhibition is a dead end, though. It might be safe in small or intermittent doses—that worked in mice to extend lifespan[12] and in humans dramatically improved the immune responses of elderly people to a flu vaccine.[13]

There are hundreds of researchers from the TOR inhibition side of the family working in universities and biotech companies to identify "rapalogs," which are compounds that act on TOR in ways similar to rapamycin but have greater specificity and less toxicity.[14]

The quality of the people involved in this line of research and development makes it hard to bet against TOR inhibition as a pathway to greater human health and vitality. But even if rapalogs don't pan out, there's another pharmaceutical pathway to prolonged vitality that has already proven to be both effective and relatively safe.

PENNIES FOR PROLONGED VITALITY

Galega officinalis is a lovely flower, with stacks of delicate purple petals that seem locked in a reverent bow to the world.

Also known as goat's rue, a rather unfortunate name, and French lilac, a far more charming sobriquet, it has been used as an herbal medicine in Europe for centuries, owing to a chemical composition rich in guanidine, a small chemical in human urine that serves as an indicator of healthy protein metabolism. In the 1920s, doctors began to prescribe guanidine as a way to lower blood glucose levels in patients with diabetes.

In 1922, a 14-year-old boy named Leonard Thompson, who was dying in a Toronto hospital, became the first diabetic patient to be given an injection of a novel pancreatic peptide hormone that had shown great promise in animal studies. Two weeks later he was given another, and news of his exceptional improvement spread quickly around the world. Type 1 diabetes, which occurs when the pancreas doesn't produce enough of the hormones needed to alert the body to sugar, is now widely treated by supplemental insulin. But the fight was not over.

The type 2 version of the disease, so-called age-associated diabetes, occurs when the pancreas is able to make enough insulin but the body is deaf to it. The 9 percent of all adults globally with this disease need a drug that restores their body's sensitivity to insulin so cells take up and use the sugar that's coursing through their bloodstreams. That's important for at least two reasons: it gives the overworked pancreas a rest, and it prevents spikes of freely floating sugar from essentially caramelizing proteins in the body. Recent results indicate high blood sugar can also speed up the epigenetic clock.

Thanks to an increasingly sedentary lifestyle and the abundance of sugars and carbohydrates on every supermarket shelf around the globe, high blood sugar is causing the premature deaths of 3.8 million people a year. These deaths do not come quickly and compassionately but in horrific ways, with blindness, kidney failure, stroke, open foot wounds, and limb amputations.

As they considered this disease in the mid-1950s, the pharmacist Jan Aron and the physician Jean Sterne—both Frenchmen who would have been exceptionally familiar with the purple-flowering plant so ubiquitous in their native land—decided to reinvestigate the potential of French lilac derivatives to fight type 2 diabetes in ways insulin doesn't.[15]

In 1957, Sterne published a paper demonstrating the effectiveness of oral dimethyl biguanide to treat type 2 diabetes. The drug, now most commonly called metformin, has since become one of the most widely taken and effective medicines on the globe. It's among the medications on the World Health Organization's Model List of Essential Medicines, a catalog of the most effective, safe, and cost-effective therapies for the world's most prevalent medical conditions. As a generic medication, it costs patients less than $5 a month in most of the world. Except for an extremely rare condition called lactic acidosis, the most common of the side effects is some stomach discomfort. Many people mitigate that side effect by taking the medication as a coated tablet or with a glass of milk or a meal, but even when that doesn't work, the mild upset feeling comes with a bit of a side benefit: it tends to discourage overeating.

What place does a diabetes medication have in a conversation about prolonging vitality? Perhaps it would have no place at all if not for the fact that, a few years ago, researchers noticed a curious phenomenon: people taking metformin were living notably healthier lives—independent, it seemed, of its effect on diabetes.[16]

In mice, even a very low dose of metformin has been shown by Rafael de Cabo's lab at the National Institutes of Health to increase lifespan by nearly 6 percent, though some have argued that the effect is due mostly to weight loss.[17] Either way, that amounts to the equivalent of five extra healthy years for humans, with an emphasis on healthy—the mice showed reduced LDL and cholesterol levels and improved physical performance.[18] As the years have gone by, the evidence has mounted. In twenty-six studies of rodents treated with metformin, twenty-five showed protection from cancer.[19]

Like rapamycin, metformin mimics aspects of calorie restriction. But instead of inhibiting TOR, it limits the metabolic reactions in mitochondria, slowing down the process by which our cellular powerhouses convert macronutrients into energy.[20] The result is the activation of AMPK, an enzyme known for its ability to respond to low energy levels and restore the function of mitochondria. It also activates SIRT1, one of my lab's favorite proteins. Among other beneficial effects, metformin inhibits cancer cell metabolism, increases mitochondrial activity, and removes misfolded proteins.[21]

A study of more than 41,000 metformin users between the ages of 68 and 81 concluded that metformin reduced the likelihood of dementia, cardiovascular disease, cancer, frailty, and depression, and not by a small amount. In one group of already frail subjects, metformin use over the course of nine years reduced dementia by 4 percent, depression by 16 percent, cardiovascular disease by 19 percent, frailty by 24 percent, and cancer by 4 percent.[22] In other studies, the protective power of metformin against cancer has been far greater than that. Though not all cancers are suppressed—prostate, bladder, renal, and esophageal cancer seem recalcitrant—more than twenty-five studies have shown a powerful protective effect, sometimes as great as a 40 percent lower risk, most notably for lung, colorectal, pancreatic, and breast cancer.[23]

These aren't just numbers. These are people whose lives were markedly improved by using a single, safe drug that costs less than a cup of bad coffee.

If all metformin could do was reduce cancer incidence, it would still be worth prescribing widely. In the United States, the lifetime risk of being diagnosed with cancer is greater than 40 percent.[24] But there's a dividend beyond preventing cancer directly, a side effect of living longer that most people don't consider: after age 90, your chances of dying of cancer drop considerably.[25] Of course, people will still die of other conditions, but the tremendous pain and costs associated with cancer would be significantly mitigated.

The beauty of metformin is that it impacts many diseases. Through the power of AMPK activation, it makes more NAD and turns on sirtuins and other defenses against aging as a whole—engaging the survival circuit upstream of these conditions, ostensibly slowing the loss of epigenetic information and keeping metabolism in check, so all organs stay younger and healthier.

Most of us assume that the effects of a pill like metformin would take years to produce any appreciable effect on aging, but maybe not. An admittedly small study of healthy volunteers claimed that the DNA methylation age of blood cells is reversed within a week and, astoundingly, only ten hours after taking a single 850 mg pill of metformin.[26] But clearly more work is needed with greater numbers of subjects to know for sure if metformin can delay the aging clock over the long run.

In most countries, metformin isn't yet prescriptible as an antiaging drug, but for the hundreds of millions of people around the world who are diabetic, it's not a hard prescription to get. In some places, such as Thailand, metformin is even available over the counter at every pharmacy—for just a few cents a pill. In the rest of the world, even if you have prediabetes, it can be challenging to convince a doctor to prescribe you metformin. If you've been good to your body, and greater than 93.5 percent of your blood's hemoglobin isn't irreversibly bound to glucose—meaning it's mostly the HbA1 type not HbA1c—you're out of luck, not just because the majority of physicians don't know the data I just shared with you, but because even if they did, aging isn't yet considered a disease.

Among the people taking metformin—and leading the charge to evaluate its long-term effects on aging in humans—is Nir Barzilai, the Israeli American physician and geneticist who, along with his colleagues at Albert Einstein College of Medicine, discovered several longevity gene variants in the insulin-like growth hormone receptor that controls FOXO3, the cholesterol gene CETP, and the sirtuin SIRT6, all of which seem to help ensure that some lucky people with Ashkenazi Jewish ancestry remain healthy beyond 100.

Yes, although genes play a back-seat role to the epigenome, it does seem that some people are genetically primed for longevity at the digital level—enjoying longer lives almost irrespective of how they live, thanks in part to gene variants that stabilize their epigenomes, preventing the loss of analog information over time. But Barzilai doesn't see these people as winners so much as markers—they represent the potential that most other humans have for long and healthy lives—and he is fond of pointing out that even if we were never to extend lives *past* 120, we know that 120 is possible. "So for most of us," he has told me, "there are 40 good years still on the table."

Barzilai is leading the charge to make metformin the first drug to be approved to delay the most common age-related diseases by addressing their root cause: aging itself. If Barzilai and his colleagues can show metformin has measurable benefits in the ongoing Targeting Aging with Metformin (TAME) study, the US Food and Drug Administration has agreed to consider aging as a treatable condition. That would be a game changer, the beginning of the end for a world in which aging is "just the way it goes."

Barzilai believes that day is coming. He has predicted that the traditional Hebrew blessing *"Ad me'ah ve-essrim shana,"* or "May you live until 120," may soon need updating, for it will be a wish not for a long life but for a very *average* one.

STAC IT UP

Back in 1999, the story of the sirtuin longevity pathway we discovered in Lenny Guarente's lab at MIT was about to get even hotter.

We had finally figured out a molecular cause of aging in yeast cells, the first for any species. We were still feeling the glow scientists get when they publish new work that shows how smart they are. In a series of prominent papers that had captured the imagination of the scientific community, we'd reported that the cause of yeast aging was the movement of Sir2 away from the mating-type genes to deal with DNA breaks and a whole lot of ensuing genome instability.[27] We'd shown that extra copies of the *SIR2* gene could stabilize the rDNA and extend lifespan. We'd linked genetic instability to epigenetic instability and found one of the world's first true longevity genes—and the yeast hadn't had to go hungry to receive its benefits.

But splicing extra copies of a gene into a single-celled organism is a much easier endeavor than putting those copies into more complex creatures. It's also far less ethically complicated. That's why a few other researchers and I entered a scientific race to find ways to ramp up sirtuin activity in mammals without inserting extra sirtuin genes.

Here is where science becomes a matter of logical guesswork and some good old-fashioned luck. Because there are more than 100 million chemicals known to science. Where do you even start?

Thankfully, Konrad Howitz was on the case. The Cornell-educated biochemist was then the director of molecular biology for Biomol, a Pennsylvania company that was a supplier of molecules for life science researchers. Howitz was looking for chemicals that would inhibit the SIRT1 enzyme, so they could be sold to the growing number of scientists who were starting to study the enzyme. In the process of evaluating different contenders, he found two chemicals that, rather than inhibiting SIRT1, stimulated or "activated" it, making it work ten times as fast. That was a serendipitous discovery, not only because he was expecting to find inhibitors but because activators are very rare in nature. They are so rare, in fact, that most drug companies don't even bother following up when one is discovered, figuring it must be a mistake.

The first SIRT1-activating compound, or STAC, was a polyphenol called fisetin, which helps gives plants such as strawberries and persimmons their color and is now known to also kill senescent cells. The second was a molecule called butein, which can be found in numerous flowering plants as well as a toxic plant known as the Chinese lacquer tree. Both had a significant effect on SIRT1, though not the sort of pedal-to-the-metal reaction that might make them ripe for further research.

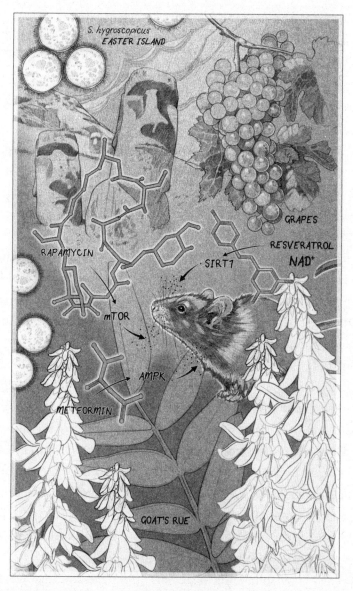

THE THREE MAIN LONGEVITY PATHWAYS, mTOR, AMPK, AND SIRTUINS, EVOLVED TO PROTECT THE BODY DURING TIMES OF ADVERSITY BY ACTIVATING SURVIVAL MECHANISMS. When they are activated, either by low-calorie or low-amino-acid diets, or by exercise, organisms become healthier, disease resistant, and longer lived. Molecules that tweak these pathways, such as rapamycin, metformin, resveratrol, and NAD boosters, can mimic the benefits of low-calorie diets and exercise and extend the lifespan of diverse organisms.

Howitz showed his initial results to Biomol's founder and scientific
director, Robert Zipkin, a brilliant chemist and entrepreneur who has
an encyclopedic knowledge of chemical structures. "Fisetin and butein,
huh?" Zipkin said. "You know what those two molecules look like?
They've got an overlapping structure: two phenolic rings connected by a
bridge. You know what else has that structure? Resveratrol."

In 2002, antioxidants were all the rage. They might not have been
the antiaging and health panaceas some believed them to be, but that
wasn't yet known. One of the antioxidants, scientists from the Karol
Marcinkowski University of Medical Sciences (now Poznan University
of Medical Sciences in Poland) had learned, was resveratrol, a natural
molecule that is found in red wine and that many plants produce in
times of stress.[28] A few researchers had suggested that resveratrol might
explain the "French paradox," the fact that the French have lower rates of
heart disease, even though their diet is relatively high in foods containing
saturated fat, such as butter and cheese.

Zipkin's guess that resveratrol might have a similar effect as fisetin
and butein was right on the money. When I studied it in my lab at Harvard, I saw that it actually far outperformed the other two molecules.

As a reminder, aging in yeast is often measured by the number of
times a mother cell divides to produce daughter cells. In most cases, a
yeast cell gets to about twenty-five divisions before it dies. Because the
experiments required a week of micromanipulation of cells while looking
down a microscope, and the fewer times you put the cells in the fridge to
get some sleep, the longer the yeast cells live, I assembled a lab at home
on my dining room table.

There, I saw something incredible: the resveratrol-fed yeast were
slightly smaller and grew slightly more slowly than untreated yeast, getting to an average of thirty-four divisions before dying, as though they
were calorie restricted. The human equivalent would be an extra 50 years
of life. We saw increases in maximum lifespan, too—on resveratrol, they
kept going past 35. We tested resveratrol in yeast cells with no *SIR2* gene,
and there was no effect. We tested it on calorie-restricted yeast, and saw

no further increase in lifespan, suggesting that the same pathway was being activated; this was how calorie restriction was working.

It seemed like a joke's punch line—not only had we found a calorie-restriction mimetic, something that could extend longevity without hunger, but we'd found it in a bottle of red wine.

Howitz and I were fascinated by the fact that resveratrol is produced in greater quantities by grapes and other plants experiencing stress. We also knew that many other health-promoting molecules, and chemical derivatives of them, are produced in abundance by stressed plants; we get resveratrol from grapes, aspirin from willow bark, metformin from lilacs, epigallocatechin gallate from green tea, quercetin from fruits, and allicin from garlic. This, we believe, is evidence of xenohormesis—the idea that stressed plants produce chemicals for themselves that tell their cells to hunker down and survive. Plants have survival circuits, too, and we think we might have evolved to sense the chemicals they produce in times of stress as an early-warning system, of sorts, to alert our bodies to hunker down as well.[29]

What this means, if it's true, is that when we search for new drugs from the natural world we should be searching the stressed-out ones: in stressed plants, in stressed fungi, and even in the stressed microbiome populations in our guts. The theory is also relevant to the foods we eat; plants that are stressed have higher concentrations of xenohormetic molecules that may help us engage our own survival circuits. Look for the most highly colored ones because xenohormetic molecules are often yellow, red, orange, or blue. One added benefit: they tend to taste better. The best wines in the world are produced in dry, sun-exposed soil or from stress-sensitive varietals such as Pinot Noir; as you might guess, they also contain the most resveratrol.[30] The most delectable strawberries are those that have been stressed by periods of limited water supply. And as anyone who has grown leaf vegetables can attest, the best heads of lettuce come when the plants are exposed to a one-two combo punch of heat and cold.[31] Ever wonder why organic foods, which are often grown under more stressful conditions, might be better for you?

Resveratrol extended the lifespan of simple yeast cells, but would it do the same for other organisms? When my fellow researcher Marc Tatar of Brown University visited me in Boston, I gave him a small vial of white, fluffy resveratrol powder—marked only with the letter *R*—to try on insects in his lab. He took it back to Rhode Island, mixed it with some yeast paste, and fed it to his fruit flies.

A few months later, I got a call from him. "David!" he said. "What is this R stuff?"

Under lab conditions, the fruit fly *Drosophila melanogaster* typically lives for an average of forty or so days. "We added a week to their lives and sometimes more than that," Tatar told me. "On average, they're living for more than fifty days."

In human terms, that's an additional fourteen years.

In my lab, resveratrol-fed roundworms also lived longer, an effect that required the worm sirtuin gene to be engaged. And when we gave resveratrol to human cells in culture dishes, they became resistant to DNA damage.

Later, when we fed resveratrol to obese mice at one year of age, something interesting happened: the mice stayed fat, causing postdoctoral fellow Joseph Baur, now a professor at the University of Pennsylvania, to conclude that I'd wasted more than a year of his time, jeopardizing his scientific career with a harebrained experiment. But when he and Rafael de Cabo, our collaborator at the NIH, opened up the mice, they were shocked. The resveratrol mice looked identical to mice on a normal diet, with healthy hearts, livers, arteries, and muscles. They also had more mitochondria, less inflammation, and lower blood sugar levels. The ones they didn't dissect wound up living about 20 percent longer than normal.[32]

Other researchers went on to show in hundreds of published studies that resveratrol protects mice against dozens of diseases, including a variety of cancers, heart disease, stroke and heart attacks, neurodegeneration, inflammatory diseases, and wound healing, and generally makes mice healthier and more resilient.[33] And in collaboration with de Cabo, we discovered that when resveratrol is combined with intermittent fasting, it can greatly extend both average and maximum lifespan even beyond what fasting alone accomplishes. Out of fifty mice, one lived more than 3 years—in human terms, that would amount to about 115 years.[34]

The first paper on resveratrol's effects on aging went on to be one of the most highly cited papers of 2006[35] and was widely circulated in the mainstream media, too. We were all over TV, and I was starting to be recognized in public. I ran off to the little German village called Burlo, where my wife was born, and the news had even made it there. Sales of red wine reportedly went up 30 percent. If you like red wine but needed a good excuse to imbibe, you can thank Rob Zipkin.[36]

On our kitchen wall hangs a variety of cartoons from the day. My favorite is one by Tom Toles. In it, a wife tries to downplay the enthusiasm of her very large husband, who covers most of a couch.

"The study said that to get the same dose as they gave the mice you'd have to drink between 750 and 1,000 glasses of red wine every day," the wife says.

"The news just keeps getting better and better," the husband replies.

As it turned out, resveratrol wasn't very potent and wasn't very soluble in the human gut, two attributes that most medicines need to be effective at treating diseases. Despite its limitations as a drug, it did serve as an important first proof that a molecule can give the benefits of calorie restriction without the subject having to go hungry, and it set off a global race to find other molecules that might delay aging. Finally, at least in scientific circles, slowing aging with a drug was no longer considered bonkers.

By studying resveratrol, we also learned that it is possible to activate sirtuins with a chemical. This prompted a flood of research into other sirtuin-activating compounds, called STACs, that are many times more potent than resveratrol at stimulating the survival circuit and extending healthy lifespans in animals. They go by names such as SRT1720 and SRT2104, both of which extend the healthy lifespan of mice when given to them late in life.[37] There are, today, hundreds of chemicals that have been demonstrated to have an effect on sirtuins that are even more effective than resveratrol's and some that have already been demonstrated in clinical trials to lower fatty acid and cholesterol levels, and to treat psoriasis in humans.[38]

Another STAC is NAD, sometimes written as NAD⁺.[39] NAD has an advantage over other STACs because it boosts the activity of all seven sirtuins.

NAD was discovered in the early twentieth century as an alcoholic fermentation enhancer. That was fortuitous: if it hadn't had the potential to improve the way we make booze, scientists might not have been so enamored by it. Instead they worked on it for decades, and in 1938 they had a breakthrough: NAD was able to cure black tongue disease in dogs, the canine equivalent of pellagra. It turned out that NAD is a product of the vitamin niacin, a severe lack of which causes inflamed skin, diarrhea, dementia, skin sores, and ultimately death. And because NAD is used by over five hundred different enzymes, without any NAD, we'd be dead in thirty seconds.

By the 1960s, however, researchers had concluded that all the interesting research on NAD had been done. For decades to come, NAD was simply a housekeeping chemical that teenage biology students had to learn about—with all the enthusiasm of a teenager doing housekeeping. That all changed in the 1990s, when we began to realize that NAD wasn't just keeping things running; it was a central regulator of many major biological processes, including aging and disease. That's because Shinichiro Imai and Lenny Guarente showed that NAD acts as fuel for sirtuins. Without sufficient NAD, the sirtuins don't work efficiently: they can't remove the acetyl groups from histones, they can't silence genes, and they can't extend lifespan. And we sure wouldn't have seen the lifespan-extending impact of the activator resveratrol. We and others also noticed that NAD levels decrease with age throughout the body, in the brain, blood, muscle, immune cells, pancreas, skin, and even the endothelial cells that coat the inside of microscopic blood vessels.

But because it's so central to so many fundamental cellular processes, no researchers in the twentieth century had any interest in testing the effects of boosting levels of NAD. "Bad stuff will happen if you mess with NAD," they thought. But not even having *tried* to manipulate it, they didn't really know what would happen if they did.

The benefit of working with yeast, though, is that the worst-case scenario in any experiment is a yeast massacre.

There was little risk in looking for ways to boost NAD in yeast. So that's what my lab members and I did. The easiest way was to identify the genes that make NAD in yeast. We first discovered a gene called *PNC1*, which turns vitamin B_3 into NAD. That led us to try boosting *PNC1* by introducing four extra copies of it into the yeast cells, giving them five copies in total. Those yeast cells lived 50 percent longer than normal, but not if we removed the *SIR2* gene. The cells were making extra NAD, and the sirtuin survival circuit was being engaged!

Could we do this in humans? Theoretically, yes. We already have the technology to do it in my lab, using viruses to deliver the human equivalent of the *PNC1* gene called NAMPT. But turning humans into transgenic organisms requires more paperwork and considerably more knowledge about safety—for the stakes are higher than a yeast massacre.

That's why we once again began searching for safe molecules that would achieve the same result.

Charles Brenner, who is now the head of biochemistry at the University of Iowa, discovered in 2004 that a form of vitamin B_3 called nicotinamide riboside, or NR, is a vital precursor of NAD. He later found that NR, which is found in trace levels in milk, can extend the lifespan of yeast cells by boosting NAD and increasing the activity of Sir2. Once a rare chemical, NR is now sold by the ton each month as a nutraceutical.

Meanwhile, on a parallel path, researchers, including us, were homing in on a chemical called nicotinamide mononucleotide, or NMN, a compound made by our cells and found in foods such as avocado, broccoli, and cabbage. In the body, NR is converted into NMN, which is then converted into NAD. Give an animal a drink with NR or NMN in it,[40] and the levels of NAD in its body go up about 25 percent over the next couple of hours, about the same as if it had been fasting or exercising a great deal.

My friend from the Guarente lab Shin-ichiro Imai demonstrated in 2011 that NMN could treat the symptoms of type 2 diabetes in old mice by restoring NAD levels. Then researchers in my lab at Harvard showed we could make the mitochondria in old mice function just like mitochondria in young mice after just a week of NMN injections.

In 2016, my other lab at the University of New South Wales collaborated with Margaret Morris to demonstrate that NMN treats a form of type 2 diabetes in obese female mice and their diabetes-prone offspring. And back at Harvard, we found that NMN could give old mice the endurance of young mice and then some, leading to the Great Mouse Treadmill Failure of 2017, when we had to reset the tracking program on our lab's miniature exercise machines because no one had expected that an elderly mouse, or *any* mouse, could run anywhere near three kilometers.

This molecule doesn't just turn old mice into ultramarathoners; we have used NMN-treated mice in studies that tested their balance, coordination, speed, strength, and memory, too. The difference between the mice that were on the molecule and the mice that were not was astounding. Were they human, those rodents would long since have been eligible for senior citizen discounts. Nicotinamide mononucleotide turned them into the equivalent of contenders on *American Ninja Warrior*. Other labs have shown that NMN can protect against kidney damage, neurodegeneration, mitochondrial diseases, and an inherited disease called Friedreich's ataxia that lands active 20-year-olds in wheelchairs.

As I write this, a group of mice that were put on NMN late in life are getting very old. In fact, only seven out of the original forty mice are still alive, but they are all healthy and still moving happily around the cage. The number of mice alive that didn't get the NMN?

Zero.

Every day I'm asked by members of the public, "Which is the superior molecule: NR or NMN?" We find NMN to be more stable than NR and see some health benefits in mouse experiments that aren't seen when NR is used. But it's NR that has been proven to extend the lifespan of mice. NMN is still being tested. So there's no definitive answer, at least not yet.

Human studies with NAD boosters are ongoing. So far, there has been no toxicity, not even a hint of it. Studies to test its effectiveness in muscle and neurological diseases are in progress or about to begin, followed by super-NAD-boosting molecules that are a couple of years behind them in development.

But a lot of people haven't been content to wait for these studies, which can take years to play out. And that has given us some interesting leads about where these molecules, or ones like them, might take us.

FERTILE GROUND

We know that NAD boosters are effective treatment for a wide variety of ailments in mice and that they extend their lifespan even when given late in life. We know that emerging research strongly suggests they could have a similar, if not duplicative, effect on human health.

We also know that the way it does this, in terms of the epigenetic landscape, is by creating the right level of stress—just enough to push our longevity genes into action to suppress epigenetic changes to maintain the youthful program. In doing so, NMN and other vitality molecules, including metformin and rapamycin, reduce the buildup of informational noise that causes aging, thus restoring the program.

How do they do this? We are still working to understand how epigenetic noise is dampened at a molecular level, but we know in principle how it works. When we give silencing proteins such as sirtuins a boost, they can maintain the youthful epigenome even with DNA damage occurring, like the long-lived yeast cells with extra copies of the *SIR2* gene. Somehow they can cope with it. Perhaps they are just superefficient at repairing DNA breaks and head home before they get lost, or if half the sirtuins head off, the remaining enzymes can hold down the fort.

Either way, the increased activity of the sirtuins may prevent Waddington's marbles from escaping their valleys. And even if they have started to head out of the valley, molecules such as NMN may push them back down, like extra gravity. In essence, this would be age reversal in some parts of the body—a small step, but age reversal nonetheless.

One of the first clues this might be true in an animal larger than a mouse came when a student who works in my lab at Harvard came into my office one afternoon.

"David," he said quietly, "do you have a moment? There is something I need to discuss. It's about my mother."

Given the expression on his face and the tone of his voice, I immediately worried that my student, who came from another country, would tell me his mother was sick. Having been half a planet away from my mother when she was dying, I very much knew how that felt.

"Whatever you need," I blurted out.

The student seemed taken aback—and I realized I hadn't yet posed the most pertinent question. "Is your mother all right?" I asked.

"Yes," he said. "Well . . . I mean, yes . . . well . . . mostly."

He reminded me that his mother had been taking supplemental NMN, as some of my students and their family members do. "The thing is, well"—his voice lowered to a whisper—"she has started her, um . . . cycle again."

It took me a few seconds to realize what cycle he was talking about.

As women approach and go through menopause, the menstrual cycle can become quite irregular, which is why a year with no periods must go by before most doctors will confirm that menopause has occurred.

After that, such bleeding can be a cause for concern, as it could be a sign of cancer, fibroid tumors, infections, or an adverse reaction to a medication.

"Has she been seen by a doctor?" I asked.

"Yes," my student said again. "The doctors say there is nothing wrong. They said this just looks like a normal period."

I was intrigued. "Okay," I said. "What we really need is more information. Can you give your mom a call to ask her some more questions?"

I've never seen the color so quickly wash away from someone's face.

"Oh, David," he pleaded, "please, please, pleeease don't make me ask my mother any more questions about that!"

Since that conversation, which took place in the fall of 2017, I have known a couple of other women and read the accounts of others claiming to have had similar experiences. These cases could, perhaps, be the result of a placebo effect. But a trial in 2018 to test whether an NAD booster could restore the fertility of old horses was successful, surprising the skeptical supervising veterinarian. As far as I know, horses don't experience the placebo effect.

Still, these stories and clinical results could be random chance. These matters will be studied in much greater detail. If, however, it turns out that mares and women can become fertile again, it will completely overturn our understanding of reproductive biology.

In school, our teachers taught us that women were born with a set number of eggs (perhaps as many as 2 million). Most of the eggs die off before puberty. Almost all the rest are either released during menstruation throughout the course of a woman's life or just die off along the way, until there are no more. And then, we were told, a woman is no longer fertile. Period.

These anecdotal reports of restored menstruation and fertile horses are early but interesting indicators that NAD boosters might restore failing or failed ovaries. We also see that NMN is able to restore the fertility of old mice that have had *all* their eggs killed off by chemotherapy or have gone through "mousopause." These results, by the way, even though they were done multiple times and reproduced in two different labs by different people, are so controversial that almost no one on the team voted to publish them. I was the exception. They remain unpublished, for now.

To me it is clear that we biologists are missing something. Something big.

In 2004, Jonathan Tilly—a highly controversial figure in the reproductive biology community—claimed that human stem cells that can give rise to new eggs, late in life, exist in the ovaries. Controversial though this theory is, it would explain how it is possible to restore fertility even in mice that are old or have undergone chemotherapy.[41,42]

Whether or not "egg precursor" cells exist in the ovary, there's no doubt in my mind that we are moving with staggering speed toward a world in which women will be able to retain fertility for a much longer portion of their lives and possibly regain it if it is lost.

All of this, of course, is good for people who wish to have a child but haven't been able to for any number of social, economic, or medical reasons. But what does it have to do with aging?

To answer that question, we need to remember what an ovary is. It's not just, as so many of us were taught in school, a slow-release mechanism for human eggs. It's an organ—just like our hearts, kidneys, or lungs—that has a day-to-day function, both holding on to eggs that were created during embryonic development and potentially being a repository for additional eggs derived from precursor cells later in life.

The ovary is also the first major organ to break down as a result of aging, in humans and animal models alike. What that means in mice is that, instead of waiting for two years for a mouse to reach "old age," we can start to see and investigate the causes and cures for aging in about 12 months, at the age female mice typically lose their ability to reproduce.

We also have to remember what NMN does: it boosts NAD, and this boosts the activity of the SIRT2 enzyme, a human form of yeast Sir2 found in the cytoplasm. SIRT2, we've found, controls the process by which an immature egg divides so that only one copy of the mother's chromosomes remain in the final egg in order to make way for the father's chromosomes. Without NMN or additional SIRT2 in old mice, their eggs were toast. Pairs of chromosomes were ripped apart from numerous directions, instead of exactly two. But if the old female mice were pretreated with NMN for a few weeks, their eggs looked pristine, identical to those of young mice.[43]

All of this is why early indicators of restored ovarian function in humans, anecdotal as they may be, are so fascinating. If true, the mechanisms that work to prolong, rejuvenate, and reverse aging in ovaries are pathways we can use to do the same thing in other organs.

One more thing that is important to bear in mind: NMN is hardly the only longevity molecule showing promise in this area. Metformin is already widely used to improve ovulation in women with infrequent or prolonged menstrual periods as a result of polycystic ovary syndrome.[44] Meanwhile, emerging research is demonstrating that the inhibition of mammalian target of rapamycin, or mTOR, may be able to preserve ovarian function and fertility during chemotherapy,[45] while the same gene pathway plays an important role in male fertility, as a central player in the production and development of sperm.[46]

LIFE WITH FATHER

Most of the time, rodent studies come long before formal human studies. That was the case for NAD boosters. But the early indicators of the safety and effectiveness of the molecules in yeast, worms, and rodents are such that many people have already begun their own private human experiments.

My father is among them.

Though he trained as a biochemist, my dad's passion was computing. He was a computer guy at a pathology company. That meant he spent a lot of his time sitting in front of a screen and on his behind—another thing experts say is devastatingly bad for our health. Some researchers have even suggested it could be as bad for us as smoking.

By the time my mother died in 2014, my father's health had also begun its seemingly inexorable decline. He had retired at 67 and was in his mid-70s, still fairly active. He liked to travel and garden. But he had passed the type 2 diabetes threshold, was losing his hearing, and his eyes were starting to go bad. He would tire fast. He repeated himself. He was grumpy. He was hardly a picture of exuberant life.

He started taking metformin for his borderline type 2 diabetes. The next year he started taking NMN.

My father has always been a skeptic. But he is also insatiably curious and was fascinated by what he heard from me about what was happening to the mice in my lab. NMN isn't a regulated substance; it's available as a supplement. So he tried it out, starting with small doses.

He knew quite well, though, that there are very big differences between mice and humans. At first he would say to me and to anyone else who asked, "Nothing has changed. How would I know?"

So the statement that came about six months into his NMN tryout was telling.

"I don't want to get carried away," he said, "but something is happening."

He was feeling less tired, he told me. Less sore. More mentally aware. "I'm outpacing my friends," he said. "They're complaining about feeling old. They can't even come for bushwalks with me anymore. I'm no longer feeling that way. I don't have aches or pains. I'm beating much younger people at rowing exercises at the gym." His doctor, meanwhile, was struck by the fact that his liver enzymes normalized after twenty years of being abnormal.

Upon his next visit to the United States, I noticed that something else was different, something very subtle. It dawned on me: for the first time since my mother's death, the smile had returned to his face.

These days, he runs around like a teenager. Hiking for six days through wind and snow to reach the peak of the highest mountain in Tasmania. Riding three-wheelers through the Aussie bush. Hunting remote waterfalls in the American West. Zip-line touring through the forest in northern Germany. Whitewater rafting in Montana. Ice cave exploring in Austria.

He's "aging in place," but he's rarely at his place.[47]

And because he missed working, he took on a new career at one of Australia's largest universities, where he sits on the ethics committee that approves human research studies, taking full advantage of his knowledge about scientific rigor, medical practice, and data security.

You might expect this sort of behavior from someone who had lived his whole life this way, but he is definitely *not* a guy who has lived his whole life this way. Dad used to say he wasn't looking forward to getting old. He isn't outgoing or optimistic by nature; he's more like Eeyore from *Winnie-the-Pooh*. He expected to have a decent ten years of retirement, then go into a nursing home. The future was clear. He had seen what had happened to his mother. He had watched helplessly as her health had declined in her 70s and 80s and as she had suffered from pain and dementia in the final decade of her life.

With all of that fresh in mind, the idea of living much past his 70s wasn't very interesting to him. In fact, it was pretty scary. But he's pretty happy with how it's turning out and wakes up every morning with a deep-seated desire to fill his life with new, exciting experiences. To that end, he faithfully takes his metformin and his NMN each morning and gets nervous when they start to run low. The turnaround in his energy, enjoyment of life, and perspective on growing old has been remarkable. It could all be unrelated to the molecules he's taking. I suppose his physical and mental transformation may just be how some people age. But it sure wasn't that way for any of my other relatives.

My father is also wondering what to think. We are a family of scientists, after all. "I can't be sure that the NMN is responsible," he told me recently. He thought about his life for a moment, then smiled and shrugged his shoulders, "but there's really no other explanation."

Recently, after touring much of the East Coast of the United States, Dad was heading home to Australia. I sheepishly asked him if he could fly back to the United States for an event being held the following month. I had been named an Officer of the Order of Australia, an honor bestowed "for distinguished service to medical research into the biology of ageing, to biosecurity initiatives, and as an advocate for the study of science," and there was going to be a ceremony at the Australian Embassy in Washington, DC.

"Sandra says it's not fair of me to ask for you to come back," I told him. "It's only four weeks from now, and you're almost eighty, and it's a long journey back, and—"

"I would love to come," he said, "but I'm just not sure I can fit it into my schedule."

He canceled some meetings and did fit the trip into his schedule, and having him there, along with Sandra and the kids, ensured that it was one of the best days of my life. As I looked at Dad, standing with my family, I thought, "*This* is what longer life is all about—having your parents there for life's important moments."

And as he stood there, he later told me, he thought, "*This* is what longer life is all about—being around for your children's important moments."

My father's story of reinvigoration is, of course, completely anecdotal. I won't be publishing it in a scientific journal anytime soon—a placebo can be a powerful drug, after all. There's simply no way to know if the combo of NMN and metformin is the reason he's feeling better or is simply what he started taking at the time he decided, subconsciously, that it was time for a big change in his approach to life.

Compelling evidence that the clock of aging is reversible will come when well-planned double-blind human clinical studies are completed. Until then, I remain very proud of my father, an average guy who grabbed life by the horns in his late 70s to start his life anew—a shining example of what life can be like if we don't accept aging as "just the way it goes."

Still, it's hard for me and anyone else who has seen what has happened to my father to not suspect that something special might be going on.

It's also hard to know what I know, to see what I've seen—the results of experiments and other clinical trials around the world years before the rest of the world learns about them—and not believe that something profound is about to happen to humanity.

COME WHAT MAY

By engaging our bodies' survival mechanisms in the absence of real adversity, will we push our lifespans far beyond what we can today? And what will be the best way to do this? Could it be a souped-up AMPK activator? A TOR inhibitor? A STAC or NAD booster? Or a combination of them with intermittent fasting and high-intensity interval training? The potential permutations are virtually endless.

Maybe the research under way on any one of these molecular approaches to battling aging will provide half a decade of additional good health. Maybe a combination of these compounds and an optimal lifestyle will be the elixir that gets us a couple of extra decades. Or maybe, as time goes by, our enthusiasm for these molecules will be dwarfed by what we discover next.

The discovery of the molecules I have described here can be credited to a lot of serendipity. But imagine what the world will discover now that we're actively and intentionally looking for molecules that engage our inbuilt defenses. Armies of chemists are now working to create and analyze natural and synthetic molecules that have the potential to be even better at suppressing epigenomic noise and resetting our epigenetic landscape.

There are hundreds of compounds that have already shown potential in this area and hundreds of thousands more that are waiting to be researched. And it's very possible that there is an as-yet-undiscovered chemical out there, hiding in a microorganism such as *S. hygroscopicus* or in a flower such as *G. officinalis*, that is just waiting to show us another way to help our bodies stay healthier longer. And that's just the natural chemicals—which are typically many times less effective than the synthetic drugs they inspire. Indeed, the emerging analogs of the molecules I've already described are demonstrating tremendous potential in early-stage human clinical trials.

It will take some time to sort out which of these molecules are best, when, and for whom. But we're getting closer every day. There will come a time in which significantly prolonged vitality is indeed only a few pills away; there are too many promising leads, too many talented researchers, and too much momentum for it to be otherwise.

Will any of these be a "cure" for aging? No. What's likely is that researchers will continue to identify molecules that are better and better at promoting both a reduction of epigenetic noise and a rejuvenation of cellular tissue. As we do, we'll be buying time for other advances that will also lead to significantly prolonged vitality.

But let's say that doesn't happen. For the sake of argument, not to mention emphasis, let's pretend we live in a world in which *none* of these molecules had ever been discovered and no one had ever thought to address aging with a pharmaceutical.

That would not change the inevitability of longer and healthier lives. Not at all. For drugs that engage the ancient survival mechanisms within us are just one of the many ways that scientists, engineers, and entrepreneurs are setting the stage for the most significant shift in the evolution of our species since . . .

. . . well, since . . .

. . . forever.

SIX

BIG STEPS AHEAD

TO THE EXTENT WE THOUGHT ABOUT IT—AND WE SELDOM DID—WE USED TO THINK aging would be a very complicated thing to change, if we could change it at all.

For most of human history, of course, we simply saw aging like the coming of the seasons; indeed, the shift from spring to summer to fall to winter was a common analogy we used to describe the movement from childhood to young adulthood to middle age to our "golden years." More recently, we figured that aging was inexorable but we *might* be able to deal with some of the diseases that made it a less appealing process. Later still, we figured that we might be able to attack each of the hallmarks and perhaps we could treat a few of the symptoms at a time. Even then, it seemed as though it would be a huge endeavor.

But here's the thing: it's really not.

Once you recognize that there are universal regulators of aging in everything from yeast to roundworms to mice to humans . . .

. . . and once you understand that those regulators can be changed with a molecule such as NMN or a few hours of vigorous exercise or a few less meals . . .

. . . and once you realize that it's all just one disease . . .

. . . it all becomes clear:

Aging is going to be remarkably easy to tackle.

Easier than cancer.

I know how that sounds. It sounds crazy.

But so did the idea of microorganisms before an amateur scientist named Antonie van Leeuwenhoek first described the world of the "small little animals" he saw under his homemade microscope in 1671; for hundreds of years to come, doctors rebelled against the idea that they needed to wash their hands before surgery. Now infections, one of the chief reasons patients used to die after surgery, have become the very thing hospital personnel are most fastidiously attentive to preventing in the operating room. Just by washing up before surgery, we have profoundly improved the rates at which patients survive. Once we understood what the problem was, it was an easy problem to solve.

For goodness' sake, we solved it with soap.

The idea of vaccines would also have sounded crazy to most people before the English physician Edward Jenner successfully used fluid he had gathered from a cowpox blister to inoculate an eight-year-old boy named James Phipps in what today would be an egregiously unethical experiment but at the time sparked a new era in immunological medicine. Indeed, the idea of giving a patient a little bit of a disease in order to prevent a lot of disease would have been seen as insane—even potentially homicidal—to many people until Jenner did it in 1796. We now know that vaccines are the single most effective medical intervention in human history in terms of saving and extending lifespans. So again, once we understood what the problem was, it was an easy one to solve.

The successes of STACs, AMPK activators, and mTOR inhibitors are a tremendously powerful indicator that we're working in an area of our biology that is upstream of every major aging-associated disease. The fact that these molecules have been shown to extend the lifespan of virtually every organism they've been tested on is further evidence that we're engaging with an ancient and powerful program to prolong life.[1]

But there is another pharmaceutical target that could increase our longevity, just a bit downstream from the processes we believe longevity molecules are impacting but still upstream of a lot of the symptoms of aging.

You might recall that one of the key hallmarks of aging is the accumulation of senescent cells. These are cells that have permanently ceased reproduction.

Young human cells taken out of the body and grown in a petri dish divide about forty to sixty times until their telomeres become critically short, a point discovered by the anatomist Leonard Hayflick that we now call the Hayflick limit. Although the enzyme known as telomerase can extend telomeres—the discovery of which afforded Elizabeth Blackburn, Carol Greider, and Jack Szostak a Nobel Prize in 2009—it is switched off to protect us from cancer, except in stem cells. In 1997, it was a remarkable finding that if you put telomerase into cultured skin cells, they don't ever senesce.

Why short telomeres cause senescence has been mostly worked out. A very short telomere will lose its histone packaging, and, like a shoelace that's lost an aglet, the DNA at the end of the chromosome becomes exposed. The cell detects the DNA end and thinks it's a DNA break. It goes to work to try to repair the DNA end, sometimes fusing two ends of different chromosomes together, which leads to hypergenome instability as chromosomes are shredded during cell division and fused again, over and over, potentially becoming a cancer.

The other, safer solution to a short telomere is to shut the cell down. This happens, I believe, by permanently engaging the survival circuit. The exposed telomere, seen as a DNA break, causes epigenetic factors such as the sirtuins to leave their posts permanently in an attempt to repair the damage, but there is no other DNA end to ligate it to. This shuts cell replication down, similar to the way that broken DNA in old yeast distracts Sir2 from the mating genes and shuts down fertility.

Triggering of the DNA damage response and major alterations to the epigenome are well known to occur in human senescent cells—and when we introduce epigenetic noise into the ICE cells they go on to senesce earlier than untreated cells, so maybe this idea has merit. I suspect that senescence in nerve and muscle cells, which don't divide much or at all, is the result of epigenetic noise that causes cells to lose their identity and shut down. This once-beneficial response, which evolved to help cells survive DNA damage, has a dark side: the permanently panicked cell sends out signals to surrounding cells, causing them to panic, too.

Senescent cells are often referred to as "zombie cells," because even though they should be dead, they refuse to die. In the petri dish and in frozen, thinly sliced tissue sections, we can stain zombie cells blue because they make a rare enzyme called beta-galactosidase, and when we do that, they light up clearly. The older the cells, the more blue we see. For example, a sample of white fat looks white when we are in our 20s, pale blue in middle age, and dark royal blue in old age. And that's scary, because when we have lots of these senescent cells in our bodies, it's a clear sign that aging is getting a strong grip on us.

Small numbers of senescent cells can cause widespread havoc. Even though they stop dividing, they continue to release tiny proteins called cytokines that cause inflammation and attract immune cells called macrophages that then attack the tissue. Being chronically inflamed is unhealthy: just ask someone with multiple sclerosis, inflammatory bowel disease, or psoriasis. All these diseases are associated with excess cytokine proteins.[2] Inflammation is also a driving force in heart disease, diabetes, and dementia. It is so central to the development of age-related diseases that scientists often refer to the process as "inflammaging." And cytokines don't just cause inflammation; they also cause other cells to become zombies, like a biological apocalypse. When this happens, they can even stimulate surrounding cells to become a tumor and spread.

We already know that destroying senescent cells in mice can give them substantially healthier and significantly longer lives. It keeps their kidneys functioning better for longer. It makes their hearts more

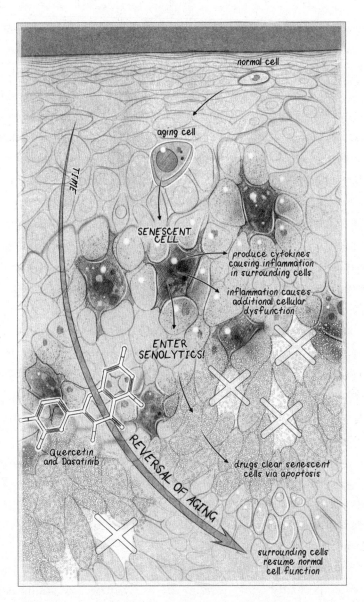

normal cell

aging cell

TIME

SENESCENT CELL

produce cytokines causing inflammation in surrounding cells

inflammation causes additional cellular dysfunction

ENTER SENOLYTICS!

Quercetin and Dasatinib

REVERSAL OF AGING

drugs clear senescent cells via apoptosis

surrounding cells resume normal cell function

DELETING THE ZOMBIE SENESCENT CELLS IN OLD TISSUES. Thanks to the primordial survival circuit we've inherited from our ancestors, our cells eventually lose their identities and cease to divide, in some cases sitting in our tissues for decades. Zombie cells secrete factors that accelerate cancer, inflammation, and help turn other cells into zombies. Senescent cells are hard to reverse aging in, so the best thing to do is to kill them off. Drugs called senolytics are in development to do just that, and they could rapidly rejuvenate us.

resistance to stress. Their lifespans, as a result, are 20 to 30 percent longer, according to research led by Mayo Clinic molecular biologists Darren Baker and Jan van Deursen.[3] In animal models of disease, killing of senescent cells makes fibrotic lungs more pliable, slows the progression of glaucoma and osteoarthritis, and reduces the size of all sorts of tumors.

Understanding why senescence evolved is not just an academic exercise; it could help us design better ways to prevent or kill senescent cells. Cellular senescence is a consequence of our inherited primordial survival circuits, which evolved to stop cell division and reproduction when DNA breaks were detected. Just as in old yeast cells, if DNA breaks happen too frequently or they overwhelm the circuit, human cells will stop dividing, then sit there in a panic, trying to repair the damage, messing up their epigenome, and secreting cytokines. This is the final stage of cellular aging—and it's not pretty.

If zombie cells are so bad for our health, why doesn't our body just kill them off? Why are senescent cells allowed to cause trouble for decades? Back in the 1950s, the evolutionary biologist George Williams was already on the case. His work, built upon by Judith Campisi from the Buck Institute for Research on Aging in California, proposes that we evolved senescence as a rather clever trick to *prevent* cancer when we are in our 30s and 40s. Senescent cells, after all, don't divide, which means that cells with mutations aren't able to spread and form tumors. But if senescence evolved to prevent cancer, why would it eventually *promote* cancer in adjacent tissue, not to mention a host of other aging-related symptoms?

This is where "antagonistic pleiotropy" comes into play: the idea that a survival mechanism that is good for us when we are young is kept through evolution because this far outweighs any problems it might cause when we get older. Yes, natural selection is callous, but it works.

Consider the 15-million-year history of hominids, the great apes. In the vast majority of our family's evolutionary journey, the forces of predation, starvation, disease, maternal mortality, infection, catastrophic weather events, and intraspecies violence meant that very few individuals saw more than a decade or two of life. Even in the relatively recent era of the *Homo* genus, what we now think of as "middle age" is an exceptionally new phenomenon.

A life expectancy of 50 and beyond was simply not a reality for most of our evolutionary history. Therefore, it didn't matter if a mechanism for slowing the spread of cancer would eventually cause more cancer and other diseases, because it generally worked, as long as it allowed people to breed and rear some children. The saber-toothed tigers took things from there.

These days, of course, few people have to worry about being picked off by hungry predators. Hunger and malnutrition are still far too common, but abject starvation is increasingly rare. We're getting better and better at staving off childhood diseases and have eliminated some of them almost entirely. Childbirth is an increasingly safe affair (although that, too, is something that can be vastly improved upon, especially in the developing world). Modern sanitation has resulted in tremendous improvements in the rates at which we die of infectious diseases. Modern technology is helping to warn us of impending catastrophes such as hurricanes and volcanic eruptions. And although the world often seems to be a vicious and violent place, the worldwide homicide rate and the numbers of wars globally have been falling for decades.

So we live longer—and evolution hasn't had a chance to catch up. We're plagued by senescent cells, which might as well be radioactive waste. If you put a tiny dab of these cells under a young mouse's skin, it won't be long before inflammation spreads and the entire mouse is filled with zombie cells that cause premature signs of aging.

A class of pharmaceuticals called senolytics may be the zombie killers we need to fight the battle against aging on this front. These small-molecule drugs are designed to specifically kill senescent cells by inducing the death program that should have happened in the first place.

That's what the Mayo Clinic's James Kirkland has done. He needed only a quick course of two senolytic molecules—quercetin, which is found in capers, kale, and red onions, and a drug called dasatinib, which is a standard chemotherapy treatment for leukemia—to eliminate the senescent cells in lab mice and extend their lifespan by 36 percent.[4] The implications of this work cannot be overstated. If senolytics work, you could take a course of a medicine for a week, be rejuvenated, and come back ten years later for another course. Meanwhile, the same medicines could be injected into an osteoarthritic joint or an eye going blind, or inhaled into lungs made fibrotic and inflexible by chemotherapy, to give them an age-reversal boost, too. (Rapamycin, the Easter Island longevity molecule, is what's known as a "senomorphic" molecule, in that it doesn't kill senescent cells but does prevent them from releasing inflammatory molecules, which may be almost as good.[5])

The first human trials of senolytics were started in 2018 to treat osteoarthritis and glaucoma, conditions in which senescent cells can accumulate. It will be a few more years before we know enough about the effects and safety of these drugs to provide them to everyone, but if they work, the potential is vast.

But there is another option, just a bit further upstream, that could be even better.

THE HITCHHIKER'S GUIDE

The selfish genes we discussed earlier, called LINE-1 retrotransposons, and their fossil remnants, make up about half of the human genome, what is often referred to as "junk DNA."

It's a lot of genetic baggage, and they are sneaky buggers. In young cells, these ancient "mobile DNA elements," also known as retrotransposons, are prevented by chromatin from jumping out of the genome, then breaking DNA to reinsert themselves elsewhere. We and others have shown that LINE-1 genes are bundled up and rendered silent by sirtuins.[6] But as mice age, and possibly as we do as well, these sirtuins become scattered all over the genome, having been recruited away to repair DNA breaks elsewhere, and many of them never find their way home. This loss is exacerbated by a drop in NAD levels—the same thing we first saw in old yeast. Without sirtuins to spool the chromatin and silence the transposon DNA, cells start to transcribe these endogenous viruses.

This is bad. And it only gets worse.

Over time, as mice age, the once silent LINE-1 prisoners are turned into RNA and the RNA is turned into DNA, which is reinserted into the genome at a different place. Besides creating genome instability and epigenomic noise that causes inflammation, LINE-1 DNA leaks from the nucleus into the cytoplasm, where it is recognized as a foreign invader. In response, the cells release even more immunostimulatory cytokines that cause inflammation throughout the body.

New work by John Sedivy at Brown University and Vera Gorbunova from the University of Rochester raises the possibility that one of the main reasons *SIRT6* mutant mice age so rapidly is that these retroviral hellhounds have no leash, causing numerous DNA breaks and the epigenome to degrade rapidly instead of slowly. Convincing evidence has come from experiments showing that antiretrovirals, the same kinds used to fight HIV, extend the lifespan of *SIRT6* mutant mice about twofold. It may turn out that, as NAD levels decline with age, sirtuins are rendered unable to silence retrotransposon DNA. Perhaps one day, safe antiretroviral drugs or NAD boosters will be used to keep these jumping genes silent.[7] We would not have stopped aging completely at its source, but we would be fighting the battle before total anarchy ensues and the genie that is aging becomes even harder to put back in the bottle.

VAX TO THE FUTURE

In 2018, scientists at Stanford University reported that they had developed an inoculation that significantly lowered the rates at which mice suffered from breast, lung, and skin cancer. By injecting the mice with stem cells inactivated by radiation and later adding a booster shot like those humans use for tetanus, hepatitis B, and whooping cough, the stem cells primed the immune system to attack cancers that normally would be invisible to the immune system.[8] Other immuno-oncological approaches are making even greater strides. Therapies such as PD-1 and PD-L1 inhibitors, which expose cancer cells so they can be killed, and chimeric antigen receptors T-cell (CAR-T) therapies, which modify the patient's own immune T-cells and reinject them to go kill cancer cells, are saving lives of people who, just a few years before, have been told to go home and make funeral arrangements. Now, some of these patients are being given a new lease on life.

If we can use the immune system to kill cancer cells, it stands to reason that we can do that for senescent cells, too. And some scientists are on the case. Judith Campisi from the Buck Institute for Research on Aging and Manuel Serrano from Barcelona University believe that senescent cells, like cancers, remain invisible to the immune system by waving little protein signs that say, "No zombie cells here."

If Campisi and Serrano are right, we should be able to take away those signs and give the immune system permission to go kill senescent cells. Perhaps a few decades from now a typical vaccine schedule that currently protects babies against polio, measles, mumps, and rubella might also include a shot to prevent senescence when they reach middle age.

When people first hear that it may be possible to vaccinate against aging, rather than just treat its symptoms or slow it down, it's not uncommon for them to immediately express worries that we are "playing God" or "interfering with Mother Nature." Maybe we are, but if so, that's not unique to people involved in the fight against aging. We fight diseases of all kinds that God or Mother Nature gave us. We've been doing so for a long time, and we're going to keep doing so for a long time to come.

The world rightfully celebrated the eradication of smallpox in 1980. When malaria is likewise eradicated—and I believe it will be sometime in the coming decades—our global community will rejoice once again. And if I could offer the world a vaccine for HIV, right now, there wouldn't be many people—no decent ones, at least—who would say that we should just "let nature run its course." These are ailments we've long considered diseases, though, and I accept that it will take some time to convince people that aging is no different.

To this end, I've found this thought experiment to be helpful: imagine an Airbus A380, a double-decker "superjumbo" filled with six hundred people on board, on approach to Los Angeles. The plane does not have landing gear, only parachutes. And all but one of the doors is stuck, so when the passengers evacuate, one by one, they'll be scattered across the most densely populated area of the country.

Oh, and one more thing: the passengers are sick. Really sick. The disease they carry is highly contagious; it starts with lethargy and sore joints, then develops into hearing and vision loss, bones as brittle as century-old teacups, excruciatingly painful heart failure, and brain signals so badly interrupted that many victims won't even be able to remember who they are. No one survives this disease, and death is almost always agonizing.

After a life of faithful service to the United States, you have found yourself behind the Resolute Desk in the Oval Office of the White House. The phone rings. The deputy director for infectious diseases from the Centers for Disease Control and Prevention tells you that if even one of the passengers is permitted to parachute into the greater Los Angeles area, tens of thousands of people will catch the disease and die. Each additional parachuter will increase the projected death toll exponentially.

The moment you put the receiver down, the phone rings again. The chairperson of the Joint Chiefs of Staff tells you that six US Air Force F-22 Raptor fighters are tracking the plane as it circles over the Pacific Ocean. The pilots have it locked in; their missiles are ready. The plane is running out of gas. The fate of the passengers, and the entire United States, rests upon your orders.

What do you do?

This, of course, is a "trolley problem," an ethical thought experiment, of the type popularized by the philosopher Philippa Foot, that pits our moral duty not to inflict harm on others against our social responsibility to save a greater number of lives. It's also, however, a handy metaphor, because the highly contagious disease the passengers are carrying is, as you doubtless have noticed, nothing more than a faster-acting version of aging.

When presented with the idea of a disease that could infect and kill legions of people—with horrendous symptoms, no less—very few of us would not make the horrible but necessary call to shoot down the plane, taking the lives of hundreds of people to protect the lives of millions.

With that in mind, consider this question: If you would sacrifice hundreds of human lives to stop a fast-acting version of aging from infecting millions, what would you be willing to do to prevent the disease as it actually occurs in the lives of everyone on the planet?

Worry not: what I'm about to suggest won't actually come at the cost of human lives. Not hundreds. Not dozens. Not even one. But it would require us to confront an idea that many people would find alarming: infecting ourselves with a virus that would quickly move into every cell in our body, turning us into genetically modified organisms. The virus wouldn't kill; it would do the opposite.

GET WITH THE REPROGRAM

Vaccines against senescent cells, CR mimetics, and retrotransposon suppressors are possible pathways to prolonged vitality, and work is under way already in labs and clinics around the world. But what if we didn't need any of that? What if we could reset the aging clock and prevent cells from ever losing their identity and becoming senescent in the first place?

Yes, the solution to aging could be cellular reprogramming, a resetting of the landscape—the way, for instance, that jellyfish have been shown to do by using small body fragments to regenerate polyps that spawn a dozen new jellies.

The DNA blueprint to be young, after all, is *always* there, even when we are old. So how can we make the cell reread the blueprint? Here it's helpful to return to the DVD metaphor. Over time, thanks to use and perhaps misuse, the digital information encoded as pits in the top layer of aluminum becomes obscured by some deep and some fine scratches, making it hard for the DVD player to read the disk. A DVD has thirty miles of data spiraled around the disk from the edge to the center, so if the disc is scratched, finding the start of a particular song becomes extremely difficult.

It's the same situation for old cells, but far worse. The DNA in our cells holds about the same amount of data as a DVD, but in six feet of DNA that's packed into a cell a tenth the size of a speck of dust. Together, all the DNA in our body, if laid end to end, would stretch twice the diameter of the solar system. Unlike a simple DVD, though, the DNA in our cells is wet and vibrating in three dimensions. And there aren't 50 songs, there are more than 20,000. No wonder gene reading becomes difficult the older we get; it's miraculous that any cell finds the right genes in the first place.

There are two ways to play an old, scratched DVD with fidelity. You could buy a better DVD player, one with a more powerful laser that could reveal the data under the scratches. Or you could polish the disc to expose the information again, making the DVD as good as new. I've heard that a rag with toothpaste on it works just fine.

Restoring youth in an organism is never going to be as simple as polishing a disk with toothpaste, but the first approach, putting a scratched DVD into a new player, was. Oxford University professor John Gurdon first did this in 1958, when he removed the chromosomes from a frog's egg and replaced them with some chromosomes from an adult frog and obtained living tadpoles. Then, in 1996, Ian Wilmut and his colleagues at the University of Edinburgh replaced the chromosomes of a sheep's egg with those from an udder cell. The result was Dolly, whose birth was met with a heated public debate about the purported dangers of cloning.

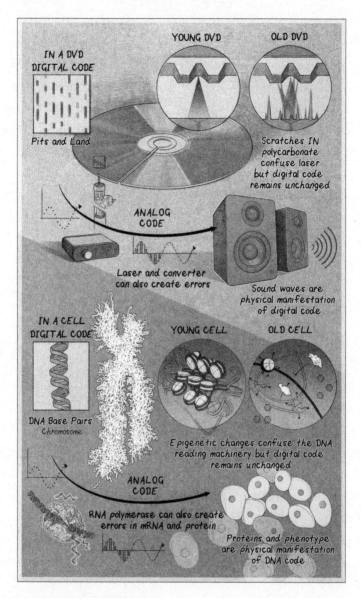

WE ARE ANALOG, THEREFORE WE AGE. According to the Information Theory of Aging, we become old and susceptible to diseases because our cells lose youthful information. DNA stores information digitally, a robust format, whereas the epigenome stores it in analog format, and is therefore prone to the introduction of epigenetic "noise." An apt metaphor is a DVD player from the 1990s. The information is digital; the reader that moves around is analog. Aging is similar to the accumulation of scratches on the disc so the information can no longer be read correctly. Where's the polish?

The debate overshadowed the most important point: that old DNA retains the information needed to be young again.

That debate has since died down; the world today has other concerns. Cloning is now routinely done to produce farm animals, racehorses, and even pets. In 2017, you could order up a dog clone for the "bargain" price of $40,000—or two of them, as Barbra Streisand did to replace her beloved Sammie, a curly-haired Coton de Tulear.[9] The fact that Sammie was 14 when she died and donated cells—that's somewhere in the range of 75 in dog years—didn't impact the clones one bit.

The implications of these experiments are profound. What they show is that *aging can be reset*. The scratches on the DVD can be removed, and the original information can be recovered. Epigenomic noise is not a one-way street.

But how might we reset the body without becoming a clone?

In his 1948 publications about the preservation of information during data transmissions, Claude Shannon provided a valuable clue.[10]

In an abstract sense, he proposed that information loss is simply an increase in entropy, or the uncertainty of resolving a message, and provided brilliant equations to back his ideas up. His work stemmed from the mathematics of Harry Nyquist and Ralph Hartley, two other engineers at Bell Labs who, in the 1920s, revolutionized our understanding of information transmission. Their notions of an "ideal code" were important for Shannon's development of his communication theory.

In the 1940s, Shannon became obsessed with communications over a noisy channel, in which information is simply a set of possible messages that needs to be reconstructed by the recipient of the message—the receiver.

As Shannon brilliantly showed in his "noisy-channel coding theorem," it is possible to communicate information nearly error free as long as you don't exceed the channel capacity. But if the data exceeds the channel capacity or is subject to noise, which is often the case with analog data, the best way to ensure it makes it to the receiver is to store a backup set of data. That way, even if some primary data are lost, an "observer" can send this "correcting data" to a "correcting device" to recover the original message. This is how the internet works. If data packets are lost, they are recovered and resent moments later, all thanks to Transmission Control Protocol/Internet Protocol (TCP/IP).

As Shannon put it, "This observer notes the errors in the recovered message and transmits data to the receiving point over a 'correction channel' to enable the receiver to correct the errors."

Though it may sound like esoteric language from the 1940s, what dawned on me in 2014 is that Shannon's "A Mathematical Theory of Communication" is relevant to the Information Theory of Aging.

In Shannon's drawing, there are three different components that have analogs in biology:

- The "source" of the information is the egg and sperm, from your parents.
- The "transmitter" is the epigenome, transmitting analog information through space and time.
- The "receiver" is your body in the future.

When an egg is fertilized, epigenetic information—biological "radio signals"—is sent out. It travels between dividing cells and across time. If all goes well, the egg develops into a healthy baby and eventually a healthy teenager. But with successive cell divisions and the overreaction of the survival circuit to DNA damage, the signal becomes increasingly noisy. Eventually, the receiver, your body when it is 80, has lost a lot of the original information.

We know that cloning a new tadpole or a mammal from an old one is possible. So even if a lot of the epigenetic information is lost in old age, obscured by epigenetic noise, there must be information that tells the cell how to reset. This fundamental information, laid down early in life, is able to tell the body how to be young again—the equivalent of a backup of the original data.

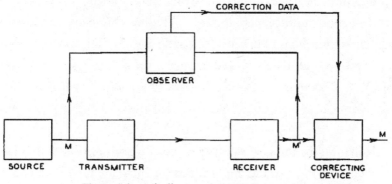

Fig. 8—Schematic diagram of a correction system.

CLAUDE SHANNON'S 1948 SOLUTION TO RECOVERING LOST INFORMATION DURING DATA TRANSMISSIONS LED TO CELL PHONES AND THE INTERNET. It may also be the solution to reversing aging.
Source: C. E. Shannon, "A Mathematical Theory of Communication," *Bell System Technical Journal* 27, no. 3 (July 1948): 379–423 and 27, no. 4 (October 1948): 623–66.

To end aging as we know it, we need to find three more things that Shannon knew were essential for a signal to be restored even if it is obscured by noise:

- An "observer" who records the original data
- The original "correction data"
- And a "correcting device" to restore the original signal

I believe we may have finally found the biological correcting device.

In 2006, the Japanese stem cell researcher Shinya Yamanaka announced to the world that after testing dozens of combinations of genes, he had discovered that a set of four—Oct4, Klf4, Sox2, and c-Myc— could induce adult cells to become pluripotent stem cells, or iPSCs, which are immature cells that can be coaxed into becoming any other cell type. These four genes code for powerful transcription factors that each controls entire sets of other genes that move cells around on the

Waddington landscape during embryonic development. These genes are found in most multicellular species, including chimpanzees, monkeys, dogs, cows, mice, rats, chickens, fish, and frogs. For his discovery, essentially showing that complete cellular age reversal was possible in a petri dish, Yamanaka won the Nobel Prize in Physiology or Medicine along with John Gurdon in 2012. We now call these four genes Yamanaka factors.

At first blush, Yamanaka's experiments might sound like a nifty laboratory parlor trick. But the implications for aging are profound, and not only because he paved the way for us to grow entirely new populations of blood cells, tissues, and organs in the dish that can be and are being transplanted into patients.

What he identified, I believe, is the reset switch responsible for Gurdon's tadpoles—the biological correcting device.

I predict, and my students are now showing in the lab, that we can use these and other switches not just to reset our cells in petri dishes but to reset an entire body's epigenetic landscape—to get the marbles back into the valleys where they belong—sending sirtuins back to where they came from, for instance. Cells that have lost their identity during aging can be led back to their true selves. This is the DVD polish we've been looking for.

We are making progress every week in restoring the youthful epigenome of mice by delivering reprogramming factors. The pace of discovery is mind spinning. A full night of sleep for me and my lab members is increasingly rare.

In the 1990s, there were major concerns about the safety of delivering genes to humans. But there are a rapidly increasing number of approved gene therapy products and hundreds of clinical trials under way. Patients with an *RPE65* mutation that causes blindness, for example, can now be cured with a simple injection of a safe virus that infects the retina and delivers, forever, the functional *RPR65* gene.

I predict that cellular reprogramming in the body will first be used to treat age-related diseases in the eye, such as glaucoma and macular degeneration (the eye is the organ of choice to trial gene therapies because it is immunologically isolated). But if the therapy is safe enough to deliver into the entire body—as the long-term mouse studies in my lab suggest they might one day be—this may be in our future:

At age 30, you would get a week's course of three injections that introduce a specially engineered adeno-associated virus, or AAV, which causes a very mild immune response, less even than what is commonly caused by a flu shot. The virus, which has been known to scientists since the 1960s, has been modified so it doesn't spread or cause illness. What this theoretical version of the virus would carry would be a small number of genes—some combination of Yamanaka factors, perhaps—and a failsafe switch that could be turned on with a well-tolerated molecule such as doxycycline, an antibiotic that can be taken as a tablet, or, even better, one that's completely inert.

Nothing, at that point, would change in the way your genes work. But when you began to see and feel the effects of aging, likely sometime in your mid-40s, you would be prescribed a month's course of doxycycline. With that, the reprogramming genes would be switched on.

During the process, you'd likely place a drop of blood in a home biotracker or pay a visit to the doctor to make sure the system was working as expected, but that's about it. Over the next month, your body would undergo a rejuvenation process as Waddington's marbles were sent back to where they once were when you were young. Gray hair would disappear. Wounds would heal faster. Wrinkles would fade. Organs would regenerate. You would think faster, hear higher-pitched sounds, and no longer need glasses to read a menu. Your body would feel young again.

Like Benjamin Button, you would feel 35 again. Then 30. Then 25.

But unlike Benjamin Button, that's where you would stop. The prescription would be discontinued. The AAV would switch off. The Yamanaka factors would fall silent. Biologically, physically, and mentally, you would be a couple of decades younger, but you'd retain all your knowledge, wisdom, and memories.

You would be young again, not just looking young but actually young, free to spend the next few decades of your life without the aches and pains of middle age, untroubled by the prospects of cancer and heart disease. Then, a few more decades down the road, when those gray hairs begin showing up again, you'd start another cycle of the prescribed trigger.

What's more, with the pace at which biotech is advancing, and as we learn how to manipulate the factors that reset our cells, we may be able to move away from using viruses and simply take a month's course of pills.

Does that sound like science fiction? Something that is very far out in the future? Let me be clear: it's not.

Manuel Serrano, the leader of the Cellular Plasticity and Disease laboratory at the Institute for Research in Biomedicine in Barcelona, and Juan Carlos Izpisua Belmonte, at the Salk Institute for Biological Studies in San Diego, have already engineered mice that have all of the Yamanaka factors from birth; these can be turned on by injecting the mice with doxycycline. In a now-famous study from 2016, when Belmonte triggered the Yamanaka factors for just two days a week throughout the lifespan of a prematurely aging mouse breed called LMNA, the mice remained young compared to their untreated siblings and lived 40 percent longer.[11] He's shown that the skin and kidneys of regular old mice heal more quickly, too.

The Yamanaka treatment, however, was highly toxic. If Belmonte overdid it by giving the mice the antibiotic for a few more days, the mice died. Serrano had also shown that by pushing the marbles too far up the landscape, the four-gene combo could induce teratomas, which are particularly disgusting tumors made up of several types of tissue, such as hair, muscle, or bone. Clearly, this tech is not ready for prime time. At least not yet. But we're getting closer every day to being able to control the Waddington marbles safely, making sure they land back precisely in their original valleys and not at the top of the mountain, where they could cause cancer.

While all this was going on, guided by the success of the ICE mouse experiments, my lab had been looking for ways to delay and reverse epigenetic aging. We'd tried many different approaches: the Notch gene, Wnt, the four Yamanaka factors. Some had worked a little, but most were turning into tumor cells.

One day in 2016, after failing consistently for two years to get old cells to age in reverse without turning into tumor cells, a brilliant graduate student named Yuancheng Lu came into my office to say he was close to quitting. As a final effort, he suggested he try leaving out the c-Myc gene that was the likely cause of the teratomas, and I encouraged him to do so.

He delivered a viral package to mice, but this time with only three of the Yamanaka factors, then turned them on using doxycycline and waited for all the mice to get sick or die. But none of them did. They were totally fine. And after months of monitoring, no tumors arose, either. It was a surprise to both of us—a great surprise.

Instead of waiting for another year to see if the mice lived longer, Yuancheng suggested he use a mouse's optic nerve as a way to test age reversal and rejuvenation. I was skeptical.

"I'm not superoptimistic this will work," I told him. "Optic nerves just don't regenerate, unless you are a newborn."

The intricate network of cells and fibers that transmit nervous signals across our bodies is divided into two parts: the peripheral system and the central system. We've known for a long time that peripheral nerves, like those in our arms and legs, can grow back, albeit very, very slowly. The nerves of the central system, though—optic nerves and the nerves of the spinal cord—never grow back. Even those scientists who bucked convention, proposing novel therapies that could regenerate some aspect of the central system, have generally been circumspect about the potential for significant regrowth. Decades of work aimed at reversing glaucoma in the eye and spinal cord injury has had almost no positive momentum.

"You've picked the hardest problem in biology to solve," I told Yuancheng.

"But," he replied, "if we *could* solve that problem . . ."

There might have been a thousand ways to measure the impact of age reversal in mice, but buoyed by his recent successes, he decided to "go big or go home." I liked that.

"No one changes the world by not taking risks," I told him. "Go test it."

The images that came to me in a text message a few months later took my breath away—so much so that I needed to make sure that what I was seeing was real.

I called Yuancheng immediately. "Am I seeing what I think I am seeing?"

"Maybe," he said. "What are you seeing?"

"The future," I said.

Yuancheng let out a tremendous sigh of relief. "David," he said, "an hour ago I thought I was going to fail."

For researchers, doubt is no vice. Doubt is the very normal and very human consequence of pushing yourself to do audacious things without knowing how those things are going to work out.

But on that day, things sure did seem to be working out. The image Yuancheng first texted to me looked like an orange, glowing jellyfish; its head was at the top, where the eye of the mouse sits, with long tentacles flowing down toward the brain. Two weeks earlier, Yuancheng and our collaborators had squeezed the optic nerve a few millimeters from the back of the eye with a set of tweezers, causing almost all the nerve cell axons, the tentacles, to die back toward the brain. They injected an orange fluorescent dye into the eye that is taken up by living neurons. So when Yuancheng took a microscope and looked below the crush site, there were no glowing nerves, just a mass of dead cell remnants.

The next picture he sent was an example of one where the reprogramming virus had been turned on after the crush. Instead of dead cells, a network of long, healthy spindly tentacles was making its way to connect up with the brain. It was the greatest example of nerve generation in history, and Yuancheng was only just getting started.

No one had *really* expected the reprogramming to work so well.

One-month-old mice were initially chosen for these experiments to give us the greatest chance of success and because that's what everyone else does. But Yuancheng and our skilled collaborators in Professor Zhi-gang He's lab at Children's Hospital at Harvard Medical School have now tested our reprogramming regimen on the damaged optic nerves of middle-aged mice aged twelve months. Their nerves also regenerate.

As I write this, we have restored vision in regular old mice.

Vision declines dramatically in a mouse by 12 months of age. Bruce Ksander and Meredith Gregory-Ksander, from Massachusetts Eye and Ear at Harvard, know this well. There is a loss of the nerve impulses in the retina, and old mice don't move their heads as often when moving lines are displayed in front of them, because they simply don't see them.

"David, I must admit," Bruce said, "I never expected this reprogramming stuff to work on normal aged eyes. I was only testing your virus because you were so excited to try it."

The result he had seen the morning before had been the most exciting day in his research life: our OSK reprogramming virus had restored vision.

A few weeks later, Meredith showed that reprogramming also works to reverse vision loss caused by internal eye pressure known as glaucoma.

"Do you know what we've discovered?" Bruce remarked. "Everyone else has been working to slow the progression of glaucoma. This treatment *restores* vision!"

If adult cells in the body, even old nerves, can be reprogrammed to regain a youthful epigenome, the information to be young cannot *all* be lost. There must be a repository of correction data, a backup set of data or molecular beacons, that is retained through adulthood and can be accessed by the Yamanaka factors to reset the epigenome using the cellular equivalent of TCP/IP.

What those youth markers are, we're still not sure. They are likely to involve methyl tags on DNA, which are used to estimate an organism's age, the so-called Horvath clock. They likely also involve something else: a protein, an RNA, or even a novel chemical attached to DNA that we

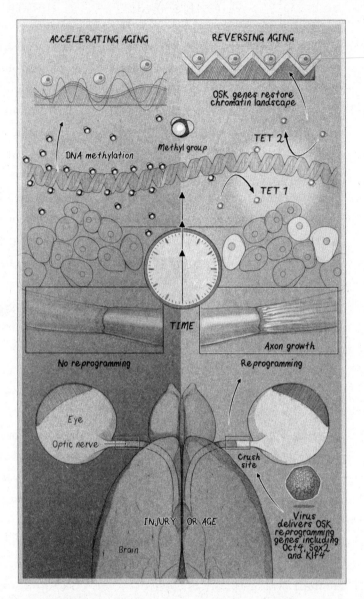

EPIGENETIC REPROGRAMMING REGROWS OPTIC NERVES AND RESTORES EYESIGHT IN OLD MICE. The Information Theory of Aging predicts that it is a loss of epigenetic rather than genetic information in the form of mutations. By infecting mice with reprogramming genes called Oct4, Sox2, and Klf4, the age of cells is reversed by the TET enzymes, which remove just the right methyl tags on DNA, reversing the clock of aging and allowing the cells to survive and grow like a newborn's. How the enzymes know which tags are the youthful ones is a mystery. Solving that mystery would be the equivalent of finding Claude Shannon's "observer," the person who holds the the original data.

haven't yet discovered. But whatever they are made of, they are important, for they would be the fundamental correcting data that cells retain over a lifetime that somehow direct a reboot.

We also need to find the observer, the one who records what the original signal is when we are young. It can't just be DNA methylation, because that doesn't explain how the reprogrammed cells know to focus on some of the youthful methyl marks and strip off the ones that accumulated during aging, the cell equivalent of the scratches on the DVD. Perhaps it is a specialized histone, or a transcription factor, or a protein that latches onto methylated DNA when we are developing in utero and stays there for eighty years waiting until a signal comes from the correcting device to restore the original information.

In Claude Shannon's parlance, when the correcting device is switched on by infecting cells with OSK genes, the cell somehow knows how to contact the observer and use the correction data to restore the original signal to that of a young cell.

Growing new nerves and restoring eyesight wasn't enough for Yuancheng. When the DNA of the damaged neurons was examined, they seemed to be going through a very rapid aging program, one that was countered by the reprogramming factors. The neurons that received the reprogramming factors didn't age, and they didn't die. This is a radical idea but one that makes a lot of sense: severe cellular injury overwhelms the survival circuit and accelerates aging of the cell, leading to death, unless the clock is somehow reversed.

With these discoveries, we may be on the verge of understanding what makes biological time tick and how to wind it back. We know from our experiments that the biological information correcting device requires enzymes called ten-eleven translocation enzymes, or TETs, which clip off methyl tags from DNA, the same chemical tags that mark the passage of the Horvath aging clock. This is no coincidence, and points to the DNA methylation clock as not just an indicator of age but a controller of it. It's the difference between a wristwatch and physical time.

In their role as a component of the correcting device, the TETs cannot just strip off all the methyls from the genome, for that would turn a cell into a primordial stem cell. We would not have old mice that can see better: we would have blind mice with tumors. How the TETs know to remove only the more recent methyls while preserving the original ones is a complete mystery.

It will likely take another decade and many other labs' work to know precisely what the biological equivalent of the TCP/IP information recovery system is. But whatever it is, eyesight that should not be able to be restored is being restored and cells that should not be able to regrow are regrowing.

Compared to the decades of research into how to slow down aging and age-related disease by a few percent, the reprogramming work has been relatively quick and easy. All it took was an intrepid idea and the courage to buck convention.

The future looks interesting, to say the least. If we can fix the toughest-to-fix and regenerate the toughest-to-regenerate cells in our body, there's really no reason to suspect we cannot regrow any type of cells our bodies need. Yes, that could mean fixing fresh spinal cord injuries, but it also means regrowing any other kind of tissue in our body that has been damaged by age: from the liver to the kidney, from the heart to the brain. Nothing is off the table.

So far, the three-Yamanaka-gene combination seems safe in mice even when turned on for a year, but there is still plenty of work to be done. There are a lot of unanswered questions: Can we deliver the combination to all cells? Will it eventually cause cancer? Should we keep the genes on or turn them off to let the cells rest? Will this work in some tissues better than others? Can it be given to middle-aged people, before they become sick, the same way we take statins to keep cholesterol in check to prevent heart disease?

I have little doubt that cellular reprogramming is the next frontier in aging research. One day it might be possible to reprogram cells via pills that stimulate the activity of the OSK factors or the TETs. This may be simpler than it sounds. Natural molecules stimulate the TET enzymes, including vitamin C and alpha-ketoglutarate, a molecule made in mitochondria that is boosted by CR and, when given to nematode worms, extends their lifespan, too.

For now, though, the best bet is gene therapy.

Because it could be so impactful, we should start debating the ethics of this technology now, before it arrives on our doorstep. The first question is who should be allowed to use this technology. A select few? The rich? The very sick? Should doctors let people who have terminal illnesses try it for so-called compassionate use? How about people over 100? Or 80? Or 60? When does the reward outweigh the risk?

There is an army of people willing to "boldly go," sound-minded volunteers in their 90s and 100s whose bodies have been broken by the disease of aging. I can assure you that there is no shortage of those who, having peered up the road at perhaps a few more years of life that is defined by ever-increasing frailty and pain, are ready to take a chance at a few more good years, if not for that, then for the chance to give their children, grandchildren, and great-grandchildren a longer, healthier life. What do they have to lose, after all?

The ethics of the technology become more difficult, though, if reprogramming becomes safe enough to use in a way that is preventive. At what age should it be given? Does a disease have to appear before an antibiotic activator of reprogramming is prescribed? If mainstream doctors refuse to help, will people head overseas? If the technology could significantly cut health care costs, should it be mandated?

And if we can help children live longer, healthier lives, do we have a moral obligation to do so? If reprogramming technology can help a child repair an eye or recover from a spinal injury, should the genes be delivered before an accident happens so they are ready to be switched on at a moment's notice, starting perhaps with an antibiotic drip in the ambulance?

If smallpox were to return to our planet, after all, parents who refused to vaccinate their children would be pariahs of the lowest order. When safe and effective treatments are available for a common childhood disease, parents who refuse to save their children's lives can have their guardianship overridden by the doctrine of *parens patriae*.

Should every human have a choice to suffer from aging? Or should that choice be made, as vaccine decisions are in most cases, for the good of both individuals and humankind? Will those who elect to be rejuvenated still have to pay for those who have decided not to? Is it morally wrong not to do so, knowing you will prematurely become a burden on family members?

These are theoretical questions today, but they probably won't remain theoretical for long.

In late 2018, a Chinese researcher, He Jiankui, reported that he had helped create the world's first genetically altered children—twin girls whose births sparked a debate in scientific circles about the ethics of using gene editing to make "designer babies."

The side effects of inducing DNA damage in embryos and the accuracy of gene editing are not well understood yet, which is why the scientific community has had such a violent negative reaction. There is also a tacit reason: scientists are fearful that gene-editing technologies, if abused, will go the way of GMOs and be outlawed for political or irrational reasons before their true potential can be realized.

These fears may be unfounded. If news of the first genetically modified children had broken in the 2000s, it would have sparked global debate and dominated the news for months. Protesters would have stormed labs, and presidents would have banned this use of the technology on embryos. But how times have changed. With a news cycle of hours and politics dished out over the internet, the story lasted a few days; then the world moved on to other, more interesting topics.

He's stated intention was to give the twins the ability to resist HIV. That may sound admirable, but if I do the numbers, the risk wasn't worth it. The chance of contracting HIV in China is less than one in a thousand. If He was going to maximize health benefits to offset the risks of the procedure, why not edit a gene that causes heart disease, which has an almost one-in-two chance of killing them?[12] Or aging, which has a

90 percent chance of killing them? HIV immunity was just the simplest edit, not the most impactful.

As these technologies become commonplace and parents ponder how to get the biggest bang for the buck, how long will it be before another rogue scientist teams up with the world's most driven helicopter parent to create a genetically modified family with the capacity to resist the effects of aging?

It may not be long at all.

SEVEN

THE AGE OF INNOVATION

THE FOUR PRESCRIPTION MEDICINES KUHN LAWAN WAS TAKING WERE PRECISELY right for the cancer with which she had been diagnosed. But the drugs weren't working. Not even a bit. The elderly Thai woman's lung cancer persisted. And with it, it seemed, the end of her life was growing ever nearer.

Her children were understandably distraught. The doctors had told them that Lawan's cancer was likely treatable. They seemed to have caught it early, after all. The fear and uncertainty they'd felt when their mother was first diagnosed had been replaced with hope, only to give way, once again, to fear and uncertainty.

Dr. Mark Boguski has spent a long time thinking about people such as Lawan and about how modern medicine has long failed so many people like her, especially later in life.

"In the most common manner of medical thinking, Lawan was getting the right care," he told me one day. "Her doctors in Thailand were top notch. But that's the thing about how we do medicine."

Most doctors, he said, still rely on early-twentieth-century technology to diagnose and treat life-threatening diseases. Take a swab and grow it in a petri dish. Bang the knee and wait for a kick. Breathe in, breathe out. Look to the left and cough.

When it comes to cancer, doctors note where a tumor is growing and cut out a tissue sample. Then they send it to a lab, where it is put into wax, cut into thin slices, stained with red and blue dyes, and looked at under the microscope. That works—sometimes. Sometimes the correct medicine is given.

But sometimes it isn't. That's because, the way I see it, looking at a tumor in this way is the equivalent of a mechanic trying to diagnose a car's faulty engine without plugging into the vehicle's computer. It's an educated guess. Most of us accept this sort of approach when it comes to potentially life-and-death decisions. Yet in the United States alone, with one of the better health care systems in the world, about 5 percent of cancer patients, or 86,500 people, are misdiagnosed every year.[1]

From the time he began studying computational biology in the early 1980s, Boguski has been driven by the idea of making medical care more exacting. He is a luminary in the field of genomics—and one of the first scientists engaged in the Human Genome Project.

"What we call 'good medicine' is doing what works for most of the people most of the time," Boguski told me. "But not everyone is most people."

And so there was a chance, and not a small one, that Kuhn Lawan was getting the wrong care. And that might have actually been making her worse.

But Boguski believes there is hope in a new way of doing medicine. A better way. A way that uses new technologies, many that are already here but simply not being utilized to their fullest potential, to refocus our medical system on individuals—upending centuries of deeply entrenched medical culture and philosophy. He coined the term *precision medicine* to describe the promise of next-generation health monitoring, genome sequencing, and analytics for treating patients based on personal data, not diagnostic manuals.

Thanks to the plummeting prices of DNA sequencing, wearable devices, massive computing power, and artificial intelligence, we're moving into a world in which treatment decisions no longer have to be based on what is best for most people most of the time. These technologies are available to some patients now and will be available to most people on the planet in the next couple of decades. That's going to save millions of lives—and it's going to extend average healthy lifespans irrespective of whether we extend maximum lifespans.

But for millions of people like Lawan, these advances cannot come soon enough. When her family sought a second opinion in the form of precision DNA sequencing of her lung tumor biopsy, the totality of the danger she was in became crystal clear. Lawan did have an aggressive cancer but not the kind of cancer for which she was being treated. She didn't have lung cancer; she had a solid form of leukemia growing in her lung.

In the vast majority of cases in which cancer is found where it was found in Lawan's body, it is indeed lung cancer. But now that we can detect the genetic signature of specific forms of cancer, using the place where you find the cancer as the only guide for what treatment to use is as ridiculous as categorizing an animal species based on where you've located it. It is like saying a whale is a fish because they both live in water.

Once we have a better idea of what kind of cancer we're dealing with, we can better apply emerging techniques for dealing with it. We can even design a therapy tailored to a patient's specific tumor—killing it before it has a chance to grow or spread to another place in the body.

That's the idea behind one of the cancer-fighting innovations we discussed earlier, CAR T-cell therapy, in which doctors remove immune system cells from a patient's blood and add a gene that allows the cells to bind to proteins on the patient's tumor. Grown en masse in a lab and then reinfused into the patient's body, the CAR T-cells go to work, hunting down cancer cells and killing them by using the body's own defenses.

Another immuno-oncology approach we discussed earlier, checkpoint blockade therapy, quashes the ability of cancerous cells to evade detection by our immune systems. Much of the early work on this technique was completed by Arlene Sharpe, whose lab is located on the floor above mine at Harvard Medical School. In this approach, drugs are used to block the ability of cancer cells to present themselves as regular cells, essentially confiscating their fake passports and thus making it easier for T-cells to discriminate between friend and foe. This is the approach that was used, along with radiation therapy, by former president Jimmy Carter's doctors to help his immune system fight off the melanoma in his brain and liver. Prior to this innovation, a diagnosis like his was, without exception, fatal.

CAR-T therapy and checkpoint inhibition are less than a decade old. And there are hundreds of other immuno-oncology clinical trials under way. The results thus far are promising, with remission rates of greater than 80 percent in some studies. Doctors who have spent their entire careers fighting cancer say this is the revolution they've been waiting for.

DNA-sequencing technology has also offered us an opportunity to understand the evolution of a specific patient's cancer. We can take single cells from a slice of a tumor, read every letter of the DNA in those cells, and look at the cells' three-dimensional chromatin architecture. In doing so, we can see the ages of different parts of the tumor. We can see how it has grown, how it has continued to mutate, and how it has lost its identity over time. That's important, because if you look at only one part of a tumor—an older part, for instance—you could be missing the most aggressive part. Accordingly, you might treat it with a less effective therapy.

Through sequencing, we can even see what kinds of bacteria have managed to make their way into a tumor. Bacteria, it turns out, can protect tumors from anticancer drugs. Using genomics, we can identify which bacteria are present and predict which antibiotics will work against those single-celled tumor protectors.

We can do all of this. Right now. Yet in many hospitals around the world the if-it-is-here-it-must-be-this and if-the-symptoms-are-this-it-must-be-that modes of diagnosis are still practiced. And so, procedurally speaking, the doctors who treated Lawan had done nothing wrong. They had simply done what doctors all over the world do, following an empirical process of diagnosis and intervention that leads to positive outcomes in most people most of the time.

If you accept that this is simply the way we care for people—and that it usually produces the right results—you could call this an understandable medical approach. But if you picture your own mother accidentally receiving a cancer treatment she doesn't need while the medicine that will save her life sits on a shelf nearby, you'll probably come to a different conclusion about what is, in fact, "understandable."

The hardworking, ethical doctors, nurses, and medical professionals who go to war with death every day, while navigating the overarching standard-of-care stipulations of governmental programs and insurance companies, should never be expected to be perfect. But we can prevent a lot of unnecessary deaths by giving medical staff more information, just as Lawan's doctors were able to get her onto a new treatment regimen once they better understood what they were dealing with.

Indeed, it wasn't long after Lawan's DNA-based diagnosis that she was on a new treatment regimen—one that was specific to the actual cancer in her body. Months later she was doing much better. Hope had returned.

There is hope for all of us. We know that humans, both male and female, are capable of living past the age of 115. It has been done, and it can be done again. Even for those who reach only their 100th year, their 80s and 90s could be among their best.

Helping more people reach that potential is a matter of bringing costs down and using emerging treatments, therapies, and technologies in a way that truly puts individuals at the center of their own care. And that's not just about diagnosing people right when something does go wrong—it's also about knowing what to do for us, as individuals, even before a diagnosis has been made.

KNOWING THYSELF

Since the new millennium, we've been told that "knowing our genes" will help us understand what diseases we are most susceptible to later in life and give us the information we need to take preventive actions to live longer. That is true, but it is only a small part of the DNA-sequencing revolution that is under way.

There are 3.234 billion base pairs, or letters, in the human genome. In 1990, when the Human Genome Project was launched, it cost about $10 to read just one *letter* in the genome, an A, G, C, or T. The entire project took ten years, thousands of scientists, and cost a few billion dollars. And that was for one genome.

Today, I can read an entire human genome of 25,000 genes in a few days for less than a hundred dollars on a candy bar–sized DNA sequencer called a MinION that I plug into my laptop. And that's for a fairly complete readout of a human genome, plus the DNA methyl marks that tell you your biological age.[2] Targeted sequencing aimed at answering a specific question—such as "What kind of cancer is this?" or "What infection do I have?"—can now be done in less than twenty-four hours. Within ten years, it will be done in a few minutes, and the most expensive part will be the lancet that pricks your finger.[3]

But those aren't the only questions that our DNA can answer. Increasingly, it can also tell you what foods to eat, what microbiomes to cultivate in your gut and on your skin, and what therapies will work best to ensure that you reach your maximum potential lifespan. And it can give you guidance for how to treat your body as the unique machine it is.

It's common knowledge that we don't all respond to drugs in the same way. Sometimes these aren't small differences in small numbers of people. G6PD genetic deficiency, for instance, affects 300 million people of primarily Asian and African descent. It's the most common genetic disease of humanity. After ingesting recommended doses of medicines for headaches and malaria and certain antibiotics, G6PD carriers can be caught unawares by hemolysis—what amounts to red blood cell mass suicide.[4]

Some mutations sensitize people to particular foods. If you're a G6PD carrier, for example, fava beans can kill you. And while gluten is usually a harmless protein that comes in foods rich in the fiber, vitamins, and minerals we need, for those with celiac disease, it's a poison.

The same is true of medical interventions: our genes can tell us which are better for us and which could do more harm than good. That's changing the game for many breast cancer patients. Those who score in a certain range on a genetic test called Oncotype DX, it has been discovered, respond every bit as well to hormone treatments as they do to chemotherapy, the latter of which has far more side effects.[5] The tragedy of this discovery is that it didn't come until 2015. The Oncotype DX test has been in use since 2004, but it wasn't until a team of researchers decided to take another look at possible treatment options and outcomes that it became clear that the medical community had been subjecting tens of thousands of women to treatments that were more harmful and no more effective.

What Lawan's case and this study demonstrate is that we can't simply rest on "this is how we do it" as a strategy for treating patients. We need to be constantly challenging the assumptions upon which medical manuals are based.

One of these assumptions is that males and females are essentially the same. We're all too slowly coming around to the shameful recognition that, for most of medical history, our treatments and therapies have been based on what was best for males,[6] thus hindering healthy clinical outcomes for females. Males don't just differ from females at a few sites in the genome; they have a whole other chromosome.

The bias begins early in the drug development process. Until recently, it was perfectly fine to study male mice only. Scientists generally aren't rodent sexists, but they are always trying to reduce statistical noise and save precious grant money. Ever since female mice have been regularly included in lifespan experiments, thanks largely to NIH stipulations, large gender differences in the effects of longevity genes and molecules have been seen.[7] Treatments that work through insulin or mTOR signaling typically favor females, whereas chemical therapies typically favor males, and no one really knows why.[8]

If females and males are in the same environment, in general, females will live longer. It's a common theme throughout the animal kingdom. Scientists have tested whether it is the X chromsome or the ovary that is important. Using a genetic trick, they created mice with one or two Xs, with either ovaries or testes.[9] Those with a double dose of the X lived longer, even if they had testes and especially if they didn't, thus proving once and for all that female is the stronger sex.

Besides the X, there are dozens of other genetic factors at play. One of the most promising uses of genomics is predicting how drugs will be metabolized. That's why an increasing number of drugs now have pharmacogenetic labels—information about how the medication is known to act differently among people of different genotypes.[10] Examples include the blood-thinning drugs Coumadin and Plavix, the chemotherapy drugs Erbitux and Vecitibix, and the depression drug Celexa. In the future, a patient's epigenetic age will also be determined and used to predict drug responses, a new field called pharmacoepigenetics. It's a rapidly advancing technology but some pharmacogenetic tests can't come soon enough.

For more than two hundred years, the drug digoxin from the digitalis family of plants has been used in small doses by doctors to treat failing hearts (and in larger doses by murderers).[11] Even under a doctor's supervision, your chance of death if you are on digoxin increases by 29 percent, according to one study.[12]

To help reduce fluid buildup owing to her weakening heart, my mother was prescribed digoxin. I had no idea of the risk and I suspect neither did my mother, who was sensitive to the drug. She steadily declined from living a reasonably normal life to being barely able to walk. Fortunately, my father, a biochemist and a pretty smart guy, diagnosed the problem: the amount of the drug prescribed was superlow, but it had been accumulating in my mom's heart. He told the doctor to test for drug levels, which she reluctantly agreed to do, and the test came back positive for an overdose.

The drug was immediately discontinued, and my mom recovered to her original self in a matter of weeks. Yes, the doctor should have done regular blood tests for drug levels, but if a test for sensitivity to digoxin prior to prescription existed, the doctor could have been on high alert. How close are we to a test? Not close enough. A few studies have identified genetic variants that predict digoxin blood levels and risk of death, but they haven't been repeated.[13] Hopefully, there will be a pharmacogenetic test for this drug soon, as well as for many more. They are badly needed. We cannot keep prescribing medicines as though we all respond to them the same way, because we don't.

Drug developers have figured this out. They are using genomic information to find new and revive failed drugs that work for people with specific genetic variations. One of these drugs is Bayer's Vitrakvi, known generically as larotrectinib, which is the first of many drugs to be designed from the beginning to treat cancers with a specific genetic mutation, not where in the body the cancer came from. A similar story is being written about the failed blood pressure drug Gencaro. It worked well on a subset of the population and, if revived by the FDA, would become the first heart drug to require a genetic test.

This is the future. Eventually, every drug will be included in a huge and ever-expanding database of pharmacogenetic effects. It won't be long before prescribing a drug without first knowing a patient's genome will seem medieval.

And vitally, with genomic information aiding in our doctors' decisions, we won't have to wait to become sick to know what treatments will work best to prevent those diseases from developing in the first place.

As Julie Johnson, the director of the University of Florida's Personalized Medicine Program, has pointed out, we are about to enter a world in which our genomes will be sequenced, stored, and already red-lighted for treatments that have been demonstrated to have adverse effects on people with similar gene types and combinations as we have.[14] Likewise, we'll be green-lighted for treatments that are known to work for people with similar genes, even if those treatments don't work for most other people most of the time. This will be particularly important in developing countries, where the local genetics and gut flora are wildly different from the population the drug was tested on.[15] These differences are rarely talked about in medical circles, but they can have a marked effect on drug efficacy and patient survival, including the efficacy of what are thought of as well-understood cancer chemotherapics.[16]

We are also learning to read the entire human proteome—all of the proteins that can be expressed by every type of cell. Researchers in my lab and others have discovered hundreds of new proteins in human blood, and each protein can tell a story about the kind of cell from which it came, a story we can use to understand what diseases are in our bodies long before they are detectable any other way. That will offer a faster, better view of the problems we're facing, giving doctors the ability to target those problems with far greater precision.

Right now, when people fall ill, especially older people, they often wait to see if things just "work themselves out" before making an appointment to see a doctor. Only when symptoms persist do they make the call. Then they have to wait—nearly a month, according to one 2017 study—before they are able to see a physician. That wait time has been growing in recent years, owing to the combination of a doctor shortage and an increase in baby boomer patients. And in some places, it's much worse. In the city in which I live, Boston—home to twenty-four of the best hospitals in the world—the wait is fifty-two days.[17] That's atrocious.

Long wait times aren't just in the United States, which has a largely private medical system; Canada's socialized system has notoriously long wait times, too. The problem isn't how we pay for care; the problem is that we've set up doctors as the only conduits to diagnosis and often, in the case of primary care physicians, as the only people who can refer a patient to a specialist.

The backlog could clear soon, thanks to technologies that give doctors the ability to conduct video home visits. Within a decade, using a device the size of a package of gum and possibly even disposable, it will be technically feasible to collect the samples your doctor needs at home, plug the device into your computer, and look together at a readout of your metabolites and your genes.

There are more than a hundred companies just in the United States pursuing lightning-fast, superfocused DNA testing that can offer us early and accurate diagnoses of a vast range of ailments and even estimate our rate of biological aging.[18] A few are aimed at detecting the genetic signature of cancer and other illnesses years before they can normally be detected. Soon, we will no longer have to wait for tumors to grow so big and so heterogeneously mutated that their spread is no longer controllable. With a simple blood test, doctors will be able to scan for circulating cell-free DNA, or cfDNA, and diagnose cancers that would be impossible to spot without the aid of computer algorithms optimized by machine learning processes trained on thousands of cancer patient samples. These circulating genetic clues will tell you not just if you have cancer but what kind of cancer you have and how to kill it. They will even tell you where in your body an otherwise undetectable tumor is growing, since the genetic (and epigenetic) signatures of tumors in one part of the body can be vastly different from those from other parts.[19]

All of this means we're on the way to a fundamental shift in the way we search for, diagnose, and treat disease. Our flawed, symptom-first approach to medicine is about to change. We're going to get ahead of symptoms. Way ahead. We're even going to get ahead of "feeling bad." Many diseases, after all, are genetically detectable long before they are symptomatic. In the very near future, proactive personal DNA scanning is going to be as routine as brushing our teeth. Doctors will find themselves saying the words "I just wish we'd caught this earlier" less and less—and eventually not at all.

But the coming age of genomics is just the start.

THE RIGHT TRACK

The dashboard on a car equipped with intelligent vehicle technologies is a marvelous thing. It can tell you how fast you're going, of course, and how many miles the vehicle has remaining before it needs a refill—adjusted second by second based on the conditions of the road and the way in which you are driving. It can tell you the temperature outside, inside, and under the hood. It can tell you what cars, bicycles, and pedestrians are around you and warn you if they're getting a little close for comfort. When something is wrong—a tire with too little air or a transmission that isn't shifting perfectly—it can let you know that, too. And if you get a bit distracted and begin to veer over the line, it will take control of the wheel and pull you back on course or drive autonomously down the highway, with no more than a bit of resistance from a hand on the steering wheel to tell it there's a human there, just in case.

Back in the 1980s, there were very few sensors in cars. But by 2017, there were nearly 100 sensors in each new vehicle—a number that had doubled in the prior couple of years.[20] Car buyers increasingly expect features such as tire sensors, passenger sensors, climate sensors, nighttime pedestrian warning sensors, steering angle guides, proximity alerts, ambient light sensors, washer fluid sensors, automatic high-beam, rain sensors, blind spot detection sensors, automatic suspension lift, voice recognition, automatic reverse parking, active cruise control, auto emergency braking, and autopilot.

Perhaps there are people out there who'd be happy to drive without any dashboard at all, relying solely on their intuition and experience to tell them how fast they are going, when their car needs fuel or recharging, and what to fix when something goes wrong. The vast majority of us, however, would never drive a car that wasn't giving us at least some quantitative feedback, and, through our purchasing decisions, we have made it clear to car companies that we want more and more intelligent cars.

Of course we do. We want them to protect us, and we want them to last.

Surprisingly, we've never demanded the same for our own bodies. Indeed, we know more about the health of our cars than we know about our own health. That's farcical. And it's about to change.

We've already taken some pretty big steps into the age of personal biosensors. Our watches monitor our heart rate, measure our sleep cycles, and can even provide suggestions for food intake and activity. Athletes and health conscious individuals are increasingly wearing sensors twenty-four hours a day that monitor the ways in which their vital signs and major chemicals are rising and falling in response to diet, stress, training, and competition.

As just about anyone with diabetes or HIV can attest, blood sugar and blood cell monitoring are exceptionally easy and increasingly pain-less affairs these days, with noninvasive and minimally invasive monitoring technologies ever more available, affordable, and accurate.

In 2017, the US Food and Drug Administration approved a glucose sensor, first launched in Europe in 2014, that you stick on your skin to provide a constant readout of blood sugar levels on your phone or watch. In thirty countries, a finger prick for diabetics is becoming a distant memory.

Rhonda Patrick, a longevity scientist turned health and fitness expert, has been using a continual blood glucose–sensing device to see what foods give her body a major sugar spike, something many of us believe is to be avoided if we are to give ourselves the greatest chance of a long life. She's seen that, at least for her, white rice is bad and potatoes aren't so bad. When I asked her what food had been the most surprising, she didn't hesitate.

"Grapes!" she exclaimed. "Avoid grapes."

Researchers at MIT are working on scanners, straight out of *Star Trek*, that can give readouts of thousands of biomarkers. Meanwhile, researchers at the University of Cincinnati have been working with the US military to develop sensors that can identify diseases, diet changes, injuries, and stress through sweat.[21] A few companies are developing hand-held breath analyzers that can diagnose cancer, infectious diseases, and inflammatory diseases. Their mission: to save 100,000 lives and $1.5 billion in health care costs.[22] Numerous other companies are working on designing clothing with sensors that can track biomarkers, and automotive engineers are exploring putting biosensors in car seats that would send an alert to your dashboard or doctor if there's something amiss in your heart rate or breathing pattern.

As I write this, I am wearing a regular-sized ring that is monitoring my heart rate, body temperature, and movements. It tells me each morning if I slept well, how much I dreamed, and how alert I will be during the day. Technology like this has been around for some time, I suppose, for people such as Bruce Wayne and James Bond. Now it costs a few hundred dollars and can be ordered by anyone online.[23]

Recently, my wife and eldest child came home with matching ear piercings, which got me thinking: there's really no reason that an even smaller piece of body jewelry—particularly one that pierces the skin—couldn't be used to track thousands of biomarkers. Every member of the family could be measured: grandparents, parents, and children. Even infants and four-legged family members will have monitors on them, because they are the ones who are least able to tell us what they are feeling.

Eventually, I suspect, very few people will want to live without tech like this. We won't leave home without it, the same way we feel about our smartphones. The next iteration will be innocuous skin patches, eventually giving way to under-skin implants. Future generations of sensors will measure and track not only a person's glucose but his or her basic vital signs, the level of oxygen in the blood, vitamin balance, and thousands of chemicals and hormones.

Combined with technologies that coalesce data from your day-to-day movements and even the tone of your voice,[24] your biometric vitals will be the bellwether for your body. If you are a man who has been spending more time in the bathroom than usual, your AI guardian will check for prostate-specific antigens and prostate DNA in your blood, then book you an appointment for a prostate exam. Changes in how you move your hands while speaking, and even the manner in which your strike the keys on your computer,[25] will be used to diagnose neurodegenerative diseases years before symptoms would be noticed by you or your doctor.

One biotechnological advancement at a time, this world is coming, and fast. Real-time monitoring of our bodies, the likes of which we could hardly have imagined a generation ago, will be as inherent to the experience of living as dashboards are to the experience of driving. And for the first time in history, that will permit us to make data-driven day-to-day health decisions.[26]

The most critical daily decisions that affect how long we live are centered around the foods we eat. If your blood sugar is high at breakfast, you'll know to avoid sugar in your morning coffee. If your body is low on iron at lunch, you'll know it and can order a spinach salad to compensate. When you get home from work, if you've failed to go outside for your daily dose of vitamin D from the sun, you'll know that, too, and you'll be able to mix up a smoothie that will address the deficiency. If you're on the road and you need X vitamin or Y mineral, you'll know not only what you need but where to get it. Your personal virtual assistant—the same AI-driven being who answers your internet search queries and reminds you about your next meeting—will point you to the nearest restaurant that has what you need or offer to have it delivered by a drone to wherever you are. It could, quite literally, be dropped into your hands from the sky.

Biometrics and analytics already tell us when and how much to exercise, but increasingly they will also help us monitor the effects of our exercise—or lack thereof. And our levels of stress. And even how the fluids we drink and the air we're breathing are impacting our body's chemistry and functionality. Increasingly, our devices will offer recommendations on what to do to mitigate suboptimal blood biomarkers: to take a walk, meditate, drink a green tea, or change the filter on the air conditioner. This will help us make better decisions about our bodies and our lifestyles.

All of this is coming soon. There are companies that are crunching data from hundreds of thousands of blood tests, comparing them to customers' genomes and providing feedback to them on what to eat and how to truly optimize their particular bodies, and looking to roll out new generations of these technologies every year.

I am fortunate to have been one of the first people to get an early look at what this sort of technology can offer us. I am a scientific adviser to a local company, spun out of MIT, called InsideTracker.[27] By signing up for regular tests, I have been able to follow a few dozen blood biomarkers over the past seven years, including vitamins D and B$_{12}$, hemoglobin, zinc, glucose, testosterone, inflammatory markers, liver function, muscle health markers, cholesterol, and triglycerides. My tests are taken every few months instead of every few seconds—as they will soon be in our future—but the reports, adjusted to my specific age, sex, race, and DNA, have been instrumental in helping me choose what to order when I sit down at a restaurant and what to pick up when I stop at the market on my way home. I can even have daily text messages, based on my most recent results, that remind me what my body needs.

Along the way, I am creating data specific to *my* body. And over time, that data is helping me identify negative and positive trends that may be subtly different for me than for other people. We know, of course, that our genetic heritage can have a significant impact on the sorts of food our bodies need, tolerate, or reject, but everyone's genetic heritage is different. What you need, what your partner needs, and what your children need can likely be found in the meals you put on your table, but the particulars may be quite different.

Biotracking will also help us stop acute and traumatic preventable deaths—by the millions. In 2018, a peer-reviewed study published by the team at InsideTracker and me, showed that biotracking and computer-generated food recommendations reduce blood sugar levels as efficiently as the leading diabetes drug, while optimizing other health biomarkers, too.

The signs of an increasingly blocked carotid artery might be hard to notice in our day-to-day lives, or even in periodic visits to the doctor, but they will be almost impossible to miss when our bodies are being

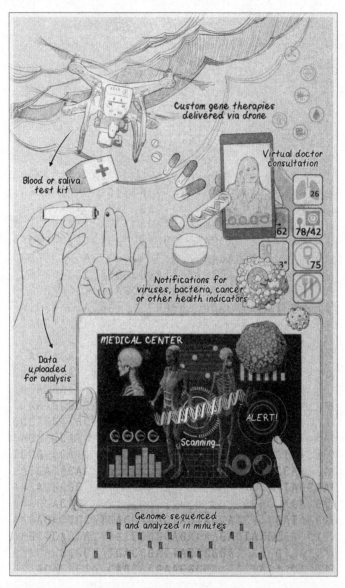

Custom gene therapies
delivered via drone

Virtual doctor
consultation

Blood or saliva
test kit

Notifications for
viruses, bacteria, cancer
or other health indicators

Data
uploaded
for analysis

MEDICAL CENTER

ALERT!

Scanning...

Genome sequenced
and analyzed in minutes

TECHNOLOGIES TO EXTEND OUR LIVES. In the near future, families will be monitored by biosensing wearables, small devices at home, and implants that will optimize our health and save lives by suggesting meals and detecting falls, infections, and diseases. When an anomaly is found, an AI-assisted, videoconferenced doctor will send an ambulance, a nurse, or medicines to your door.

measured and monitored all the time. Same, too, for heartbeat irregularities, minor strokes, venous blockages during air medical transport, and many other medical problems that currently are almost always treated in critical care conditions—when it is too late. Before, if you suspected your heart was malfunctioning, and even if you didn't, it would take a visit to a couple of doctors to get an electrocardiogram. Now millions of people can conduct their own accurate ECG in 30 seconds, wherever they are, just by pressing their finger to the dial on their watch.

Of course, I use the term *watch* loosely, given that today's wrist devices don't just tell you the time and date. They are also calendars, audiobooks, fitness trackers, email and text programs, newsstands, timers, alarms, weather stations, heart rate and body temperature monitors, voice recorders, photo albums, music players, personal assistants, and phones. If these devices can do all of that, there's no reason we should not expect them to help us avoid traumatic health incidents, too.

In the future, if you are experiencing a heart attack—even if it's perceptible only as a slight pain in your arm—or a ministroke, which so often goes undiagnosed until it's identified on a brain scan years later, you'll be alerted, and so will those around you who need to know. In an emergency, a trusted neighbor, a best friend, or whatever doctor happens to be closest to you can also be alerted. An ambulance will be dispatched to your door. This time, the doctors at the nearest hospital will know *exactly* why you are coming in before you even arrive.

Do you know an emergency room doctor? Ask her about the value of a single minute of additional treatment time. Or a single blood test's worth of additional information. Or a recent electrocardiogram. Or a patient who is still conscious, not in pain, and not suffering from a loss of blood to their brain when they arrive—a person who is able to help in the process of making appropriate emergency health care choices. It may not be long before medics routinely ask for a download of your most recent biotracking data to aid them in making what could be life-and-death decisions.

Biotracking is already helping us identify diseases faster than ever before. That is what happened in the summer of 2017 to a woman named Suzanne. After a time of subtle shifts in her menstrual cycle, changes her doctor very reasonably attributed to a shift into menopause, the 52-year-old woman downloaded an app that helped her track her periods. Three months later, the app sent her an email alerting her to the possibility that her data might be "outside the norm" for women of her age. Armed with that data, Suzanne returned to her doctor. She was immediately sent for blood tests and an ultrasound that revealed mixed Müllerian tumors, a malignant form of cancer found predominantly in postmenopausal women over the age of 65. It took a radical hysterectomy to remove the cancer before it could spread further, but Suzanne's life was spared.[28]

The app she used was relatively simple compared to those that are on the way. It required proactive data entry and tracked only a few metrics. Yet it saved her life. Imagine, then, what "hands-off" trackers that collect millions of daily data points can offer us. Now imagine coupling that data with what we learn from routine DNA sequencing.

And don't stop imagining there. Because biotracking won't just tell you when your heart rate is up, your vitamin levels are low, or your cortisol level is spiking, it will also tell us when our bodies are under attack—and that could save everyone on this planet.

READY FOR THE WORST

In 1918—long before our modern, superfast, hyperconnected global transportation network took shape—a worldwide influenza pandemic that some historians believe originated in the United States killed more people in absolute numbers than any other disease outbreak in history.[29] It was a violent death, with hemorrhage from mucous membranes, especially from the nose, stomach, eyes, ears, skin, and intestines.[30] At a time in which the era of human flight was in its infancy and most people had never ridden in a car, the H1N1 virus found its way to some of the furthest reaches of our globe. It killed people on remote islands and in arctic villages. It killed without regard to race or national boundaries.

It killed like a new Black Death. Average life expectancy in the United States plummeted from 55 to 40 years. It recovered, but not until more than 100 million people of all ages globally had had their lives cut short.

This could happen again. And given how much more humans and animals are in contact and how much more interconnected our planet is now than it was a century ago, it could happen quite easily.

The gains in life expectancy we've witnessed over the past 120 years, and those to come, could be wiped out for a generation unless we address the greatest threat to our lives: other life-forms that seek to prey on us. It doesn't matter if we live decades upon decades longer if a pandemic quickly snuffs out hundreds of millions of lives—negating and even rolling back the gains in average lifespan we will have achieved. Global warming is a long-term, critical issue to deal with, but one could also argue that, at least within our lifetimes, infections are our greatest threat.

Ensuring the next big outbreak never happens could be the greatest gift of the biotracking revolution. Individually, of course, real-time monitoring of vitals and body chemicals offers incredible benefits for

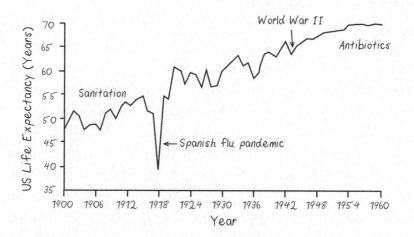

CHANGE IN LIFE EXPECTANCY DURING THE 1918 FLU EPIDEMIC.
Source: S. L. Knobler, A. Mack, A. Mahmoud, and S. M. Lemon, eds., The Threat of Pandemic Influenza: Are We Ready? Workshop Summary, Institute of Medicine (Washington, DC: National Academies Press, 2005), https://doi.org/10.17226/11150, PMID: 20669448.

optimizing health and preventing emergencies. Collectively, though, it could help us get ahead of a global pandemic.

Thanks to wearables, we already have the technology in place to monitor the body temperature, pulse, and other biometric reactions of more than a hundred million people in real time. The only things separating us from doing so are a recognized need and a cultural response.

The need is already here. It has been for quite some time. It took about twenty years for the deadly mosquito-transmitted Zika virus to spread from Central Africa, where it was first documented, to South Asia and about forty-five more years to reach French Polynesia in the Central Pacific in 2013. In the span of those sixty-five years, it affected just a small part of the world. In the next four years, though—four years, that's all—the virus spread like wildfire across South America, through Central America, into North America, and back across the Atlantic Ocean to Europe.

Zika, at least, is somewhat limited in the way it can be spread—mostly through mosquito bites but also from mother to child and from sexual partner to sexual partner. It cannot, as far as we know, be transmitted by doorknobs, via food, or in the air-recirculating climate control systems on airplanes.

But influenza can, as can other, potentially deadlier viruses.

On March 23, 2014, the World Health Organization reported cases of Ebola virus disease in the forested rural region of southeastern Guinea, and from there it spread rapidly to three neighboring countries, causing widespread panic. Even the richest country in the world, where eleven people were treated for Ebola, was caught without a unified plan.

That October, people in hazmat suits boarded American Airlines flight 45 when it landed in New Jersey to shine infrared heat detectors on people's foreheads in an attempt to detect a fever. Kaci Hickox, who worked for Doctors Without Borders, later won a lawsuit that led to a "quarantine bill of rights," after she was placed in Governor Chris Christie's "private prison." On that occasion, and a few since then, the deadly virus has been contained, but humanity may not always be so lucky.

"Whether it occurs by a quirk of nature or at the hand of a terrorist, epidemiologists say a fast-moving airborne pathogen could kill more than 30 million people in less than a year," Bill Gates told a crowd at the Munich Security Conference in 2017, "and they say there is a reasonable probability the world will experience such an outbreak in the next 10–15 years."[31]

If that happens, 30 million could be a very conservative estimate.

As our transportation networks continue to expand in reach and speed, as more people travel to more corners of our world faster than our ancestors could possibly have imagined, pathogens of all sorts are traveling faster than ever, too. But with the right data in the right hands, we can move faster, especially if we combine mass "biocloud" data with superfast DNA sequencing to detect pathogens as they spread through cities and along transportation corridors. In doing so, we can get ahead of a killer pathogen with emergency travel restrictions and medical resources. In this fight, every minute will matter. And every minute that passes without action will be measured in human lives.

Not everyone is ready for the biotracking world. That makes sense. To many, clearly, it will feel like a step too far. Maybe several steps too far.

In order to get to a world in which hundreds of millions of humans—all being tracked in real time for hormone levels, chemicals, body temperature, and heart rate—are standing as sentinels to warn us of public health crises as they happen, someone is going to have to have the data. Who will that be? One government? A coalition of governments? Any and every government?

Maybe a computer company. Or maybe a pharmaceutical manufacturer. Or an internet shopping company. Or an insurer. Or a pharmacy. Or a supplement company. Or a hospital network.

Most likely, it will be a combination of these companies, all under one roof. Consolidation has already started and will continue as these companies set their sights on the largest and fastest-growing sector of the global economy, health care, which now exceeds 10 percent of global GNP and is increasing at an annual rate of 4.1 percent.

Whom do you trust to know your every move? To listen to your every heartbeat? To see you when you're sleeping and know when you're awake, like a certain benevolent mythical being of wintertime lore? To be able to identify, through the data, when you are feeling sad, driving too fast, having sex, or had too much to drink?

There's no sense in trying to convince people that there is nothing to worry about. Of course there are things to worry about. Think having your credit card data stolen is bad? That's nothing. You can always call the bank and get a new credit card, but your medical records are permanent—and far more personal. More than 110 million medical records were breached in the United States between 2010 and 2018.[32] Jean-Frédéric Karcher, the head of security at Maintel, a UK communications provider, predicts that attacks will become far more common.

"Medical information can be worth ten times more than credit card numbers on the deep web. Fraudsters can use this data to create fake IDs to buy medical equipment or drugs," he has warned.[33]

We already trade a tremendous amount of privacy for technological services. We do it all the time. We do it every time we start a bank account or sign up for a credit card. We do it often when we point our internet browsers to a new web page. We do it when we sign up for school. We do it when we get onto an airplane. And we do it—a lot—when we use our mobile phones. Have these been good trade-offs for everyone? That's a matter of personal opinion, of course. But when most people imagine not being able to use a credit card, surf the web, sign up for school, travel by air, or use their phones and smart watches, they quickly conclude that the trade is tolerable.

Will people trade a little more privacy to stop a global disease pandemic? Sadly, probably not. The tragedy of the commons is that humans are not very good at taking personal action to solve collective problems. The trick to revolutionary change is finding ways to make self-interest align with the common good. For people to accept widespread biometric tracking in a way that could help us get ahead of fast-moving deadly viruses, they'll need to be offered something they have a hard time seeing themselves without.

How to get ready for this world is a conversation that needs to be had. And soon.

I'm there already. Before I began having my biomarkers checked on a regular basis, I did worry about what the ever-changing chemical signals in my body could disclose about me to someone with access to my data. All the data are held on health care– or HIPAA-compliant servers, and the data are encrypted. But there's always the fear that the data will be hacked. There's always a way.

But after I began, the information I received was worth far more than the concerns I carry. It's a personal choice, no question. Now, having seen the changes on my dashboard, I cannot imagine living without it. Just as I now wonder how I ever managed to drive without a GPS, I wonder how I ever made decisions about what I should be eating and how much I should be exercising before I received regular updates from my biosensor ring and blood biomarker reports. Indeed, I am eager for the day in which the data about my health are processed in real time. And if that can help protect others, all the better.

MOVING FASTER

While I was doing my PhD, I had a night job. For about eight dollars an hour, I tested body fluids—urine, feces, spinal fluid, blood, and genital swabs horribly twisted in hair—for the presence of deadly bacteria, parasites, and fungi. It was glamorous work.

At my disposal, I had all the trappings of nineteenth-century technology: microscopes, petri dishes, sterile water. A lab technician transported from 1895 to that 1980s microbiology lab would have felt right at home. Today, this is still how many microbiology labs operate.

Making life-and-death calls this way was frustrating. In every other branch of medicine, we've made enormous strides technologically with robotics, nanotechnology, scanners, and spectrometers.

These days, though, I'm no longer frustrated. I'm furious.

Antibiotic-resistant strains of bacteria continue to spread, and new studies implicate bacteria as causal agents in cancer, heart disease, and Alzheimer's disease.[34]

But I wasn't working to solve this problem, until recently. A brush with Lyme disease has a way of intensifying a person's feelings about these sorts of things.

Our daughter Natalie was 11 years old when it happened. In New England, where we live, there is an epidemic of ticks that carry the bacterial spirochete *Borrelia burgdorferi*, which causes Lyme disease. Recent estimates suggest that approximately 300,000 people in the United States may contract the disease each year. Left untreated, *Borrelia* hides out in skin cells and lymph nodes, causing facial paralysis, heart problems, nerve pain, memory loss, and arthritis. It hides in a protective biofilm, making it extremely difficult to kill.

Natalie never had a red ring of skin around a tick bite—a sure sign you've contracted the parasite. She had been complaining about a headache and sore back, typical signs of flu. But quickly it became clear that this wasn't flu—it was something much worse.

She was unable to turn her head. She was losing her eyesight. She was terrified. My wife and I were, too—we'd never felt so helpless in our lives. We began searching online for answers. Potential diseases included leukemia and a viral infection of the brain.

Doctors at Boston's Children's Hospital began poring over her. The first test lit up Lyme disease proteins, but the insurance company needed confirmation because the first test occasionally gives a false positive. The second test failed, putting the course of treatment into limbo pending more lab results.

I asked for a microliter sample of Natalie's spinal fluid to test. My lab was across the street, and I could sequence the DNA of the pathogen. The hospital refused.

Given the state of her symptoms at that point, I've since learned, she had a 50 percent chance of survival. Her life came down to a coin flip. At a time when every second counted, doctors were waiting on lab results.

It took three days to confirm that it was a Lyme disease infection, and finally the doctors gave Natalie intravenous antibiotics directly into the large vein next to her heart. She received that treatment every day for nearly a month.

She is okay now, but it was clear to everyone involved, especially Natalie, that we desperately need to be applying twenty-first-century technologies to diagnosing infectious diseases. In Cambridge, Massachusetts, and Menlo Park, California, I've helped gather a group of very smart folks—infectious disease doctors, microbiologists, geneticists, mathematicians, and software engineers—to develop tests that can rapidly and unambiguously tell physicians what an infection is and how best to kill it, using "high-throughput sequencing."

The first step in this process is the extraction of nucleic acids from blood samples, saliva, feces, or spinal fluid. Because it adds cost and reduces sensitivity, the patient's DNA is removed using innovative methods honed by the same scientists who extract ancient DNA from mummies—one of countless cases of one field of science benefiting another. Next, the samples are processed through agnostic DNA-sequencing technologies, meaning that the system is not looking for any one specific infectious agent but rather reading the genomes in the entire sample. That list is then scanned against a database of all known human pathogens at the strain level. The computer spits out a highly detailed report about what invaders are present and how best to kill them. The tests are as accurate as the standard ones, but they provide strain-level information and are pathogen agnostic. In other words, soon doctors won't have to guess what to look for when ordering a test or what treatment will work best. They will know.

Just a few years ago, this wouldn't just have been a slow process, it wouldn't even have been possible. Now it can be done in days. Soon it will be able to be done in hours and eventually minutes.

But there's another way to deal with such diseases: we could prevent them altogether.

THE NEW AGE OF INOCULATION

There is no rational debate over the immensely positive impact of vaccines on life expectancy and healthy lifespans over the past century. Childhood mortality around the world has plummeted, in no small part because we've wiped out diseases such as smallpox. The number of healthy children in the world has risen because we've destroyed polio. The number of healthy adults has, too. Within fifty years, postpolio syndrome, which causes fatigue, muscle weakness, abnormal spinal curvature, and speech defects in adults, will be extinct.

And, of course, the more diseases we can vaccinate for, especially those that claim elderly people's lives, such as flu and pneumonia, the more life expectancy will rise in the coming years.

When we inoculate the herd, it doesn't just protect us individually, it protects the weakest among us: the young and the old. Chickenpox once claimed thousands of lives each year around the world—mostly among the very young and the very old—and accounted for hundreds of thousands of hospitalizations and millions of days of missed work. Those days are over.

A shining example of the power of vaccines to extend lifespan came in the years after the introduction of vaccines for *Streptococcus pneumoniae*, a major source of illness in older persons and the most common cause of death by respiratory infection. After the Prevnar vaccine for infants was launched in 2000, hospitalization and deaths from pneumonia fell across the board, according to a study published in the *New England Journal of Medicine*.

"The protective effect we saw in older adults, who do not receive the vaccine but benefit from vaccination of infants, is quite remarkable," the study's first author, Marie Griffin, explained. "It is one of the most dramatic examples of indirect protection, or herd immunity, we have seen in recent years."[35]

In the first three years alone, deaths from pneumonia were halved, averting more than 30,000 cases and 3,000 deaths in the United States alone, according to another study.[36]

We can snuff out a lot of killers with vaccines like this.

Yet for several decades, the promise of vaccines to improve the lives of billions of people around the world has been slowed, not only by a distrust of vaccines promulgated by debunked science, but by plain old market forces. The golden era of vaccine research was in the mid–twentieth century, a time that saw the quick development of a succession of exceptionally effective inoculations against whooping cough, polio, mumps, measles, rubella, and meningitis.

But by the latter part of the century, the business model that long sustained vaccine research and development was badly broken. The cost of testing new vaccines had risen exponentially, thanks in large part to increasing public concerns about safety and risk-averse regulatory bodies. The "low-hanging fruit" of the inoculation world had already been picked. Now a simple vaccine can take more than a decade to produce and cost more than half a billion dollars, and there is still the chance it won't be approved for sale. Even some vaccines that have worked well and been critical for the prevention of epidemics, such as GlaxoSmithKline's Lyme disease vaccine, have been taken off the market because the unfounded backlash against vaccines made continuation of the product "just not worth it."[37]

Governments don't make vaccines; companies do. So when the market forces are not conducive, we don't get the medicines we so badly need. Funding gaps are sometimes made up by charitable organizations, but not nearly enough. And economic downturns such as the global recession of the late 2000s and early 2010s left foundations—many of which base their giving on market-tied endowment earnings—unable or unwilling to invest as much in these lifesaving interventions.[38]

The good news is that we are experiencing a minirenaissance in vaccine research and development, which has tripled between 2005 and 2015, now accounting for about a quarter of all biotechnology products being developed.[39]

The big one is malaria, infecting 219 million people and claiming 435,000 people in 2017.[40] Thanks to Bill and Melinda Gates, GlaxoSmithKline, and Program for Appropriate Technology in Health (PATH), a partially effective vaccine against malaria known as Mosquirix was deployed for the first time in 2017, giving hope that the malaria parasite will one day be pushed to extinction.[41]

We are also learning how to quickly grow vaccines in human cells, mosquito cells, and bacteria, avoiding the time and expense of infecting the millions of fertilized chicken eggs we currently use, a remarkably antiquated process. One Boston-based research consortium was able to get a vaccine for Lassa fever, a disease similar to Ebola, all the way to the animal-testing stage in just four months and for about $1 million, cutting many years and many millions of dollars from the usual process.[42] That's nothing short of astounding.

At this moment, researchers are starting the final sprint toward the end of a very long race to develop vaccines that will inoculate us against diseases that are so ubiquitous that we simply accept them as part of life. Many thought leaders predict, though with some trepidation, that it won't be long before we're no longer throwing Hail Mary passes such as the annual influenza vaccine, which in some years protects less than a third of its recipients, which is still far better than nothing. (If you don't get flu vaccines or vaccinate your kids, please do. We are privileged to live in an age in which we can protect ourselves and our children from potentially deadly diseases.)

The ability to quickly detect, diagnose, treat, and even prevent diseases that aren't related to aging but that claim many millions of lives each year will allow us to continue to push our average life expectancy higher and higher, closing the gap between the mean and the maximum.

Even then, organs will fail and body parts will wear out. What will we do when all other technologies fail? There's a revolution happening there, too.

THE ORGAN GRIND

The Great Ocean Road, which runs along the Australian coast west of Melbourne, is among the most beautiful stretches of highway in the world. But whenever I'm on it, I can't help but remember one of the most frightening days of my life—the day I got a call telling me that my brother, Nick, had been in a motorcycle accident.

He was 23 years old at the time and touring the country by motorbike. He was an expert rider, but he hit an oil patch, flew from his bike, and slid under a metal barrier that crushed his ribs and ruptured his spleen.

Fortunately, he pulled through, but to save his life, the emergency room doctors had to remove his spleen, which is involved in the production of blood cells and is an important part of the immune system. For the rest of his life, he has to be careful not to get a major infection, and he certainly seems to get sicker more often and take longer to get better. People without a spleen are also at higher risk of dying of pneumonia later in life.

It doesn't take age or disease to do a number on our organs. Sometimes life does that to us in other ways, and we're lucky if it's just our spleen that we lose. Hearts, livers, kidneys, and lungs are a lot harder to live without.

The same kind of cellular reprogramming we can use to restore optic nerves and eyesight may one day offer us the ability to restore function in damaged organs. But what can we do for organs that have completely failed or need to be removed because of a tumor?

Right now, there's only one way to effectively replace damaged and diseased organs. It's a morbid truth, but it's a truth nonetheless: when people pray for an organ to become available for a loved one who needs one, part of what they're praying for is a deadly car accident.

There's a lot of irony, or some would say logic, in the fact that the Department of Motor Vehicles is the organization that asks people whether they want to be organ donors: each year in the United States alone, more than 35,000 people are killed in motor vehicle accidents, making this mode of death one of the most reliable sources of tissues and organs. If you haven't signed up to be an organ donor, I hope you consider it. Between 1988 and 2006, the number of people waiting for a new organ grew sixfold. As I write this sentence, there are 114,271 people on the

US online registry waiting for organ transplants, and every ten minutes, someone new is added to the transplant waitlist.[43]

It's even worse for patients in Japan, where the ability to get an organ transplant remains far below those of Western countries. The reasons are both cultural and legal. In 1968, the Buddhist belief that the body should not be divided after death fueled an emotion-laden firestorm in the media about whether the first Japanese heart donor had truly been "brain dead" when the heart was removed by Dr. Juro Wada. A strict law was immediately enacted that banned the removal of organs from a cadaver until the heart had stopped beating. The law was relaxed thirty years later, but the Japanese remain divided on the issue and good organs remain hard to come by.

My brother also suffers from an eye disease called keratoconus, which caused the corneas covering his lenses to wrinkle like a finger pushed into plastic wrap. To treat this, he had two separate corneal transplant surgeries, one in his twenties, the other in his thirties, that swapped two other people's corneas for his. Both times, he suffered through six months of corneal stitches that felt like "branches" in his eyes, but his vision was saved. The fact that Nick now literally sees the world through others' eyes is an amusing topic for dinner conversation that belies the true depth of our family's gratefulness to his deceased donors.

Now, as we rapidly approach the era of self-driving cars—a technological and social paradigm shift that almost every expert expects will rapidly reduce car crashes—we need to confront an important question: Where will the organs come from?

The geneticist Luhan Yang and her former mentor Professor George Church in my department at Harvard Medical School had just discovered how to gene edit mammalian cells when they began working to edit out genes in pigs. To what end? They envisioned a world in which pig farmers raise animals specifically designed to produce organs for the millions of people who are on transplant waiting lists. Though scientists have had dreams of widespread "xenotransplantation" for many decades, Yang took one of the biggest steps toward that goal when she and her colleagues demonstrated that they could use gene editing to eliminate dozens of retroviral genes from pigs that currently prevent them from donating organs. That's not the only obstacle to xenotransplantation, but it's a big one—and one that Yang figured out how to overcome before her 32nd birthday.

That's not the only way we'll be getting organs in the future. Ever since researchers discovered in the early 2000s that they could modify inkjet printers to lay down 3D layers of living cells, scientists around the world have been working toward the goal of printing living tissue. Today scientists have implanted printed ovaries into mice and spliced printed arteries into monkeys. Others are working on printing skeletal tissue to fix broken bones. And printed skin is likely to start being used for grafts in the next few years, with livers and kidneys coming soon after that and hearts—which are a bit more complicated—a few years behind.

Soon it won't matter if the morbid pipeline for human organ transplantation ends. That pipeline never met the demand anyway. In the future, when we need body parts, we might very well print them, perhaps by using our own stem cells, which will be harvested and stored for just such an occasion, or even using reprogrammed cells taken from blood or a mouth swab. And because there won't be competition for these organs, we won't have to wait for things to go catastrophically wrong for someone else to get one—we'll only have to wait for the printer to do its job.

JUST IMAGINE

Is all this hard to imagine? That's understandable. We've spent a long time building up our expectations of what medical care should look like—and indeed what human life should look like. For a lot of people, it's simply easier to say, "I just don't believe that will happen," and leave things at that.

But we're actually quite a bit better at changing our minds about what we expect out of life, and what age actually means, than many of us think we are.

Consider Tom Cruise. As the *Top Gun* actor entered his late 50s, with bulging muscles and a straight line of dark hair sprouting from a minimally wrinkled forehead, he was still at work. Not just acting, but doing the sort of acting that has long been the purview of much younger actors. He was still doing many of his own dangerous stunts, too: riding motorcycles at high speed through alleys, being strapped to the outside of a plane as it takes off, hanging off the top of the world's tallest building, skydiving from the upper reaches of the atmosphere.

How easily do the words "Fifty is the new thirty" slide from our lips these days? We forget what we used to expect life past 50 to look like, not hundreds of years in the past but just a few decades ago.

It didn't look like Tom Cruise jumping out of airplanes. It looked like Wilford Brimley. In the 1980s, Brimley was one of Cruise's costars in the movie *The Firm*. Cruise was 39 and Brimley 58, already a gray-haired old man with a walrus mustache.

A few years earlier, Brimley had starred in *Cocoon*, a movie about a group of senior citizens who stumble upon an alien "fountain of youth" that gives them the energy—although not the looks—of their youth. The image of old folks running around like teenagers was played to great comedic effect.

It was audacious to think of someone that age acting so youthful. At the time the movie was released, though, Brimley was five or six years younger than Cruise is now. According to Ian Crouch of the *New Yorker*, Cruise has easily blasted through what he calls "the Brimley Barrier."[44]

Barriers fall. And they will fall again. In another generation, we'll be well accustomed to seeing movie stars in their 60s and 70s riding motorcycles at high speeds, jumping from great heights, and delivering kung fu kicks high into the air. Because 60 will be the new 40. Then 70 will be the new 40. And on it will go.

When will this happen? It's already happening. It is not fanciful to say that if you are reading these words, you are likely to benefit from this revolution; you will look younger, act younger, and *be* younger—both physically and mentally. You will live longer, and those extra years will be better.

Yes, it is true that any one technology might lead to a dead end. But there is simply no way that all of them will fail. Taken separately, any of these innovations in pharmaceuticals, precision medicine, emergency care, and public health would save lives, providing extra years that would otherwise have been lost. When we take them together, though, we are staring up the road at decades of longer, healthier life.

Each new discovery creates new potential. Each minute saved in the quest for faster and more accurate gene sequencing can help save lives. Even if it doesn't move the needle much on the maximum number of years we live, this age of innovation will ensure that we stay much healthier much longer.

Not many of us much of the time, but all of us.

PART III

WHERE WE'RE GOING

(THE FUTURE)

EIGHT

THE SHAPE OF THINGS
TO COME

LET'S DO A LITTLE MATH.

And let's make it conservative math. Let's assume that each of these vastly different technologies emerging over the next fifty years independently contributes to a longer, healthier lifespan.

DNA monitoring will soon be alerting doctors to diseases long before they become acute. We will identify and begin to fight cancer years earlier. If you have an infection, it will be diagnosed within minutes. If your heartbeat is irregular, your car seat will let you know. A breath analyzer will detect an immune disease beginning to develop. Keystrokes on the keyboard will signal early Parkinson's disease or multiple sclerosis. Doctors will have far more information about their patients—and they will have access to it long before patients arrive at a clinic or hospital. Medical errors and misdiagnoses will be slashed. The result of *any one* of these innovations could be decades of prolonged healthy life.

Let's say, though, that all of these developments *together* will give us a decade.

Once people begin to accept that aging is not an inevitable part of life, will they take better care of themselves? I certainly have. So, too, it seems, have most of my friends and family members. Even as we have all stepped forward to be early adopters of biomedical and technological interventions that reduce the noise in our epigenomes and keep watch over the biochemical systems that keep us alive and healthy, I've noticed a definite tendency to eat fewer calories, reduce animal-based aminos, engage in more exercise, and stoke the development of brown fat by embracing a life outside the thermoneutral zone.

These are remedies available to most people regardless of socioeconomic status, and the impact on vitality has been exceptionally well studied. Ten additional healthy years is not an unreasonable expectation for people who eat well and stay active. But let's cut that by half. Let's call it five.

That's fifteen years.

Molecules that bolster our survival circuit, putting our longevity genes to work, have offered between 10 and 40 percent more healthy years in animal studies. But let's go with 10 percent, which gives us another eight years.

That's twenty-three years total.

How long will it be before we are able to reset our epigenome, either with molecules we ingest or by genetically modifying our bodies, as my student now does in mice? How long until we can destroy senescent cells, either by drugs or outright vaccination? How long until we can replace parts of organs, grow entire ones in genetically altered farm animals, or create them in a 3D printer? A couple of decades, perhaps. Maybe three. One or all of those innovations is coming well within the ever-increasing lifespans of most of us, though. And when that happens, how many more years will we get? The maximum potential could be centuries, but let's say it's only ten years.

That's thirty-three years.

At the moment, life expectancy in the developed world is a tad over 80 years. Add 33 to that.

That's 113 years, a conservative estimate of life expectancy in the future, as long as most people come along for the ride. And recall that this number means that over half the population will exceed that number. It's true that not all of these advances will be additive, and not everyone will eat well and exercise. But also consider that the longer we live, the greater chance we have of benefiting from radical medical advances that we cannot foresee. And the advances we've already made are not going away.

That's why, as we move faster and faster toward a *Star Trek* world, for every month you manage to stay alive, you gain another week of life. Forty years from now, it could be another two weeks. Eighty years from now, another three. Things could get really interesting around the end of the century if, for every month you are alive, you live another four weeks.

This is why I say that Jean Calment, who may have had the longest lifespan of any person on our planet, will eventually fall off the list of the top ten oldest humans in history. And it won't be more than a few decades after that that she will leave the top 100. After that she will leave the top million. Imagine if people who have lived beyond 110 had had access to all these technologies. Could they have made it to 120 or 130? Perhaps.

Fellow scientists often warn me not to be so publicly optimistic. "It's not a good look," one well-meaning colleague recently told me.

"Why?" I asked.

"Because the public isn't ready for these numbers."

I disagree.

Ten years ago, I was a pariah to many of my colleagues for even talking about making medicines to help patients. One scientist told me that our job as researchers is to "just show a molecule extends lifespan in mice, and the public will take it from there." Sadly, I wish that were true.

Today, many of my colleagues are just as optimistic as I am, even if they don't admit it publicly. I'd wager that about a third of them take metformin or an NAD booster. A few of them even take low doses of rapamycin intermittently. International conferences specifically about longevity interventions are now held every few weeks, the participants not charlatans but renowned scientists from the world's most prestigious universities and research centers. In these gatherings it is no longer unusual to hear chatter about how raising the average human lifespan by a decade, if not more, will change our world. Mind you, the debate is not about *whether* this will happen; it is about what we should do *when* it happens.

The same is increasingly true among the political, business, and religious leaders with whom I spend more and more of my time these days, talking not just about new technologies but about their implications. Slowly but surely, these individuals—legislators, heads of state, CEOs, and thought leaders—are coming to recognize the world-changing potential of the work being done in the field of aging, and they want to be ready.

All these people might be wrong. I might be wrong. But I expect to be around long enough to know one way or the other.

If I am wrong, it might be that I was too conservative in my predictions. Though there are many examples of false predictions—who can forget nuclear-powered vacuum cleaners and flying cars?—it is far more common for people *not* to see something coming. All of us are guilty of it. We extrapolate linearly. More people, more horses, more horse manure. More cars, more air pollution, always more climate change. But that's not how it works.

When technologies go exponential, even experts can be blindsided. The American physicist Albert Michelson, who won a Nobel Prize for measuring the speed of light, gave a speech at the University of Chicago in 1894, declaring that there was probably little else to discover in physics besides additional decimal places.[1] He died in 1931, as quantum mechanics was in full swing. And in his 1995 book, *The Road Ahead*, Bill Gates made no mention of the internet, though he substantially revised it about a year later, humbly admitting that he had "vastly underestimated how important and how quickly" the internet would come to prominence.[2]

Kevin Kelly, the founding editor of *Wired* magazine, who has a better track record than most at predicting the future, has a golden rule: "Embrace things rather than try and fight them. Work with things rather than try and run from them or prohibit them."[3]

We often fail to acknowledge that knowledge is multiplicative and technologies are synergistic. Humankind is far more innovative than we give it credit for. Over the past two centuries, generation after generation has witnessed the sudden appearance of new and strange technologies: steam engines, metal ships, horseless carriages, skyscrapers, planes, personal computers, the internet, flat-screen TVs, mobile devices, and gene-edited babies. At first we are shocked; then we barely notice. When the human brain was evolving, the only things to change in a lifetime were the seasons. It should come as no surprise that we find it hard to predict what will happen when millions of people work on complex technologies that suddenly merge.

No matter if I'm right or wrong about the pace of change, barring a war or an epidemic, our lifespan will continue to rise. And the more thought leaders I speak to around the globe, the more I realize how vast the implications are. And yes, some of those people have allowed me to think and plan for events well beyond the initial scope of my research. But the people who push me to think even harder are the younger people I teach at Harvard and other universities, and the often even younger people I hear from via email and social media nearly every day. They push me to think about how my work will impact the future workforce, global health care, and the very fabric of our moral universe—and to better understand the changes that must take place if we are to meet a world of significantly prolonged human healthspans and lifespans with equity, equality, and human decency.

If the medical revolution happens and we continue on the linear path we're already on, some estimates suggest half of all children born in Japan today will live past 107.[4] In the United States the age is 104. Many researchers believe that those estimates are overly generous, but I don't. They might be conservative. I have long said that if even a few of the therapies and treatments that are most promising come to fruition, it is not an unreasonable expectation for anyone who is alive and healthy today to reach 100 in good health—active and engaged at levels we'd expect of healthy 50-year-olds today. One hundred twenty is our known potential, but there is no reason to think that it needs to be for the outliers. And I am on record as saying, in part to make a statement and in part because I have a front-row seat on what's around the corner, that we could be living with the world's first sesquicentenarian. If cellular reprogramming reaches its potential, by century's end 150 may not be out of reach.

At the moment I write this, there is no one on our planet—no one whose age can be verified, at least—who is over the age of 120. So it will be several decades, at least, before we know if I'm right about this, and it could take 150 years before someone steps over that threshold.

But as for the next century? And the next? It is not at all extravagant to expect that someday living to 150 will be standard. And if the Information Theory of Aging is sound, there may be no upward limit; we could potentially reset the epigenome *in perpetuity.*

This is terrifying to a lot of people—and understandably so. We're on the cusp of upending nearly every idea we've ever had about what it means to be human. And that has a lot of people saying not just that it *can't* be done but that it *shouldn't* be done—for it will surely lead to our doom.

The critics of my life's work aren't nameless, faceless social media trolls. Sometimes they are my colleagues. Sometimes they are close personal friends.

And sometimes they are my own flesh and blood.

Our oldest child, Alex, who at 16 hopes for a career in politics and social justice, has often struggled to see the future with the same optimism I do. Especially when you're young, it is hard to see much of an arc to the moral universe, let alone one that bends toward justice.[5]

Alex grew up, after all, in a world that is quickly and disastrously warming; in a nation that has been at war for the better part of two decades; and in a city that suffered a terrorist attack on the people participating in one of its most cherished traditions, the Boston Marathon. And like so many other young people, Alex lives in a hyperconnected universe where news of one humanitarian crisis after the next, from Syria to South Sudan, is never far from the screen of a smartphone.

So I understand. Or I try to, at least. But it was disappointing to learn, one recent night, that Alex didn't share the optimism I've always had about the future. Of course I'm proud that our kid has such a strong moral compass, but it was saddening to realize this more pessimistic view of the world casts a significant shadow over the way Alex sees my life's work.

"Your generation, just like all the ones that came before, didn't do anything about the destruction that is being done to this planet," Alex told me that evening. "And now you want to help people live longer? So they can do even more damage to the world?"

I went to bed that night troubled. Not by our firstborn's denouncement of me; of that, I admit, I was a little proud. We'll never destroy the global patriarchy if our children don't first practice on their fathers. No, what I was troubled by—what kept me up that night and has done so many since—were the questions that I simply could not answer.

Most people, upon coming to the realization that longer human lives are imminent, also quickly recognize that such a transition cannot possibly occur without significant social, political, and economic change. And they are right; there can be no evolution without disruption. So what if the way I see the future isn't at all what we're headed toward? What if giving billions of people longer and healthier lives enables our species to do greater harm to this planet and to one another? Greater longevity is inevitable; I'm sure of it. What if it inevitably leads to our self-destruction?

What if what I do makes the world worse?

There are plenty of people out there—some of them very smart and very informed—who think that's the case. But I'm still optimistic about our shared future. I don't agree with the naysayers. But that doesn't mean I do not listen to them. I do. And we all should. That's why, in this chapter, I'm going to explain some of their concerns—indeed, concerns

I share in many cases—but I'll also present a different way of thinking about our future.

You can take it from there.

THE HUNDRED YEARS' WARNING

The number of *Homo sapiens* grew slowly over the first few hundred thousand years of our history, and at least on one occasion, we almost went extinct. While there are many young skeletons from the Late Archaic and Paleolithic periods, there is only a handful of skeletons of individuals over the age of 40. It was rare for individuals to make it to the point we now have the luxury of calling middle age.[6] Recall, this was a time when teenage girls were mothers and teenage boys were warriors. Generations turned over quickly. Only the fastest, smartest, strongest, and most resilient tended to survive. We rapidly evolved superior bipedal and analytical skills but at the expense of millions of brutal lives and early deaths.

Our ancestors bred as fast as biology allowed, which was only slightly faster than the death rate. But that was enough. Humanity endured and scattered to all ends of the planet. It wasn't until right around the time Christopher Columbus rediscovered the New World that we reached the 500-million-individuals mark, but it took just three hundred more years for that population to double. And today, with each new human life, our planet becomes more crowded, hurtling us toward, or perhaps further beyond, what it can sustain.

How many is too many? One report, which examined sixty-five different scientific projections, concluded that the most common estimated "carrying capacity" of our planet is 8 billion.[7] That's just about where we're at *right now*. And barring a nuclear holocaust or a global pandemic of historically deadly proportions—nothing anyone in his or her right mind would ever wish for—that's not where our population is going to peak.

When the Pew Research Center polled members of the largest

association of scientists in the world, 82 percent said that there isn't enough food and other resources on this planet for its fast-growing human population.[8] Among those who held that opinion was Frank Fenner, an eminent Australian scientist who helped bring an end to one of the world's deadliest diseases as the chairman of the Global Commission for the Certification of Smallpox Eradication. It was Fenner, in fact, who had the distinct honor of announcing the eradication of the disease to the World Health Organization's governing body in 1980. Having helped millions of people avoid a virus that killed nearly a third of those who contracted the disease it caused, Fenner would have been justified in indulging in a little exuberant optimism about the ways in which people can come together to save themselves.

He had planned a quiet retirement.[9] But his mind wouldn't stop working. He couldn't stop trying to identify and solve big problems. He spent the next twenty years of his life writing about other threats to humankind, many of which had been virtually ignored by the same world health leaders who had banded together to stop smallpox.

His final act of forewarning came just a few months before his death in 2010, when he told the *Australian* newspaper that the human population explosion and "unbridled consumption" had already sealed our species' fate. Humanity would be gone in the next hundred years, he said. "There are too many people here already."[10]

We've heard this song before, of course. At the turn of the nineteenth century, as the global human population was screaming past the 1 billion mark, the English scholar Thomas Malthus warned that advances in food production inevitably led to population growth, placing increasing numbers of poor people at greater risk of starvation and disease. Viewed from the developed world, it often looks as though a Malthusian catastrophe has largely been avoided; agricultural advances have kept us one step ahead of disaster. Viewed globally, though, Malthus's warnings were little short of prophetic. About the same number of people who lived on the planet in Malthus's time go hungry in our time.[11]

In 1968, as the global population approached 3.5 billion, Stanford

University professor Paul Ehrlich and his wife, associate director of Stanford's Center for Conservation Biology Anne Ehrlich, sounded the Malthusian alarm once again in a best-selling book called *The Population Bomb*. When I was young, that book had a rather prominent place on my father's bookshelf—right at eye level for a young boy. The cover was disturbing: a plump, smiling baby sitting inside a bomb with a lit fuse. I had nightmares about that.

What was inside the cover was worse, though. In the book, Ehrlich described his "awakening" to the horrors to come, a revelation he had during a cab ride in New Delhi. "The streets seemed alive with people," he wrote. "People eating, people washing, people sleeping. People visiting, arguing, and screaming. People thrusting their hands through the taxi window, begging. People defecating and urinating. People clinging to buses. People herding animals. People, people, people, people."[12]

With every new year, the Ehrlichs wrote, global food production "falls a bit further behind burgeoning population growth, and people go to bed a little hungrier. While there are temporary or local reversals of this trend, it now seems inevitable that it will continue to its logical conclusion: mass starvation."[13] It's horrifyingly clear, of course, that millions of people have indeed died of starvation in the decades that have passed since *The Population Bomb* was first published, but not nearly at the levels the Ehrlichs predicted and not typically because of a lack of food production but rather as a result of political crises and military conflicts. When a child starves, though, it doesn't much matter to them or their family how it came to happen.

Though the direst of their predictions did not come to pass, in focusing so intently on the food production–population relationship Malthus and the Ehrlichs may actually have underestimated the greater and longer-term risk—not mass starvation that might claim hundreds of millions of lives but a planetary rebellion that will kill us all.

In November 2016, the late physicist Stephen Hawking predicted that humanity had less than 1,000 years left on "our fragile planet." A few months of contemplation later, he revised his estimate downward by

90 percent. Echoing Fenner's warnings, Hawking believed that humanity would have 100 years to find a new place to live. "We are running out of space on Earth," he said. A lot of good that will do; the Earth-like planet that is nearest to our solar system is 4.2 light-years away. Barring major advances in warp speed or wormhole-transit technology, it would take us ten thousand years to get there.

The problem is not just population, it's consumption. And it's not just consumption, it's waste. In comes the food; out goes the effluent. In come the fossil fuels; out go the carbon emissions. In come the petrochemicals; out goes the plastic. On average, Americans consume more than three times the amount of food they need to survive and about 250 times as much water.[14] In return, they produce 4.4 pounds of trash each day, recycling or composting only about of a third of it.[15] Thanks to things such as cars, planes, big homes, and power-hungry clothes dryers,[16] the annual carbon dioxide emissions of an average American are five times as high as the global average. Even the "floor"—below which even monks living in American monasteries typically do not go—is twice the global average.[17]

It isn't just that Americans consume and waste so much, it's that hundreds of millions of other people consume and waste as much and in some cases more,[18] and billions of other people are moving in that same direction. If everyone in the world consumed as Americans do for one year, the nonprofit Global Footprint Network estimates it would take the Earth four years to regenerate what has been used and absorb what has been wasted.[19] This is textbook unsustainability; we use and use and use, and return little of value to our natural world.

The increasing number of scientists making hundred-year warnings has formed around a terrifying environmental reality: even with "very stringent and unrealistically ambitious abatement strategies,"[20] we likely will not be able to prevent global temperature changes that will be greater than 2°C, a "tipping point" that many scientists believe will be catastrophic for humanity.[21] Indeed, as Fenner said, it might truly be "too late."

We are not yet at that two-degree tipping point, and nonetheless the consequences are already quite staggering. Human-caused climate change is destroying food webs around the globe, and by some estimates, one in six species is now at risk of extinction. Warming temperatures have "cooked the life out of the corals" of our oceans,[22] including the Great Barrier Reef, which is roughly the size of California and the most diverse ecosystem on our planet. More than 90 percent of that Australian natural wonder has suffered from bleaching, meaning it is being starved of the algae it needs to survive. In 2018, the Australian government released a report acknowledging what scientists had been saying for many years: that the reef is headed toward "collapse."[23] And in the same year, Australian researchers said that global warming had claimed its first mammalian victim, a long-tailed marsupial mouse called the Bramble Cay melomys, which was sent into extinction when its island ecosystem was destroyed by surging seawater.

There can also be no debate, at this point, that the melting of the Antarctic and Greenland ice caps is driving a rise in sea levels, which the National Oceanic and Atmospheric Association and others have warned will worsen coastal flooding in the coming years, threatening cities such as New York, Miami, Philadelphia, Houston, Fort Lauderdale, Galveston, Boston, Rio de Janeiro, Amsterdam, Mumbai, Osaka, Guangzhou, and Shanghai. A billion people or more live in areas likely to be affected by rising sea levels.[24] Meanwhile, we're facing more—and more severe—hurricanes, floods, and droughts; the World Health Organization estimates that 150,000 people are already dying each year as a direct result of climate change, and that number is likely to at least double in coming years.[25]

All of these dire warnings are predicated upon a world in which humans live for an average of about 75 or 80 years. Thus even the most pessimistic of assertions about the future of our environment are actually underestimating the extent of the problem. There is simply no model in which more years of life does not equate to more people and in which that does not lead to more crowding, more environmental degradation, more consumption, and more waste. As we live longer, these environmental crises will be exacerbated.

And that could be only part of our woes.

THE HUNDRED-YEAR POLITICIAN

If there has been a consistent driving force that has made our world a kinder, more tolerant, more inclusive, and more just place, it is that humans don't last long. Social, legal, and scientific revolutions, after all, are waged, as the economist Paul Samuelson often noted, "one funeral at a time."

The quantum physicist Max Planck also knew this to be true.

"A new scientific truth does not triumph by convincing its opponents and making them see the light," Planck wrote shortly before his death in 1947, "but rather because its opponents eventually die, and a new generation grows up that is familiar with it."[26]

Having witnessed a few different sorts of revolutions during my life—from the fall of the Berlin Wall in Europe to the rise of LGBTQ rights in the United States to the strengthening of national gun laws in Australia and New Zealand—I can vouch for these insights. People can change their minds about things. Compassion and common sense can move nations. And yes, the market of ideas has certainly had an impact on the way we vote when it comes to issues such as civil rights, animal rights, the ways we treat the sick and people with special needs, and death with dignity. But it is the mortal attrition of those who steadfastly hold on to old views that most permits new values to flourish in a democratic world.

Death by death, the world sheds ideas that need to be shed. Ipso facto, birth by birth, the world is offered an opportunity to do things better. Alas, we don't always get it right. And it's often a slow and uneven sort of progress. With a generation time of twenty minutes, bacteria evolve rapidly to survive a new challenge. With a generation time of twenty years, human culture and ideas can take decades to evolve. Sometimes they devolve.

In recent years, nationalism has moved from being the purview of angry fringe groups to being the force behind powerful political movements around the world. There is no one single factor that can explain all of these movements, but the economist Harun Onder is among those who have made a demographic observation: nationalist arguments tend to resonate with older people.[27] Therefore, it is likely that the antiglobalist wave will be with us for some time to come. "Virtually every country in the world," the United Nations reported in 2015, "is experiencing growth in the number and proportion of older persons in their population." Europe and North America already have the largest per capita share of older persons; by 2030, according to the report, those over the age of 60 will account for more than a quarter of the population on both of these continents, and that proportion will continue to grow for decades to come. Once again, these are estimates based on ridiculously low projections for lengthened lifespans.[28]

Older constituencies support older politicians. As it is now, politicians seem steadfastly opposed to stepping down in their 70s and 80s. More than half of the US senators running for reelection in 2018 were 65 or older. Democratic leader Nancy Pelosi was 78 that year. Dianne Feinstein and Chuck Grassley, two powerful senators, were 85. On average, members of the US Congress are 20 years older than their constituents.

At the time of his death in 2003, Strom Thurmond was 100 years old and had served 48 years as a US senator. That Thurmond was a centenarian in Congress is no vice—we want our leaders to have experience and wisdom, as long as they aren't stuck in the past. The travesty was that Thurmond somehow managed to keep his seat in spite of a long record of supporting segregation and opposing civil rights, including basic voting rights. At the age of 99, he voted to use military force in Iraq, opposed legislation to make pharmaceuticals more affordable, and helped kill a bill that would have added sexual orientation, gender, and disability to a list of categories covered by hate crimes legislation.[29] After his death, the "family values" politician was revealed to have had a daughter with his family's teenage African American housekeeper when he was 22, which was almost certainly an act of statutory rape under South Carolina law. Though he knew about the child, he never publicly acknowledged her.[30] Thurmond lived in retirement only six months; those who were too young to vote then will have to live with the consequences of his votes for the rest of their lives.

We tend to tolerate a bit of bigotry among older people as a condition of the "age in which they grew up," but perhaps also because we know we won't have to live with it for long. Consider, though, a world in which people in their 60s will be voting not for another twenty or thirty years but for another sixty or seventy. Imagine a man like Thurmond serving in Congress not for half a century but for an *entire* century. Or, if it makes it easier to envision from your place on the political spectrum, picture the politician you despise more than any other holding power longer than any other leader in history. Now consider how long despots in far less democratic nations will cling to power—and what they will do with that power.

What will this mean for our world politically? If a steadfast driving force for kindness, tolerance, inclusivity, and justice suddenly ceases to exist, what will our world look like?

And the potential problems don't stop there.

SOCIAL INSECURITY

Few people were spared the trauma inflicted by the worldwide Great Depression during the 1930s. But the impact was particularly felt by those in the last decades of their lives. Stock market crashes and bank failures claimed the life savings of millions of older Americans. With so many people out of work, the few employers who were offering jobs were reluctant to hire older workers. Destitution was rampant. About half of the elderly were poor.[31]

Those people had been deacons in churches, pillars of communities, teachers and farmers and factory workers. They were grandmothers and grandfathers, and their desperation shook the nation to its core, prompting the United States in 1935 to join about twenty other countries that had already instituted a social insurance program.

Social Security made moral sense. It made mathematical sense, too. At that time, just over half of men who reached their 21st birthday would also reach their 65th, the year at which most could begin to collect a supplemental income. Those who reached age 65 could count on about thirteen more years of life.[32] And there were a lot of younger workers paying into the system to support that short retirement; at that time only about 7 percent of Americans were over the age of 65. As the economy began to boom again in the wake of World War II, there were forty-one workers paying into the system for every beneficiary. Those are the numbers that supported the system when its first beneficiary, a legal secretary from Vermont named Ida May Fuller, began collecting her checks. Fuller had worked for three years under Social Security and paid $24.75 into the system. She lived to the age of 100 and by the time of her death in 1975 had collected $22,888.92. At that point, the poverty rate among seniors had fallen to 15 percent, and it has continued to fall ever since, owing largely to social insurance.[33]

Now about three-quarters of Americans who reach the age of 21 also see 65. And changes to the laws that govern the US social insurance safety net have prompted many to retire—and begin collecting—earlier than that. New benefits have been added over the years. Of course, people are living longer, too; individuals who make it to the age of 65 can count on about twenty more years of life.[34] And as just about every social insurance doomsdayer can tell you, the ratio of workers to beneficiaries is an unsustainable three to one.

That is not to say that Social Security is necessarily doomed. There are reasonable adjustments that can be made to keep it solvent for decades to come. But all of the most commonly recommended adjustments, as you might by now suspect, are predicated on the assumption that we will enjoy only modest gains in lifespan in coming years. There are very few policy makers in the United States—let alone the 170 other countries that now have some form of social insurance program—who have so much as considered a world in which, at the age of 65, many people will be reaching the midpoint of their lives.

Even upon considering this, it can be assured that many politicians, if not the overwhelming majority of them, will choose to bury their heads in the sand. Lyndon Johnson's landslide victory over Barry Goldwater in the 1964 US presidential race can largely be attributed to Goldwater's perceived hostility to social insurance. But by the 1980s, politicians on both sides of the political aisle had taken to calling Social Security the "third rail" of American politics: "Touch it, you're dead."[35] At that time 15 percent of Americans were collecting Social Security. Today about 20 percent are.[36] Today, people over the age of 65 make up 20 percent of the voting population and will grow by 60 percent by 2060,[37] in addition to which they are about twice as likely as 18- to 29-year-olds to go to the polls.[38]

There is a very rational argument for the resistance of the AARP (formerly the American Association of Retired Persons) to any change to social insurance. A few more years of waiting for retirement might not seem so bad to people who work in occupations with low physical impact or in a job they love, but what of those who have spent 45 years doing heavy manual labor, working on an assembly line, or toiling in a meatpacking plant? Is it fair to expect them to work even longer? Longevity drugs and healthspan therapies are very likely to help those people feel better and stay healthier for longer, but that wouldn't justify forcing people who have worked arduously for most of their lives back to the mines.

There are no easy answers, but if past is prologue—and it so often is with human behavior—politicians will watch this slow-moving disaster until it becomes a fast-moving disaster; then they will sit and watch some more. In many nations, and particularly those of western Europe, social insurance programs are relatively generous to beneficiaries and have been embraced by the political Left and Right alike. These programs have become strained in recent years under the weight of government deficits and the inability to meet long-held promises to aging workers,[39] prompting fights over which entitlements are most sacred, pitting education against health care and health care against pensions and pensions against disability compensation. These fights will only increase as the systems become further strained. And that strain is inevitable without revolutionary reforms that account for the fact that the ranks of retirees will soon be brimming with those who, when the systems were designed in the mid-1900s, were aged outliers.

At least every couple of months, I get a call from a politician for an update on the latest developments in biology, medicine, or defense. Almost always we end up discussing what will happen to the economy as people live longer and longer. I tell him or her that there is simply no economic model for a world in which people live forty years or more past the time of traditional retirement. We literally have no data whatsoever on the work patterns, retirement arrangements, spending habits, health care needs, savings, and investments of large groups of people who live, quite healthily, well into their 100s.

Working with the world-renowned economists Andrew Scott at the University of London and Martin Ellison at Oxford University, we are developing a model to predict what the future looks like. There are quite a few variables, not all of them positive. Will people continue to work? What jobs will they be able to get in a world in which the labor market will already be being upended by automation? Will they spend a half century or more in retirement? Some economists believe that economic growth is slowed when a country ages, in part because people spend less in retirement. What will happen if people spend half of their very long lives out of work, spending only enough to get by?

Will they save more? Invest more? Get bored soon after retirement and start a new career? Take long sabbaticals from work, only to return decades later when their money runs out? Spend less on health care because they are so much healthier? Spend more on health care because they are living so much longer? Invest more years and money into their educations early on?

Anyone who claims to know the answer to any of these questions is a charlatan. Anyone who says these questions aren't important is a fool. We have absolutely no idea what's going to happen. We are flying blind into one of the most economically destabilizing events in the history of the world.

Yet that is not the worst of it.

WHAT DIVIDES US GROWS GREATER

If you were a member of the American upper middle class in the 1970s, you weren't just enjoying a more affluent life, you had a longer one, too. Those in the top half of the economy were living an average of 1.2 more years than those in the bottom half.

By the early 2000s, the difference had increased dramatically. Those in the upper half of the income spectrum could expect nearly six additional years of life, and by 2018, the divide had widened, with the richest 10 percent of Americans living thirteen more years of life than the poorest 10 percent.[40]

The impact of this disparity cannot be understated. Just by living longer, the rich are getting richer. And of course, by getting richer, they are living longer. Extra years offer more time to preside over family businesses, and more time for family investments to multiply exponentially.

Riches are not just invested into companies; they provide rich people with access to the world's leading doctors (there are about five in the United States that they all seem to use), nutritionists, personal trainers, yoga instructors, and the latest medical therapies—stem cell injections, hormones, longevity drugs—which mean they stay healthier and live longer, which allows them to accumulate even more wealth during their lifetimes. The accumulation of wealth has been a virtuous cycle for families lucky enough to get onto it.

And the rich don't invest only in their health; they also invest in politics, which is no small part of the reason why a series of revisions to the US tax code has resulted in a dramatic reduction in taxes on the wealthy.

Most countries tax people when they die as a way to limit wealth accumulation over generations, but it's a little-known fact that, in the United States, estate taxes weren't initially designed to limit multigen-erational wealth; they were imposed to finance wars.[41] In 1797, a federal tax was imposed to build a navy to fend off a possible French invasion; in 1862, an inheritance tax was instituted to finance the Civil War. The 1916 estate tax, which was similar to present-day estate taxes, helped pay for World War I.

In recent times, the burden of paying for wars has shifted to the rest of the population. Thanks to tax loopholes, the percentage of rich American families who pay what were cleverly branded as "death taxes" decreased fivefold, providing the lowest cost for "dying rich" in modern times.[42]

All this means that the children of the wealthy are faring extremely well. Unless there is an upward revision to the tax code, they will con-tinue to do better, both in how much money they inherit and in how much longer they will live than others do.

Remember, too, that aging is not yet considered a disease by any nation. Insurance companies don't cover pharmaceuticals to treat dis-eases that aren't recognized by government regulators, even if it would benefit humanity and the nation's bottom line. Without such a desig-nation, unless you are already suffering from a specific disease, such as diabetes in the case of metformin, longevity drugs will have to be paid for out of pocket, for they will be elective luxuries. Unless aging is desig-nated a medical condition, initially only the wealthy will be able to afford many of these advances. The same will be true for the most advanced biotracking, DNA sequencing, and epigenome analyses to permit truly personalized health care. Eventually prices will come down, but unless governments act soon, there will be a period of major disparity between the very rich and the rest of the world.

Imagine a world of haves and have-nots unlike anything we have experienced since the dark ages: a world in which those born into a certain station in life can, by virtue of nothing more than exceptional fortune, live thirty years longer than those who were born without the means to literally buy into therapies that provide longer healthspans and enable more productive working years and greater investment returns.

We have already taken the first tenuous steps into a world that was predicted by the 1997 film *Gattaca,* a society in which technologies originally intended to assist in human reproduction are used to eliminate "prejudicial conditions," but only for those who can afford them. In the coming decades, barring a safety issue or a global backlash against the unknown, we'll likely see the increased ability and acceptance of gene editing globally, providing would-be parents with the option to limit disease susceptibility, choose physical traits, and even select intellectual and athletic abilities. Those of means who wish to give their children "the best possible start," as a doctor tells two prospective parents in *Gattaca,* will be able to do so, and with longevity genes identified, they could be given the best possible finish, too. Whatever advantages genetically enhanced people will already have, they could be multiplied by virtue of economic access to longevity drugs, organ replacements, and therapies we haven't even yet dreamed of.

Indeed, unless we act to ensure equality, we stand at the precipice of a world in which the über-rich could ensure that their children, and even their companion animals, live far longer than some poor people's children do.

That would be a world in which the rich and poor will be separated not simply by differing economic experiences but by the very ways in which human life is defined—a world in which the rich will be permitted to evolve and the poor are left behind.

Yet . . .

Notwithstanding the potential that extending human longevity has to exacerbate some of the direst problems of our world—and indeed to give us new troubles in the decades to come—I remain optimistic about the potential of this revolution to change the world for the better.

We've been here before, after all.

TO WEND OUR WAY

To understand the future, it is often helpful to travel into the past. So if we want to better understand the desperate world we are about to enter, a good place to go is to another desperate time.

In a city brimming with iconic landmarks, from the Tower of London to Trafalgar Square, from Buckingham Palace to Big Ben, it is perfectly reasonable that many people, and indeed even many Londoners, have never dedicated so much as a thought to the Cannon Street Railway Bridge.

There are no songs about it; not to my knowledge, at least. I know of no authors who have set their stories upon its rusted rails. When it appears in cityscape paintings, it is almost always an incidental character.

Granted, it is a rather unsightly thing, an uninvolving and utterly utilitarian structure of green-painted steel and concrete. And if you were to look easterly upon the River Thames from the far more charming, lamp-lined sidewalks of Southwark Bridge, you could indeed be forgiven for missing it altogether, although it is right before you, for just beyond on the right is architect Renzo Piano's famous Shard building, and just beyond that, spanning the river, is the even more famous London Bridge, among other grand sights downstream.

In 1866, the year the Cannon Street Railway bridge was opened, there were nearly 3 million people in London. More arrived in the years to come, often arriving from abroad by boat to Cannon Street Station, London's equivalent of Ellis Island, and dispersing from there by rail, across the humble bridge, to the other parts of the city as it grew more and more crowded by the day. I can scarcely imagine what someone looking upon the throngs of out-of-town arrivals must have thought in the years in which London seemed so clearly unable to sustain any more people, let alone the masses coming from other parts of the world and the many more being born into the already overcrowded city.

Even the exodus to colonies in the Americas and Australia did nothing to stem the population explosion. By 1800, approximately a million people were living in London, and by the 1860s that number had tripled, unleashing dire consequences on the capital of the British Empire.

Central London was a particularly hellish place. The mud and horse manure were often ankle deep in streets further littered by newspapers, broken glass, cigar ends, and rotting food. Dockworkers, factory workers, laundresses, and their families were packed into tiny hovels with dirt floors. The air was thick with soot in the summer and soot-drenched fog in the winter. With every breath, Londoners filled their lungs with mutagenic, acid-coated particles of sulfur, wood, metals, soil, and dust.

A sewer system intended to take human waste away from the richer neighborhoods of central London did just that—sending it into the River Thames, where it flowed east past the Isle of Dogs toward the poorer quarters, where people drew the water to wash and drink.[43,44]

In those squalid conditions, it should come as little surprise that cholera could spread with devastating speed. And it had, with three large outbreaks so far that century, in 1831, 1848, and 1853, claiming more than 30,000 lives, with thousands more lost to smaller outbreaks during the intermediate years.

The Final Catastrophe, as it came to be known, was focused almost exclusively on the inhabitants of Soho in the West End, where a contaminated well provided water to more than a thousand people. Today, the Broad Street pump is preserved on what is now Broadwick Street, surrounded by pubs, restaurants, and high-end clothing stores. The pump's granite base is often used as a seat by unsuspecting tourists. Save for the keystone plaque on the building nearby, there are no clues about the misery this site wrought.

Twenty people died in the first week of the cholera outbreak, July 7 to 14 of 1866, falling to diarrhea, nausea, vomiting, and dehydration. Doctors had only just realized that they were dealing with another outbreak when the second wave began. More than three hundred additional people had died by July 21. From there it only got worse. On no day between July 21 and August 6 did fewer than a hundred people perish, and the death toll continued to mount through November.

That was the hellscape in which a former domestic servant named Sarah Neal gave birth to her fourth child on September 21, 1866, just six miles south of the epicenter of the outbreak. She called her son "Bertie." So did her husband, Joseph Wells. But the boy would ultimately choose to go by the initials of his given name, Herbert George.

In the center of despair and squalor, in a city breaking under the weight of a population boom, in the heart of hopelessness, was born the father of utopian futurism, H. G. Wells.

Wells is most famous today for his dystopian fiction *The Time Machine*, but in stories such as *The Shape of Things to Come*, he audaciously predicted a "future history" that included genetic engineering, lasers, airplanes, audiobooks, and television.[45] He also predicted that scientists and engineers would lead us away from fighting war after war toward a world devoid of violence, poverty, hunger, and disease.[46] It was, in many ways, a blueprint for *Star Trek* creator Gene Roddenberry's vision of a future Earth that would be a utopian base for exploration of the "final frontier."[47]

How did we go from a world of such misery to one in which such dreams were even possible?

Well, as it turned out, the disease was the cure.

The Cannon Street Bridge, completed the same year that cursed London with the Final Catastrophe and blessed the world with the genius of H. G. Wells, stands as a testament to the ways in which the London of yesterday came to be the London of today, of how population and progress are intrinsically connected, and, indeed, of utopian dreams realized. For London's nineteenth-century population boom forced the city to confront its most horrific challenges. There was simply no other option. The choice was clear: adapt or perish.[48]

And so it was that the late nineteenth century brought to London some of the world's first public housing projects, replacing dirt-floored shanties with plumbed tenements that would, upon the passage of the Housing of the Working Classes Act of 1900, also have access to electric power. The same time period saw a tremendous rise in the number and quality of public institutions of education, including mandated schooling for children between the ages of 5 and 12, imperfectly but increasingly drawing legions of children away from the dangerous and exploitative conditions of life on London's streets.

Perhaps the most important of the reforms, however, was in the field of public health, beginning in 1854 with the physician John Snow's rebellion against the entrenched medical view that cholera was caused by miasma, or "bad air." By talking to residents and triangulating the problem, Snow had the Broad Street pump's handle removed. The epidemic soon ended. Government officials were quick to replace the pump handle, in part because the fecal-oral route of infection was too horrific to contemplate. Finally, in the eventful year of 1866, Snow's chief opponent, William Farr, was investigating another cholera outbreak and came to the realization that Snow was right. The resolution of that public health skirmish led to improved water delivery and sewage systems in the capital of the world's largest empire.

Those innovations were soon copied around the globe—one of the greatest global health achievements in human history. Far more than any other lifestyle change or medical intervention, clean water and working sanitation systems have led to longer and healthier lives the world over. And London, where this all began, is Exhibit A. Lifespans in the United Kingdom have more than doubled in the past 150 years, in no small part because of innovations that were made in direct response to the overcrowding in it that the early-nineteenth-century parliamentarian William Cobbett derisively called the Great Wen, a nickname that compared the city to a swelling, pus-filled, sebaceous cyst.

The movement from miasmatic theory to germ theory, meanwhile, fundamentally shifted ideas about how to combat all sorts of other diseases, setting the stage for Louis Pasteur's breakthroughs in fermentation, pasteurization, and vaccination. The ripples are manifold and can be measured, without the slightest hint of hyperbole, in hundreds of millions of human lives. If it hadn't been for the advances that came out of that period of our history, billions upon billions of people would not be alive today. You might be here. I might be here. But the chances that we would both be here would be very slim. It turned out that the population of London wasn't the problem after all.

The problem wasn't *how many* people lived in the city but *how* they lived in the city.

At 9 million residents and still growing, London today has three times as many people as it did in 1866 but far less death, disease, and despair.

Indeed, if you were to describe the London of today to Londoners of the 1860s, I submit that you would be hard-pressed to find a single soul who would not agree that their city, in the twenty-first century, would have far surpassed their most sanguine utopian dreams.

Do not get me wrong: the limitless and legitimate concerns people express about a world in which humans live twice as long as they do now—or longer—cannot be dismissed with a story about old London. The city is by no means perfect. Anyone who has ever priced a one-bedroom flat in the city knows this to be true.

But today, we can plainly see that the city is flourishing not in spite of its population but because of it, such that today the capital of and most populous city in the United Kingdom is home to a myriad of museums, restaurants, clubs, and culture. It is home to several Premier League football clubs, the world's most prestigious tennis tournament, and two of the best cricket teams on the globe. It is home to one of the world's largest stock exchanges, a booming tech sector, and many of the world's biggest and most powerful law firms. It is home to dozens of institutions of higher education and hundreds of thousands of university students.

And it is home to what is arguably the most prestigious national scientific association in the world, the Royal Society.

Founded in the 1600s during the Age of Enlightenment and formerly headed by Australia's catalyst, the botanist Sir Joseph Banks, as well as such legendary minds as Sir Isaac Newton and Thomas Henry Huxley, the society's cheeky motto is a pretty good one to live by: *"Nullius in Verba,"* it says underneath the society's coat of arms. That's Latin for "Take nobody's word for it."

So far in this chapter I have presented a case—one agreed upon by many great scientists—that even at current and very conservative population growth projections, based on lives that are extended only slightly in the coming decades, our planet is already past its carrying capacity and we, as a species, are only exacerbating that problem with the ways in which we are increasingly choosing to live. And yes, advances in healthspans and lifespans could greatly exacerbate some of the problems we already face as a society.

But there is another way of seeing our future—one in which prolonged vitality and increasing populations are every bit as inevitable but not damning to our world. In this future, the coming changes are our salvation.

But, please: don't just take my word for it.

A SPECIES WITH NO LIMITS

When he is remembered at all, the Dutch amateur scientist Antonie van Leeuwenhoek is almost always thought of as the father of microbiology. But Leeuwenhoek dabbled in great questions of all sorts, including one that may impact the world every bit as grandly. In 1679, by way of trying to convey to the Royal Society just how multitudinous the unseen microscopic world was, he embarked upon an effort to calculate—"but very roughly," he hastened to add—the number of human beings who could survive on the Earth.[49] Using the population of Holland at the time, which was roughly 1 million people, and some very round estimates for the globe's size and total land surface, he came to the conclusion that the planet could carry about 13.4 billion people.

That wasn't a bad guess for someone using what we might today call "back-of-the-napkin" math. Albeit high, it's in the ballpark of the estimates of many more contemporary scientists who have explored the same question with far more data to work with.

A United Nations Environment Programme report detailing sixty-five scientific estimates of global carrying capacity found that the majority—thirty-three—had pegged the maximum sustainable human population at 8 billion or fewer people. And yes, by these estimates, we have either already met or will soon meet the maximum number of human beings our planet can sustain.[50]

But an almost equal number of estimates—thirty-two of them—concluded that the number is somewhere above 8 billion. Eighteen of those estimates suggested that the carrying capacity is at least 16 billion. And a few estimates suggested that our planet has the potential to sustain more than 100 billion people.

Clearly, someone's numbers must be way off.

As you might imagine, these varying estimates are largely dependent on differences in the ways in which the constraining limits of population are defined. Some researchers consider only the most basic factors; not unlike Leeuwenhoek, they speculate as to a maximum population per square mile, multiply that by the roughly 25 million square miles of habitable land on Earth, and that's that.

More robust estimates have included basic constraining factors such as food and water. After all, it does not matter if we can fit tens of thousands of people into a square mile—as is the case in exceptionally dense cities such as Manilla, Mumbai, and Montrouge—if those people starve or die of thirst.

Detailed estimates of the entire globe's carrying capacity include the interaction of constraining factors and the impact of human exploitation of the global environment. Having enough land and water doesn't matter, either, if continued population growth aggravates the already dire consequences of climate change, further destroying the forests and biological diversity that sustain our existence.

But whatever the method and whatever the resulting number, the very act of engaging in the process of trying to derive a carrying capacity acknowledges that there is, in fact, a definitive uppermost limit. Indeed, my colleague at Harvard, the Pulitzer Prize–winning biologist Edward O. Wilson, wrote in *The Future of Life*, "it should be obvious to anyone not in a euphoric delirium that whatever humanity does or does not do, Earth's capacity to support our species is approaching the limit."[51] That was in 2002, when the Earth's population was a paltry 6.3 billion. In the next fifteen years, another 1.5 billion people were added.

Scientists generally pride themselves on rejecting the notion that anything "should be obvious." Evidence, not obviousness, drives our work. So at the very least, the overwhelming certainty that a limit exists deserves to be debated, as any scientific idea does.

It needs to be pointed out that very few of the global carrying capacity models account for human ingenuity. As we have discussed, it is easier *not* to see things coming than to see them, so we tend to extrapolate into the future directly from the way things are now. That's unfortunate and, in my view, scientifically wrong, for it eliminates an important factor from the equation.

Positive views about the future aren't as popular as negative ones. In rejecting well-meaning but imperfect estimates and arguing that there is no scientifically foreseeable limit to the number of people the planet can sustain, the environmental scientist Erle C. Ellis at the University of Maryland has taken a lot of heat. That, of course, is what happens when scientists challenge entrenched ideas. But Ellis has stood firm, even penning an op-ed for the *New York Times* in which he called the very notion that we might be able to identify a global carrying capacity "nonsense."[52]

"The idea that humans must live within the natural environmental limits of our planet denies the realities of our entire history, and most likely the future," he wrote. ". . . Our planet's human-carrying capacity emerges from the capabilities of our social systems and our technologies more than from any environmental limits."[53]

If there were anything like a "natural" limit, Ellis has argued, the human population probably exceeded it tens of thousands of years ago, when our hunter-gatherer ancestors began to rely upon increasingly so-phisticated water control systems and agricultural technologies to sustain and grow their numbers. From that point on, our species has grown only by the combined grace of the natural world and our ability to adapt to it technologically.

"Humans are niche creators," Ellis stated. "We transform ecosystems to sustain ourselves. This is what we do and have always done."

In this way of thinking, few of the adaptations that sustain our lives are "natural." Water delivery systems are not natural. Agriculture is not natural. Electricity is not natural. Schools and hospitals and roads and clothes are not natural. We have long since crossed all of those figurative and literal bridges.

On a plane from Boston to Tokyo recently, I introduced myself to a man sitting next to me and we chatted about our work. When I told him that I was endeavoring to extend human lives, he curled his upper lip.

"I don't know about that," he said. "It sounds unnatural."

I gestured for him to look around. "We are in reclinable chairs, flying at six hundred miles an hour seven miles above the North Pole, at night, breathing pressurized air, drinking gin and tonics, texting our partners, and watching on-demand movies," I said. "What about *any* of this is natural?"

You don't have to be in an airplane to be removed from the natural world. Look around. What about your current situation is "natural"?

We long ago left a world in which the vast majority of humans could expect a life of "no arts; no letters; no society," as Thomas Hobbes wrote in 1651, "and which is worst of all, continual fear, and danger of violent death."

If that is indeed what is natural, I have no interest in living a natural life, and I would wager that you do not wish for that, either.

So what is natural? Certainly we can agree that the impulses that compel us to live better lives—to strive for existences with less fear, dan-ger, and violence—are natural. And it is true that most of the adaptations that enable survival on this planet, including our wonderful survival circuit and the longevity genes it has created, are the products of natural selection, weeding out over billions of years those who failed to hunker down when times were tough, but a great many are skills we've accumulated over the past 500,000. When chimps use sticks to probe termite nests, birds drop rocks on mollusks to break their shells, or monkeys bathe in warm volcanic pools in Japan, it's all natural.

Humans just happen to be a species that excels at acquiring and passing on learned skills. In the past two hundred years, we have invented and utilized a process called the scientific method, which has accelerated the advancement of learning. In this way of thinking, then, culture and technology are both "natural." Innovations that permit us to feed more people, to reduce disease, and, yes, to extend our healthy lives are natural. Cars and planes. Laptop computers and mobile phones. The dogs and cats who share our homes. The beds on which we sleep. The hospitals in which we care for one another in times of sickness. All of this is natural for creatures who long ago exceeded the numbers that could be sustained in conditions Hobbes famously described as "solitary, poor, nasty, brutish, and short."

To me, the only thing that seems unnatural—in that it has *never* happened in the history of our species—is to accept limitations on what we can and cannot do to improve our lives. We have always pushed against perceived boundaries; in fact, biology compels us to.

Prolonging vitality is a mere extension of this process. And yes, it comes with consequences, challenges, and risks, one of which is increased population. But possibility is not inevitability, for as a species we are *naturally* compelled to innovate in response. The question, then, is not whether the natural and unnatural bounties of our Earth can sustain 8 billion, 16 billion, or 20 billion people. That's a moot point. The question is whether humans can continue to develop the technologies that will permit us to stay ahead of the curve in the face of population growth, and indeed make the planet a better place for all creatures.

So can we?

Absolutely. And the past century is proof.

PEOPLE, PEOPLE, GLORIOUS PEOPLE

After our species was almost driven to extinction 74,000 years ago, up until 1900, the human population grew at a rate amounting to a fraction of a percent each year as we expanded to all habitable regions on the planet, breeding with at least two other human species or subspecies. By 1930, thanks to sanitation and decreases in child-mother mortality, our species was increasing its numbers at 1 percent each year. And by 1970, due to immunization and improvements in food production globally, the rate was 2 percent each year.

Two percent might not seem like a lot, but it added up fast. It took more than 120 years for our population to move from 1 billion to 2 billion, but after reaching that mark in 1927, it took just thirty-three more years to add another billion and then fourteen years to add another.

This is how, at the end of the second decade of the twenty-first century, we came to have more than 7.7 billion people on our planet, and every year one additional person per square kilometer.[54] Stepping back, if you graph human population size over the last 10,000 years, the transition from humans being very rare creatures to being the dominant species on Earth looks like a vertical step up. That baby inside the bomb would, on the face of it, seem justified.

Over the past few decades, however, the rate of human population growth has been falling steadily—principally as women who have better economic and social opportunities, not to mention basic human rights, choose to have fewer children. Until the late 1960s, each woman on the planet had an average of more than five children. Since then that average has fallen fast, and with it the rate at which our population is increasing has fallen, too.

The annual population growth rate has plummeted, from 2 percent around 1970 to about 1 percent today. By 2100, some researchers believe, the growth rate could fall as low as one-tenth of 1 percent. As this happens, United Nations demographers anticipate that our total global population will plateau, reaching about 11 billion people by the year 2100, then stop and drop from there.[55]

This assumes, as we have discussed, that most people will continue to live longer on average but will still die in their 80s. That's not likely going to be the case. In my experience, most people tend to significantly overestimate the impact of death on population growth. Of course death keeps the human population in check, but not by much.

Bill Gates made a convincing argument for why improving human health is money well spent, and won't lead to overpopulation, in his 2018 video "Does Saving More Lives Lead to Overpopulation?"[56] The short answer is: No.

If we were to stop *all deaths*—every single one around the globe— right now, we would add about 150,000 people to our planet each day. That would be 55 million people each year. That might sound like a lot, but it would be less than a single percentage point. At that rate, we would add a billion people to our ranks every eighteen years, which is still considerably slower than the rate at which the last few billion people have come along and easily countered by the global decline in family sizes.

It's still an increase, but it's not the sort of exponential growth many people fret about when they first encounter the idea of slowing aging.

Recall, these calculations are what we'd face if we ended *all* deaths right away. And although I'm very optimistic about the prospects for prolonged vitality, I'm not *that* optimistic. I don't know any reputable scientist who is. One hundred years is a reasonable expectation for most people alive today. One hundred twenty is our known potential and one that many people could reach—again, in good health if technologies in development come to fruition. If epigenetic reprogramming reaches its potential or someone comes up with another way to convince cells to be young again, 150 might even be possible for someone living on this planet with us right now. And ultimately there is no upward biological limit, no law that says we must die at a certain age.

But these milestones will come one at a time, and slowly. Death will

The Gompertz Makeham Equation

MORTALITY RATE THROUGHOUT HISTORY

United States 1970

Rome 100 BC

Europe 15 000 years ago

Africa 50000 years ago

max has not changed

PERCENTAGE ALIVE

AGE

$$m(t) = A_0 e^{Gt} + M_0$$

THE LAW OF HUMAN MORTALITY. Benjamin Gompertz, a self-taught mathematical genius, was barred from attending university in nineteenth-century London for being a Jew yet was elected to the Royal Society in 1819. His brother-in-law, Sir Moses Montefiore, in partnership with Nathan Rothschild, founded Alliance Assurance Company in 1824, and Gompertz was appointed actuary. His tidy equation, which replaced mortality tables, tracks the exponential increase in the chance of death with age. As important as this "law" is to insurance companies, it does not mean that aging is a fact of life.

remain a part of our lives for a very long time to come, even as the time of it is pushed out in the coming decades.

That change, though, will be set against an ongoing fall in birth rates that has been under way for decades. So overall, our population might continue to grow but more slowly and not at all in the explosive ways we experienced in the past century. Rather than fearing the more moderate population increase we are likely to see, we should welcome it. Let us not forget what happened during the past century: our species not only survived in the midst of exponential population growth, it thrived.

Yes: *thrived*. No one can ignore the vast devastation we have unleashed upon our planet, not to mention the evils we have inflicted upon one another. We should rightfully focus our attention on these failures; that's the only way to learn from them. But the continual focus on the negative impacts the way we think about the state of our world today and in the future, which is likely why, when the global polling company You-Gov asked people in nine developed nations, "All things considered, do you think the world is getting better or worse, or neither getting better nor worse?" only 18 percent of people believed that things were getting better.

Oh, wait. That was 18 percent of people in Australia—which was the *most* optimistic of the Western nations included in the survey. In the United States, only 6 percent of people were similarly confident that things were getting better in our world.

It's important to note that the pollsters didn't ask about whether respondents' individual lives were getting better or worse. They asked about the *world*. And they asked people in some of the richest nations in the world.[57] And sure, these are people who might have reason to think that their individual standards of living—supported until recently by economic benefits rooted in slavery and colonialism—have been falling a bit in recent years. These are also people, however, who have tremendous access to information about the world, and thus, quite frankly, they should know better.

In much of the rest of the world, however, the future is not viewed in nearly such a dismal way. Not at all.

In China, which holds about a fifth of the global population, some 80 percent of people polled in 2014 by Ipsos MORI, a UK research company, believed the lives of younger people will be better than their own. The same survey identified similarly significant levels of optimism in Brazil, Russia, India, and Turkey—all places where standards of living have been on the rise.[58] And yes, this includes habits of increasing consumption, but it also includes lowering birth rates, falling rates of poverty, greater access to clean water and electricity, more stable access to food and shelter, and greater availability of medical care.

Pessimism, it turns out, is often indicative of exceptional privilege. When viewed globally, however, it gets a lot harder to make the case that the world is an increasingly miserable place. It's simply not.

In the past two hundred years—an era that saw the most explosive population growth in human history—we transformed from a world in which nearly everyone but monarchs and their viceroys was living in poverty to a global society in which the rate of extreme poverty is now below 10 percent and rapidly falling. Meanwhile, in a century in which we added billions of people to our planet's population, we also improved educational access for people around the world. In 1800, the global literacy rate was 12 percent, by 1900 it was 21 percent, and today it's 85 percent. We now live in a world where more than four out of five people can read, the majority of whom have instant access to essentially all the world's knowledge.

One significant reason our population grew so fast in the past century was that child mortality fell from more than 36 percent in 1900 to less than 8 percent in 2000.[59] No decent person could possibly believe that our world would be better if a third of all children were still dying before their fifth birthday.

Did these improvements to the human condition occur in spite of our population boom or because of it? I contend it is the latter, but it actually doesn't matter. They happened simultaneously. As yet, there is really no evidence in modern times that population levels correlate with, let alone cause, increases in human misery. Much to the contrary, in fact, our world is more populated today than it ever has been—and it's a better place for more people, too.

The Harvard psychologist Steven Pinker put it this way in his book *Enlightenment Now: The Case for Reason, Science, Humanism, and Progress*: "Most people agree that life is better than death. Health is better than sickness. Sustenance is better than hunger. Abundance is better than poverty. Peace is better than war. Safety is better than danger. Freedom is better than tyranny. Equal rights are better than bigotry and discrimination. Literacy is better than illiteracy."[60] We have all of those things in greater plenitude today than we did a hundred years ago, when our planet was far less populated and we lived far shorter lives.

So, when I consider the prospect of a more populated planet, it is far easier to envision one in which a greater share of the global population is living better than it ever has. The science simply compels me to dream this way.

But why? Why do we live better even though there are more of us and more of us are living longer lives?

There are a great many factors, including the good that comes from networks of human capital of all ages. But if I had to explain it in just one word, that word would be: "elders."

THE LONG RACE

It was a beautiful day in San Diego, California, in June 2014. Thousands of runners were lined up for a marathon. Among them was a woman who most people would likely have pegged as 70. That alone would have made her an outlier among the throngs of runners predominantly in their 20s, 30s, and 40s.

Except that Harriette Thompson wasn't in her 70s. She was 91. And that day, she broke the official US record for a marathon by a woman in her 90s—by nearly two hours.

When she ran the same race again the following year, she was just a tad slower but set a new record as the oldest woman known to have completed a marathon. She crossed the finish line to cheers of "Go, Harriette!" as red, white, and blue confetti rained down around her.[61]

Thompson, who raised more than $100,000 for the Leukemia & Lymphoma Society through her running, was an exceptionally special person, for her vigor and her big heart. But what she did physically doesn't need to be special. In the future, no one will do a double take upon seeing a marathoner in his or her 90s step up to the starting line among a chronologically younger crowd. The truth is that it will be hard to tell how old the veteran runners are.

That will be the case in every other facet of life, too. In our classrooms, where ninety-year-old teachers will stand before seventy-year-old students embarking on a new career, as my father did. In our homes, where great-great-grandparents will play rough-and-tumble games with their great-great-grandchildren. And in our businesses, where older workers will be revered and fought over by employers. You can already see it happening in workplaces that depend on experience.

And it's about time.

Old people were revered in traditional cultures as sources of wisdom. Of course they were: before written text—and long before the advent of digital information—elders were our *only* wellsprings of knowledge. That began to change, quickly and significantly, when a fifteenth-century goldsmith, Johannes Gutenberg, developed a press that led to the Printing Revolution. The subsequent Education Revolution, in the nineteenth and twentieth centuries, led to rates of literacy that grew to meet the availability of information. Elders were no longer the only sources of long-held information. Rather than being seen as an essential asset to a functioning society, the elderly came to be viewed as a burden.

The Nobel laureate Seamus Heaney described our complicated relationship with aging parents in his poem "The Follower," ostensibly about his own father, who had shoulders like sails, and Seamus, as a child, "tripping and falling" in his father's wake. The poem ends, "But today / It is my father who keeps stumbling / Behind me, and will not go away."

Heaney's tragic poem echoes the sentiments expressed in a *Life* magazine article from 1959 titled "Old Age: Personal Crisis, U.S. Problem."[62]

"The problem has never been so vast or the solution so inadequate," the author wrote. "Since 1900, with better medical care, life expectancy has increased an average of 20 years. Today there are five times as many aged as in 1900 . . . the problem of old age comes almost overnight—when a man retires, after a woman's husband dies."

When I came upon the musty magazine in a Cape Cod bookstore on Old King's Highway, I first marveled at how far gender equality has come since 1959, but then was struck by how little has changed in the way we fret about the calamity of the impending deluge of old people. Whatever will we do with them? Will they overwhelm our hospitals? What if they want to keep working?

The impact of this shift in the way many people view elders has been particularly hard felt in the workforce, where age discrimination is rampant. Hiring managers hardly bother to hide their prejudices. They view older workers as more likely to be sick, slow working, and incapable of handling new technologies.

Absolutely none of that is true, especially for people in management and leadership positions.

Yes, it used to be that technology was slow to catch on. But educated older people now use technology just as frequently as those under 65. Don't forget, these are the generations who sent rockets to the moon, and invented the supersonic passenger jet and personal computer.

"Every aspect of job performance gets better as we age," Peter Cappelli, the director of the Wharton Center for Human Resources, reported after he began to investigate the stereotypes that often surround older workers. "I thought the picture might be more mixed, but it isn't. The juxtaposition between the superior performance of older workers and the discrimination against them in the workplace just really makes no sense."[63]

Between 2012 and 2017, the average age of new CEOs at the largest companies in the United States increased from 45 to 50 years. Yes, it's true that older people cannot work physically the same way they did when they were 20, but when it comes to management and leadership, it's the opposite. Consider some examples of leadership: Tim Cook, Apple's CEO, is currently 58; Bill Gates, Microsoft cofounder, is 63; Indra Noori, who recently stepped down as CEO of PepsiCo and now sits on Amazon's board, is 63; and Warren Buffett, the CEO of the investment firm Berkshire Hathaway, is 87. These people are not what you'd call technophobes.

It's bad enough when companies allow themselves to be deprived of great workers because of untrue stereotypes. But this is done at a national and international scale, sidelining millions of people in the best years of their work lives—all because of old ideas about age that aren't true now and that are going to be even less true in the near future. Thanks to the Age Discrimination in Employment Act of 1967, individuals in the United States over age 40 are legally protected from employment discrimination based on age. But in Europe, most workers are forced to retire in their mid-60s, including professors, who are just getting good at what they do. The best ones move to the United States so they can keep on innovating.

It's Europe's loss, and it's completely backward.

If you were the transportation director of a large company preparing to spend hundreds of thousands of dollars to purchase some new trucks for your fleet, would it be better to invest in a model known to be reliable for about 150,000 miles or one known to last twice as long? All other things being equal, of course you'd choose the trucks that would last longer; that's simply the right investment.

We don't tend to think about *people* this way, though. It feels cold. Humans aren't products that have been rolled off assembly lines, after all. But people *are* investments. Every society in our world places a bet on each one of its individual citizens—chiefly through education and training—that pays off over the course of a taxpaying lifetime. Those investments already produce tremendous dividends to our societies—for every dollar a government spends on education, that nation's GDP grows on average by about $20.[64] And this is in an era in which age-related sickness and death rob us of years of productivity. Imagine, then, what the returns would be if we extended the best working years of people's lives.

Right now, about half of the people in the United States and Europe between the ages of 50 and 74 are suffering from a mobility impairment. About a third have hypertension. More than one in ten is fighting heart disease or diabetes. More than one in twenty is suffering from cancer or lung disease.[65] Many are fighting several of these diseases at once. Even so, they far outperform the young at most mental tasks, writing and vocabulary, and leadership.

When we extend *healthy* lives, we exponentialize this investment. The longer people stay in the workforce, the better our return. That doesn't mean people should *have* to keep working. The way I see it, once you've repaid the investment our society has made in you, and if you can support yourself, there's little reason why you shouldn't be able to do whatever you want for as long as you want. But as we continue to evolve into a species that stays healthier for a lot longer, old ideas about who "belongs" in the workforce are going to change, and fast.

A lot of people worry that young workers will be "crowded out" of jobs if no one ever retires. I don't. Countries stagnate because they don't innovate and don't utilize their human capital, not because there aren't enough jobs. This explains why countries with an earlier retirement age have a lower GDP. In the Netherlands, Sweden, the United Kingdom, and Norway, the retirement age is 66 to 68, while in Moldova, Hungary, Latvia, Russia, and Ukraine, it is 60 to 62.[66] I have nothing against young people—I teach and train them every day—but I also know science and technology is getting more and more complex, and young people can benefit greatly from learning the wisdom that decades of experience can bring.

Looking through old magazines, it's easy to see what scared previous generations. It's always the same; there are too many people and not enough resources: too many people and not enough jobs.

In another edition of *Life* magazine, this one from 1963, an article says that automation "displaces men. It has thrown hundreds of thousands of people out of work and will throw out many more."[67]

It then quotes from a then-recent study on the topic: "Within the next two decades machines will be available outside the laboratory that will do a credible job of original thinking, certainly as good as that expected of most mid-level people who are supposed to 'use their minds.'"

The foreboding article concludes, "While we are fast running out of use for people, we are at the same time ironically producing people faster than ever."

Those fears never materialized as fact, not even in the face of another tremendous disruption of the status quo. In 1950, the US labor force participation rate of women was about 33 percent; by the turn of the century, it had nearly doubled. Tens of millions of women began working during those decades; that didn't result in tens of millions of men losing their jobs.

The labor market isn't a pizza with a limited number of pieces. Each of us can have a slice of the pie. And in fact, greater labor participation by older people, men and women, may be the best antidote to concerns that we're going to bankrupt our social insurance programs. The answer to the challenge of keeping Social Security solvent is not to *force* people to work longer but to *allow* them to do so. And given the pay, respect, and advantages that will come with extra decades of vitality and the opportunity to continue to find purpose through meaningful work, many will do so.

Even as it stands, many Americans plan to work beyond the traditional age of retirement, at least on a part-time basis, not always because they have to but often because they want to.[68] And as more people recognize that working well into their so-called golden years doesn't mean feeling tired or confused at work, being treated poorly, or having to take time off to visit the doctor all the time, the number of people who will want to stay engaged in this part of their lives is certain to grow. Age-related discrimination will fall, particularly as it becomes harder to tell who is "older" in the first place.

And if you are a politician wondering how it will be possible to provide meaningful, productive work to all the people, consider the city of Boston, where I live. Since it opened the first American university in 1724 and the first American patent office in 1790, the city has been home to the invention of the telephone, razor, radar, microwave oven, the internet, Facebook, DNA sequencing, and genome editing. In 2016 alone, Boston produced 1,869 start-ups and the state of Massachusetts registered more than 7,000 patents, about twice as many per capita as California.[69] It is impossible to know how much wealth and how many jobs Boston has generated for the United States and globally, but in 2016 the robotics industry alone employed more than 4,700 people in 122 start-ups and generated more than $1.6 billion in revenue for the state.[70]

The best way to create jobs for productive people of any age, even less skilled workers, is to build and attract companies that hire highly skilled ones. If you want a country in which your citizens flourish and that others envy, don't reduce the retirement age or discourage medical treatments for the elderly, hoping to save money and make room for the young. Instead, keep your population healthy and productive, and destroy all barriers to education and innovation.

I try my best to be aware of how lucky I am to be living in Boston and working on things I love. So long as I am feeling physically and mentally fit, I don't ever want to retire. When I envision myself at 80, I see a person who doesn't feel a whole lot different than he does at 50 (and if reprogramming works, won't look a lot different, either). I imagine walking into my lab at Harvard, much as I do these days most mornings of the week, to be bombarded by the energy and optimism of a motley group of researchers working to make discoveries aimed at changing billions of people's lives for the better. I absolutely *love* the idea of applying sixty or seventy years of experience to the task of leading and mentoring other scientists.

Yes, it's true: when people choose to keep working for eighty, ninety, or a hundred years, it will fundamentally change the way our economy works. Trillions of dollars have been hidden away in virtual and quite a few literal mattresses by people dreading the prospect of running out of money at a time in their lives in which they are too frail to return to work. The option to work at any age—if and when work is wanted and needed—will offer a sort of freedom that would have been unfathomable just a few years ago. The risk of spending one's savings on fulfilling a dream, innovating, starting a business, or going on a new educational journey will not be such a risk at all; it will simply be an investment in a long and fulfilling life.

And it's an investment that will pay off in other ways, too.

UNLEASHING THE ARMY

Dana Goldman had heard from all the naysayers.

The University of Southern California economist understood—far more than most people do—that health care costs had risen dramatically over the past decades, not just in his native United States but around the world. He knew those costs were coming at a time in which human lifespans were being extended, resulting in multitudes of patients who were sicker for longer. And he was fully aware of the never-ending nightmare about the future solvency of programs such as Social Security that provide for the common welfare. The prospect of billions of people growing even older seemed like a perfect storm of economic catastrophes.

A few years back, however, Goldman began to realize that there was a difference between extending lives and extending *healthy* lives. As it stands, aging presents a double economic whammy, because adults who get sick stop making money and contributing to society at the same time they start costing a whole lot to keep alive.

But what if older people could work longer? What if they were to use fewer health care resources? What if they were able to continue to give back to society through volunteering, mentorship, and other forms of service? Perhaps—just perhaps—the value of those extra healthy years would lessen the economic blow?

So Goldman began to crunch the numbers.

As any good economist would, he sought to be both rigorous and conservative in his estimation of the benefits of delayed aging. He and his colleagues developed four different scenarios: one that simply projected spending and savings under status quo conditions, two that estimated the impact of modest improvements to delaying specific diseases, and one that evaluated the economic benefits of delaying aging and thus reducing *all* the symptoms of aging. For each scenario the researchers ran a simulation fifty times and averaged the outcomes.

When Goldman reviewed the data, something became clear: reducing the burden of any one disease, even several, wouldn't change much. "Making progress against one disease means that another will eventually emerge in its place," his team reported in *Perspectives in Medicine*. "However, evidence suggests that if aging is delayed, all fatal and disabling disease risks would be lowered simultaneously."[71]

For the record, that's precisely what I am suggesting will happen to the total disease burden as we slow and even reverse aging. The result will be an upgrade of the health care system as we know it. Treatments that once cost hundreds of thousands of dollars could be rendered obsolete by pills eventually costing pennies to make. People will spend the last days of their lives at home with their families instead of racking up huge bills in centers intended for nothing more than "aging in place." The idea that we once spent trillions of dollars trying to eke out a few more weeks of life from people who were already teetering on the edge of death will be anathema.

The "peace dividend" we will receive from ending our long war on individual diseases will be huge.[72] Over fifty years, Goldman estimated, the potential economic benefits of delayed aging would add up to more than $7 trillion in the United States alone. And that's a *conservative* estimate, based on modest improvements in the percentages of older people living without a disease or disability. Whatever the dollar figure, though, the benefits "would accrue rapidly," Goldman's team wrote, "and would extend to all future generations," because once you know how to treat aging, that knowledge isn't going away.

Even if we reinvest only a small amount of that dividend into research, we'll enter a new golden era of discovery. That discovery will be hypercharged as we unleash a vast army of brilliant people not only to continue the fight to prolong human vitality but to combat the many other challenges we currently face, such as global warming, the rise of infectious diseases, moving to clean energy, increasing access to quality education, providing food security, and preventing extinctions. Those are challenges we cannot effectively fight in a world in which we spend tens of trillions of dollars each year battling age-related diseases one by one.

Even now, while we spend so much of our intellectual capital on whack-a-mole medicine, there are thousands of labs around the world with millions of researchers. That sounds like a lot, but globally, researchers account for just one-tenth of 1 percent of the population.[73] How much faster would science move if we were to unleash even a small bit of the physical and intellectual capital that's tied up in hospitals and clinics treating diseases one at a time?

This army could be augmented by billions of additional women if they can be provided much longer windows of opportunity for pregnancy and parenting. Animal studies in my lab indicate that the window of female fertility could be extended by up to a decade. This is an exciting prospect because, in the United States, 43 percent of women step away from their careers for a period of time, almost always to shoulder the burden of child rearing. Many never return to work. As a woman's lifespan and fertility lengthens, the consequences of taking a break will be seen as relatively minor. By this century's end, we will almost certainly look back with sadness at the world we currently inhabit, in which so many people, particularly women, are forced to choose between parenting and career success.

Now add to the ranks of this army the combined intellectual power of the men and women who are currently sidelined due to age discrimination, socially enforced ideas about "the right time to retire," and diseases that rob them of the physical and intellectual capacity to engage as they once did. Many people in their 70s and 80s will reenter the workforce to do something they've always wanted to do, earning more than they ever did, or serving their communities as volunteers and helping raise their grandkids, as my father has. With the money saved by preventing expensive medical care, a retraining fellowship could be provided for a few years to allow people over 70 to go back to school and start the career they always wished they'd started but didn't because they made the wrong decisions or life simply got into the way.

With active people over 70 still in the workforce, imagine the experiences that could be shared, the institutional knowledge that could be relied upon, and the wizened leadership that would emerge. Problems that seem insurmountable today will look very different when met by the tremendous economic and intellectual resources offered by prolonged human vitality.

That could be especially true if we're all engaging in our world with the best version of ourselves.

THE GREATEST OF THESE

In the early 1970s, two psychologists decided to put the Parable of the Good Samaritan to the test.

The biblical story, as you might recall, centers around the moral obligation to help those in need, and the psychologists figured that people who had the parable on their minds would be more likely to stop to help someone in distress. So they hired an actor to pretend to be in pain and put the young man—who was doubled over and coughing—into an alley next to the doorway of the Green Hall Annex at Princeton Theological Seminary.

The psychologists had also recruited forty seminary students to present a talk at the annex. First, though, the students were asked to stop by another building on campus. Once there, some of the seminarians were told they could take their time getting to the annex, others were told they would be on time as long as they left immediately, and a final group was told that they needed to hurry to make it to the annex on time.

Just 10 percent of those in the "high-hurry" group stopped to help the man. Those were seminary students, for goodness' sake, and they ignored a brother in need. One literally stepped over the man in distress to get where he was supposed to be.

In the "low-hurry" group, though, more than 60 percent stopped to help. In that experiment, the difference between whether a person made a compassionate choice had nothing to do with personal morality or religious scholarship but whether he felt rushed.[74]

This isn't a new idea, of course. Back in the days in which Christ was first telling the Good Samaritan story, his contemporary in ancient Rome Seneca the philosopher was begging his followers to stop and smell the roses. "Life is very short and anxious for those who forget the past, neglect the present, and fear the future," he wrote.[75]

For people who don't appreciate life, time is "reckoned very cheap . . . in fact without any value," he lamented. "These people do not know how precious time is."

This might be the least considered societal advantage of prolonged vitality, and it might just be the greatest advantage of all. Perhaps when we're not all so afraid of the ticking clock, we'll slow down, we'll take a breath, we'll be stoic Samaritans.

I would like to emphasize the word "perhaps," here. I will be the first to say that this thesis is supposition more than science. But the small-sample Princeton experiment both followed and portended a lot of other research demonstrating that humans are a lot more humane when they've got more time. All of the studies, though, take stock of how people behave when they have a few more minutes, or perhaps a few more hours, to spare.

What would happen if we had a few more years? A few more decades? A few more centuries?

Maybe we would do nothing differently, even if we had two or three hundred years. In the grand scheme of the universe, after all, three hundred years is nothing. My first fifty years went by like a blink, and I suspect that a thousand years, a mere twenty blinks, would also feel short.

And so it comes down to this: When those years do come, how do we wish to spend them? Will we follow the perilous path that ultimately leads to a dystopian doom? Will we band together to create a world that exceeds our wildest utopian dreams?

The decisions we make right now will determine which of those futures we create. And this is important. Preventing disease and disability is possibly the single most impactful thing we can do to avert a global crisis precipitated by climate change, crippling economic burdens, and future social upheavals. We have to get this right.

Because there has been no more consequential choice in the history of our species.

NINE

A PATH FORWARD

IN 1908, ONLY FIVE YEARS AFTER THE WRIGHT BROTHERS BEGAN FLYING, H. G. Wells published a book titled *A War in the Air*, in which Germany starts an air war against Great Britain, France, and the United States.

To say Wells had a penchant for prescience would be a vast understatement.

In 1914, the Institute of International Law tried to ban the dropping of bombs from flying machines,[1] but it was too little, too late. Giant German "Gotha" planes began bombing Great Britain in 1917. That year, 180 miles west of London, a baby was born, named Arthur, who would come to be regarded as the preeminent science fiction writer of the twentieth century. As he became more famous, Arthur C. Clarke increasingly regarded predicting the future a "discouraging, hazardous occupation." That may be true, but Clarke was awfully good at it, anticipating satellites, home computers, email, the internet, Google, live-streaming TV, Skype, and smart watches.

Clarke had some strong opinions about scientists: a physicist in his

30s was already too old to be useful. In other scientific disciplines, a 40-year-old has likely experienced "senile decay." And scientists of over 50 are "good for nothing but board meetings, and should at all costs be kept out of the laboratory!"

Toward the end of his life, Clarke gave a series of interviews. Most of them were recorded and edited because he had halting speech caused by postpolio syndrome. In one interview, he revealed that he had a use for washed-up scientists: "When a distinguished but elderly scientist states that something is possible, he is almost certainly right. When he states that something is impossible, he is very probably wrong."[2]

I'm a scientist who is now 50. Some people might call me distinguished. And my students definitely don't want me in the laboratory. So although I can't say I'm certain about my predictions, I'm apparently well qualified to make them.

I've been asked on occasion, by members of the US Congress and the like, for predictions of technological breakthroughs and how they might be used for good or evil purposes. A few years ago, I gave an opinion on the top five future advances in the biological sciences that would be of relevance to national security. Though I can't reveal what I said, I expect that most people would have thought they were science fiction. My best estimate was that they would happen sometime before 2030. Within six months, two of them became science fact.

I don't know precisely when the first individual to cross the threshold of 125 years will be, but he or she will certainly be an outlier, as pioneers always are. In just a few years, he or she will be joined by another. Then dozens more. Then hundreds. Then the fact won't be worth noting. Even longer lifespans will become more and more common. The world may see the first sesquicentenarian sometime in the twenty-second century. (If you think that's far off, consider that some researchers believe *half* of all American children born today will celebrate New Year's Eve 2120. Not outliers—half.)[3]

Those who think all this is impossible are ignorant of the science.

Or they are denying it. Either way, they are almost certainly wrong. And because things are moving so fast, many of them might even live to realize they are wrong.

No biological law says there is a limit to how long we can live; there is no scientific mandate that the average age at death must be 80 years. And there is no God-given mandate to die after fourscore years. Indeed, in Genesis 35:28, Isaac is claimed to have lived "*one hundred* and fourscore."[4]

Thanks to the technologies I've described, a prolonged, healthier human lifespan is inevitable. How and when we'll achieve it is a bit less certain, although the general path is quite clear. The evidence of the effectiveness of AMPK activators, TOR inhibitors, and sirtuin activators is deep and wide. On top of what we already know about metformin, NAD boosters, rapalogs, and senolytics, every day the odds increase that even more effective molecule or gene therapy will be discovered, as brilliant researchers around the world join the global fight to treat aging, the mother of all diseases.

All of that comes on top of the other innovations that are on track to further lengthen our lives and strengthen our health, such as senolytics and cellular reprogramming. Add to that the power of truly personalized care to keep our bodies running, prevent disease, and get ahead of problems that could be troublesome down the road. That's not to mention the very easy steps we can all take right now to engage our longevity genes in ways that will provide us with more good years.

With significantly prolonged vitality an inexorable part of our future world, what do you want that world to look like?

Are you comfortable with a future in which the rich live much longer than the poor and in doing so get richer with every passing year? Do you want to live in a world in which an ever-increasing population greedily scraps for every last remaining resource while the world continues be less and less habitable?

If so, there is nothing left for you to do. The status quo will get us there—regardless, in fact, of whether we prolong human life or not. You can sit back, relax, and watch the world burn.

There is another potential future, though, one in which prolonged youthfulness is the torch that lights the way to greater universal prosperity, sustainability, and human decency. This is a future in which tremendous resources are freed from a medical-industrial complex that is based on battling diseases one by one, thus creating tremendous opportunities to tackle other challenges. It is a future in which people who have lived on this planet for a long time are revered for their knowledge and skills. It is a future of global Good Samaritanism.

It's also a future for which we must fight, for it is in no way guaranteed.

To get there, we have some work to do.

INVEST PUBLIC MONEY TO TACKLE AGING, NOW

I am a serial entrepreneur, a disciple of innovation, and a grateful beneficiary of people's investment in me and the teams I've brought together to solve hard problems.

I also recognize, however, that the free market doesn't magically produce good science or equitable outcomes when it comes to health care. any research endeavor, a balance of public and private funding is vit producing the conditions that encourage unbound scientific explor the investment in early discoveries, and a degree of common ow that better ensures that the benefits of newfound knowledge wi available to the greatest possible number of people.

That balance has become ever more precarious in recen ing in 2017, for the first time since World War II, the US ment was no longer the majority source of funding f research in the United States.

Federal funding for science in the United State when the Marine Hospital Service, the predecesso tutes of Health, was charged by Congress with arriving ships for clinical signs of infectious di 1901, a routine supplemental appropriation

a new building, became the founding legislation of the NIH. Congress was not convinced that the money would go to good use, so it ensured that every year, funding would be at the discretion of Congress, and so it remains. Hopefully Congress remains convinced that NIH funding, which provides hundreds of competitive grants to scientists around the country, is money well spent, because without NIH-funded research, the majority of the medicines and medical technologies we rely on would never have been discovered, not to mention the thousands of new medicines still waiting to be discovered.

At least for now, the federal government still makes up a large share of total funding for medical research at hospitals and universities, ensuring that R&D isn't driven by profit alone. This is important, so scientists like me can run with their imaginations and instinct, sometimes for a decade, before any commercial applications are apparent and long before any investor would consider supporting the work to help it survive the innovation "valley of death."

Clearly, government is essential in this ecosystem, but in a world in which there is more competition for overall research funding than ever before, good scientists investigating aging are having to seek more and more private financial support for their work; world-changing research certainly isn't cheap, and when it's funded by a company with short-term goals, it isn't free, either. That's why it's important that we reverse the decline in public funding for medical research, which fell by 11 percent in real dollars from 2003 to 2018.[6]

The situation is particularly hard for researchers studying aging. Funding to understand the "biology of aging" gets less than 1 percent of the total US medical research budget.[7] With an aging population and ever-increasing health care costs, why aren't governments dramatically increasing funding for aging research to keep people healthier for longer? The reason is, in nearly every nation in the world that has made a investment in medical research, that research is tied back, if not completely tied up, by the definition of disease.

If you are a scientist with an idea for a novel way to slow the progression of cancer or a researcher with an inventive idea for ending Alzheimer's disease, the NIH and similar national research funding agencies around the world are there to help. The NIH isn't simply a bunch of buildings in Bethesda, Maryland. It allocates more than 80 percent of its budget to almost 50,000 competitive grants to about 300,000 researchers at more than 2,500 universities and research institutions. Medical research would almost grind to a halt without that money.

It's worth drilling down into the NIH budget to see which of the 285 diseases that are being researched get the most attention.[8]

- Heart disease gets $1.7 billion for a disease that affects 11.7 percent of the population.
- Cancer gets $6.3 billion to impact 8.7 percent.
- Alzheimer's disease gets $3 billion for a disease that impacts 3 percent—at most.[9]

How much does obesity, which affects 30 percent of the population and reduces lifespan by over a decade, get? Less than a billion dollars.

Don't get me wrong. Compared to how the government spends a lot of its money—the cost of a single F-22 Raptor fighter jet is upward of $335 million, for instance—this is all money well spent. To put it into even greater perspective, though, consider this: US consumers spend more than $300 billion per year on coffee.[10]

To be fair, life without coffee might not be worth living. But if you are a researcher who wants to make life *even better*—by slowing or reversing diseases of aging—you have a bit of a problem. There just isn't that much public money being spent on that area of science.

In 2018, Congress appropriated $3.5 billion for research into aging, but if you were to dig into the budgetary documents, you'd see that the money went almost entirely to research Alzheimer's disease, perform clinical trials of hormone replacement therapy, and study the lives of the elderly. Less than 3 percent of the funding for "aging research" was actually for the study of the biology of aging.

Aging disables 93 percent of people over the age of 50, but in 2018 the NIH spent on aging less than a tenth of what was spent on cancer research.[11]

One scientist who is particularly annoyed by the budgetary focus on individual diseases is Leonard Hayflick, the scientist who first discovered that human cells in a dish have a limited capacity to divide and eventually senesce, after having reached the Hayflick Limit.

"The resolution of Alzheimer's disease as a cause of death will add about 19 days onto human life expectancy," he noted in 2016.[12] Hayflick has suggested that the name of the National Institute on Aging, a division of the NIH, might as well be changed to the National Institute on Alzheimer's Disease.

"Not that I support ending research on Alzheimer's disease, I do not," he said, "but the study of Alzheimer's disease and even its resolution will tell us nothing about the fundamental biology of aging."

The relatively tiny amount the United States spends on research into aging, however, is generous when compared to that of most other advanced nations, which invest next to nothing. There is no doubt that this situation is a direct result of the establishment view that aging is an inevitable part of life rather than what it actually is, a disease that kills about 90 percent of the population.

Aging is a disease. This is so clear that it seems almost insane that those four words need to be repeated again and again. But I'll do so anyway: aging is a disease. And not only is it a disease, but it is the mother of all diseases, the one we all suffer from.

Paradoxically, no public funding agency around the world classifies aging as a disease. Why? Because, if we are fortunate to live long enough, we *all* suffer from it. And thus for now, the pool of public funding available for research aimed at prolonged vitality is rather paltry; the biggest checks are still being written to support initiatives aimed at recognized diseases. And at the moment I am writing these words, aging isn't recognized as a disease. Not in *any* nation.

There are several ways to speed innovation to find and develop medicines and technologies that prolong healthy lifespan, but the easiest is also the simplest: define aging as a disease. Nothing else needs to change. Researchers working on aging will compete on equal footing with researchers working to cure every other disease in the world. The science-based merits of grant proposals will dictate which research efforts are funded. And private investment will continue, as it should, to drive innovation and competition.

Labs like mine, focused specifically on developing innovative therapies to treat, stop, and reverse aging, will no longer be rare. There will be one or more at every health science university in the world.

And there should be, because there is no shortage of scientists lining up to enlist in this army. Right now, I and other researchers who study aging are being besieged by eager, experienced, and absolutely brilliant youths who want nothing more than to devote their lives to the fight to stop aging. For lab heads like me, it's a virtual buyer's market. There are far more people who want to work in aging than there are labs they can work in. What this means is that there are a lot of people who, despite being wicked smart and raring to tackle the aging problem, are having to work in other fields or other professions. This will soon change.

The first nations to define aging as a disease, both in custom and on paper, will change the course of the future. The first places to provide large amounts of public funding to augment the fast-growing private investments in this field will prosper in kind. It will be their citizens who benefit first. Doctors will feel comfortable prescribing medicines, such as metformin, to their patients before they become irreversibly frail. Jobs will be created. Scientists and drug makers will flock to that country. Industries will thrive. Their national budget will see a significant return on investment. Their leaders' names will be in the history books.

And the holders of the patents, the universities and the companies, will have more money than they know what to do with.

I'm proud to say that Australia is leading the charge to define aging as a treatable disease. I recently made a trip to Canberra to meet with Greg Hunt, the minister for health, and Deputy Secretatry Professor John Skerritt of the Therapeutic Drug Authority, and about 15 of Australia's other top aging researchers. I learned that developing a drug for aging may be far easier in my native land than in the United States. While the United States expects evidence that a disease is cured or alleviated, in Australia it is possible for a drug to receive approval for "influencing, inhibiting or modifying a physiological process in persons." In the aging field, we know how to do that!

Singapore and the United States are among the nations that are also seriously considering a regulatory shift. Whichever does so first will be making a historically important decision, one that will benefit itself first and foremost.

There's a reason why the United States virtually owns the aerospace sector—exporting products worth more than $131 billion in 2017, or more than the next three national exporters combined. "First in Flight" isn't just a good slogan for North Carolina license plates; it's a statement about why being out front matters. Americans retain the pioneering spirit of their ancestors: anything is possible. More than a century after the Wright brothers flew the first planes at Kitty Hawk, and after almost losing out to the French and British, the United States is still ahead in the flight game. It has the world's most powerful air force. It got to the moon first. And it has a big lead in the development of public and private initiatives to put people on Mars.

But none of that will impact human history as much as the first nation to declare aging a disease.

At a bare minimum, governments have a vested interest in making sure the innovations we develop to protect human life are used wisely and for our collective benefit. The time to talk about ethics and how personal privacy will be impacted by these coming technologies is now, for once the bottle is opened, it will be exceedingly hard to put the genie back. DNA-based technologies that enable the detection of specific pathogens, for example, could also be used to search for specific people. Technology now exists to create humans that are stronger and longer lived. Will parents choose to give their children "the best possible start"?[13] Will the United Nations outlaw the genetic improvement of citizens and military?

To create a future worth living in, it won't be enough to simply fund research that lengthens and protects people's lives and ban its misuse. We must also ensure that everyone benefits together.

IT'S TIME TO INSIST ON THE RIGHT TO BE TREATED

The dentist looked bored. "Your teeth are fine," she told me as she peered into my mouth. "Just the normal wear and tear. I'll send the hygienist in for your cleaning, and we'll get you on your way."

It seemed as though she were turning away before her fingers were even out of my mouth.

"Doctor, if you can spare a moment," I said, "can you tell me what you mean by 'normal wear and tear'?"

"You're getting older, and your teeth are showing that," she said. "Your two front teeth are worn down. Totally normal. If you were a teenager, we would probably fix them but—"

"Well, then," I said. "I'd like to have them fixed."

Eventually the dentist relented, although not before I told her what I did for a living and explained to her that I was hoping to use my teeth for a very long time to come. I also assured her that I would be happy to pay for the procedure even if my insurance wouldn't.

Her resistance was understandable. When dentists look into the mouths of patients who are in their 40s and 50s, they have long been looking at teeth that are halfway done with their jobs. But that's no longer the case. Our teeth—like all of our other body parts—are going to have to last a lot longer now.

My experience at the dentist was a microcosm of the way middle-aged people are treated in every facet of the health care system. When a doctor looks at a 50-year-old person right now, his or her goal is to keep the patient "less sick," not to ensure that he or she will be healthy and happy for decades to come. Who among us over the age of 40 has not heard a doctor say the words "Well, you're not twenty anymore"?

There are two things that guide medical treatments more than anything else: age and economics. The first often limits what doctors are even willing to discuss in terms of treatment options, because they assume that people are *supposed* to slow down, begin dealing with a bit of pain, and gradually experience the degradation of various body parts and functions. The second dictates these discussions even more, because regardless of how much potential a procedure may have to improve a patient's life, it is pointless, and even heartless, to tell someone about care he or she can't afford.

Indeed, our medical system is built on ageism. When we are young, we don't get treatments that could keep us healthy as we grow old. When we are old, we don't get the treatments that are routinely used on the young.

This all has to change. The quality of our medical care should not be predicated on age or income. A 90-year-old and a 30-year-old should be treated with the same enthusiasm and support. There will be enough money to pay for this because of the trillions of dollars that won't have to be spent by insurance companies or the government, and hence ourselves, on treating the chronically ill.[14] Everyone should be entitled to treatments and therapies that improve quality of life, no matter what the date on his or her birth certificate is. As we move toward a world in which the number of our birth years indicates less about us than ever, we will need to adjust the assumptions, rules, and laws that govern what medical treatments people can receive.

Equitable access to medical care, no matter how long life may be, is a terrifying idea to many people, because it sounds awfully expensive. That's understandable, because as it stands, social medical programs across the globe are straining under the ever-increasing cost of treatment, especially treatments that are provided to those who are very sick, very old, and likely to get nothing more than a few extra years—if that—out of the deal.

That's not what the future of medical care needs to look like. Right now, the overwhelming majority of the money we spend on medical care is spent fighting diseases. But when we are able to treat aging, we will be tackling the biggest driver of all disease. Effective longevity drugs will cost pennies on the dollar compared to the cost of treating the diseases they will prevent.

In 2005, a study by Dana Goldman and his colleagues at RAND in Santa Monica put some numbers on this. They estimated the value that new discoveries would add to society and the cost to society to extend a human life by one year.[15] The cost of an innovative medicine to prevent diabetes: $147,199. Of a cancer treatment: $498,809. Of a pacemaker: $1,403,740. Of an "antiaging compound" that would extend healthy years by a decade: a mere $8,790. Goldman's numbers support an idea that should be common sense: that there is no cheaper way to address the health care crisis than to address aging at its core.

But what if the drugs don't keep people healthy? What if they simply prolong life, like many cancer chemotherapy drugs, which are approved based on their ability to provide a longer life, not a higher quality of life? Society should debate whether longevity medicines that don't keep us healthier should ever be approved. If they were to be allowed, there would be even more elderly people with disease and disability, and, according to Goldman, health care spending in thirty years' time would be 70 percent higher.

Fortunately, the science suggests this nightmare scenario is not going to happen. When we have safe and effective drugs to slow aging, they will also extend our healthspans. What will be left will be medical maintenance, which is exceptionally cheap; emergency medicine, which is costly but rare; and communicable diseases, which we'll be able to track, treat, and prevent with far greater efficiency and effectiveness. It's similar to making the switch from gasoline-powered cars that need oil, belts, tuning, and regular maintenance to electric cars that tell you occasionally to top up the windscreen washer fluid.

Having lived in Australia, the United Kingdom, and the United States—three countries with an intertwined history, language, culture, and trade—I've found it interesting to see how similar they are in some ways and how different they are in others.[16] One big difference is that most Australians and Britons rarely assume that their way of doing things is the best. Americans, however, often believe that their way of doing things is assuredly the best.

I'm not saying the United States doesn't do a lot well and shouldn't continue to blaze its own path in many areas of domestic and global policy, but I've long been perplexed by the American resistance to studying what actually works elsewhere.

In science we call this experimentation, and it's what propels our civilization forward. The more experiments are conducted, the better informed we are. And some experiments work really well.

Seeded as a prison colony, Australia is one of the least religious countries in the world, but when it comes to providing for its citizens, it is a city upon a hill.[17] Like the United States, Australia has its problems: traffic snarls, a high cost of living, and strict rules aimed at saving lives, even if those rules often take the fun out of life.

There is a statistic, though, of which Australians are increasingly proud: a fifty-year-long experiment to protect and preserve every citizen, regardless of social position, education, or income. Deaths from car accidents and smoking are the lowest in the world, thanks to strict laws and hefty fines. Even before these laws passed, there was bigger change afoot. In the mid-1970s, a universal health care system was enacted, one of the first ever, and life expectancy in Australia began to shoot up. Similar to the United States in the 2010s, the next government tried to limit the scope of that progressive reform but ultimately failed.

A controversial, right-wing politician, Bronwyn Bishop, helped create an independent Australian Federal Department of Health and Ageing, which lasted from 2002 to 2013 with a budget of around AU$36 billion, focusing on health promotion, disease prevention, and services and caregivers for the aged.

During this time, Australia continued on an upward trajectory, using its wealth to create more health and productivity in its workforce, and its health and productivity to create more wealth, a virtuous cycle of the highest moral order.

Between 1970 and 2018, Australian men gained an extra twelve years of life. Their *healthy* life expectancy is 73 years, ten years higher than the global average, thanks to a significant decline in the percentage of people suffering from disabling health conditions.[18]

The elderly in Australia are being less elderly, less of a burden, and much more productive than other nations. If you visit Australia, the difference between its fit, active elderly and those in the United States who are saddled with obesity, diabetes, and disability is noticeable.

My father thought he was headed for the grave. Instead, he's most often headed to concerts or the mountains. He spends several nights a week eating out with friends. He's adept at computers and new high-tech gadgets and was one of the first people in Australia to have a smart speaker with a virtual assistant at home. He's unbothered by international travel, so we get to see him frequently. He's gone back to work. Physically and mentally, he is at least thirty years younger than his mother was at his age.

His remarkable health may or may not be due to the molecules he takes—the coming years of his life will be an indicator, whereas scientific proof will come only in the form of double-blind placebo-controlled trials—but he is also helped by frequent exercise, access to excellent medical care, and a system that believes in disease prevention, not just late-stage treatment. He is a shining example of a new generation of Australians in their 70s and 80s who are not just living longer but living far better than any of their forebears. In 2018, Australia ranked seventh on the global Human Capital Index, a measure of the knowledge, skills, and health that people in a nation accumulate over their lives, just behind Singapore, Korea, Japan, Hong Kong, Finland, and Ireland. The United States ranked twenty-fourth. China ranked twenty-fifth.

The trajectory for Australia is up, and the Aussies aren't looking back.

Having seen what works, other, mostly European countries have adopted similar health care systems. Australia now has reciprocal agreements with the United Kingdom, Sweden, the Netherlands, Belgium, Finland, Italy, Ireland, New Zealand, Malta, Norway, and Slovenia, which means that citizens from those countries can receive the same medical care in Australia as they can at home, and vice versa. Imagine an entire world like that.

Meanwhile, some countries are being left behind. And one, in particular, is moving backward.

Thanks to the burgeoning addiction to calories and opioids, and a health care system that is inadequate, if not completely inaccessible to one-third of its population, the United States recently experienced a decline in life expectancy for the first time since the early 1960s. That decline may soon exceed the decline in life expectancy caused by the Spanish flu epidemic in 1918. This is happening despite the fact that the United States spends 17 percent of its GDP on health care, nearly double that of Australia.

I don't mean to disparage the country in which I live—it has been very generous to my family and me. But I am frustrated. Ever since I arrived in the country that actually put humans onto the moon, it has been a shock to see opportunities to help more people for less money wasted over and over again.

The United States has been a leader in both public and private investment into lifesaving medical research. And although it can be hard to track the origin of every drug in this increasingly interconnected world, by one estimate 57 percent of all medications are developed in the United States. Other nations, especially those that don't invest as heavily in medical research, should be grateful to the United States for discovering and developing most of the drugs that ensure their increasingly long lives.

In a just world, the citizens of the United States would be the greatest beneficiaries of the medical breakthroughs they subsidize and produce. They're not.

Australians are. Britons are. As are the Swedes, the Dutch, the Irish, and the Slovenians. They're all benefiting in terms of lifespans and health-spans, because they have the sort of universal access to health care that 15 percent of registered Democrats and half of all Republicans in the United States have come to fear.[19] That the average American lifespan is just four years shorter than Australia's[20] belies the fact that in the poorest regions of the United States citizens live a decade even shorter than that.[21]

As the Australian example proves, when *everyone* is living longer and healthier, *everyone* does better. So why isn't this a topic of discussion in the United States? Why aren't people charging Capitol Hill with protest signs and the proverbial pitchforks, demanding more investment, universal access to medicines, and the healthiest lifespan on the planet? As other countries enjoy increasingly longer, healthier lives, perhaps Americans will wake up and smell the disparity. But I suspect they won't. Though the World Health Organization ranks the United States at number 37, below Dominica, Morocco, and Costa Rica and one up from Slovenia,[22] it's still common to hear US politicians say, without any justification, that the United States has the best health care system in the world, and millions of people believe it.[23]

The alternative to a universal right to be treated—regardless of age and regardless of the ability to pay—is a world in which rich people increasingly benefit from even longer and even healthier lives than they already enjoy, while poor people suffer through short, disease-ridden existences. This is a terrible idea for rich and poor alike.

My line of work has put me into contact with some of the wealthiest people in the world, who are understandably interested in learning the secrets of longer and healthier lives. I've yet to meet a single one who wishes to see such a divide come to pass. In that direction, after all, lie the seeds of revolution—and revolt seldom goes well for the ruling class. As the venture capitalist and "very large yacht" owner Nick Hanauer wrote in a memo to "My Fellow Zillionaires" in 2014, "there is no example in human history where wealth accumulated like this and the pitchforks didn't eventually come out. You show me a highly unequal society, and

I will show you a police state. Or an uprising. There are no counterex-
amples. None. . . . We will not be able to predict when, and it will be
terrible—for everybody. But especially for us."[24]

Hanauer's warning came before longevity genes were on most people's radar and long before most people had so much as contemplated
what significantly lengthened lifespans and healthspans could do to the
rich-poor divide.

Universal access to technologies that prolong vitality won't fix every
problem associated with income inequality, but it's a crucial start.

WE SHOULD BE ABLE TO DIE WHENEVER WE WANT TO

By cosmic standards, this region of the Milky Way isn't a horribly inhos-
pitable place for life to evolve in. We're here, after all. And the outer edges
of spiral galaxies like ours seem to hold reasonably good promise for a few
life-sustaining planets to materialize,[25] far better than the dwarf galaxies
that are the most abundant type of star systems in the universe.

The way the astronomer Pratika Dayal sees it, however, the most
likely places for life to form and thrive in are the rarer, metal-rich, giant
elliptical galaxies—twice as big as the Milky Way and often much bigger,
holding as many as ten times the number of stars and perhaps 10,000
times as many habitable planets.[26] By the way, if you're under the misconception that if we screw up this planet we can just travel to a new one,
consider that the closest known habitable exoplanet is twelve light-years
away, as the crow flies. That sounds close, but barring the discovery of
a space wormhole or light sailing of tiny cargoes at near light speed, it
would take at least 10,000 years to get a few humans there[27] (which,
I've argued, is another good reason to figure out how to extend human
lifespan).

The closest giant elliptical galaxy is Maffei 1, which is about 10 million light-years away. We can assume that if explorers from Maffei 1
ever make the trip to visit us, they'll be from an exceptionally advanced

society. I expect they'll have a few questions, for they will want to know how far we have advanced, too.

First, I believe, they'll be curious about the *easy* things: Have we figured out *pi* to a million decimal places? The speed of light? The fact that mass and energy are the same thing? Quantum entanglement? The age of the universe? Evolution?

Next they'll ask us about some of the harder stuff: Have we learned to use the resources available on our planet wisely? We'll get passing marks on that one, I suppose, as long as we don't mention lead pipes, nuclear bombs, and Furbys. Have we done so sustainably? "Um, pass."

Then they'll likely want to hear about what other worlds we've visited. "We sent twelve guys to Luna," we'll say. "Where's that?" they'll ask. We'll point to the big white orb in our night sky. "Hmmm," they'll say. "Just the men of your species?" We'll nod, and they'll roll their 146 eyes.

After that, they'll want to know about our lifespan. Have we figured out how to live far beyond the time given to us by evolution? "Er, we didn't know that was a thing worth studying until a few years ago." They'll offer a bit of overly enthusiastic encouragement, as a human adult might do for a baby who is learning to eat solid food.

The next question will be a rather grave one: "How do you die?" they'll ask. And how we answer that question is going to be an important indicator of just how advanced we truly are.

Right now, as my mother's death exemplified, the way most of us die is barbaric. We go through a long period of decline, and we've come up with ways to extend that period of pain, grief, confusion, and fear so that we must experience even more pain, even more grief, even more confusion, and even more fear. The sorrow, sacrifice, and turmoil this creates for our families and friends are protracted and traumatic, so that when we finally pass on, it often comes as a relief to those who love us.

The most popular means to the end, of course, are diseases—which can strike in the prime of life. Heart disease at 50. Cancer at 55. Stroke at 60. Younger-onset Alzheimer's at 65. Way too frequently, what is said at funerals is that someone left this life "way too early." Or the diseases don't kill, and the fight to beat them back again and again is a decades-long exercise in suffering.

These are terrible answers to the question of how we die. The answer we should strive for—just as much as we strive for prolonged vitality—is "when we are ready, then quickly and painlessly."

Fortunately, the science of longevity shows that the longer we make rodents live, the faster they tend to die. They still die of the same diseases, but, perhaps because they are very old, and the animals are on the brink anyway, they tend to suffer for days rather than months, then keel over.

This is not the only way we should meet our end, though.

"Physician-assisted suicide." "Death with dignity." "Elective euthanasia." Whatever we call it, we need to end the patchwork of laws and customs that force people to travel great distances, often when they are already suffering in one way or many, to bring their lives to a peaceful end.

These are the sorts of barriers that the eminent ecologist David Goodall faced in 2018, at the age of 104, when he was forced to leave his home in Australia, where physician-assisted suicide is illegal, and journey to a clinic in Switzerland, where it is lawful and safe. No one should have to choose between dying in a foreign land and committing a crime as his or her last act on Earth.

Thus no one with a sound mind who is over the age of 40—about the age at which one has paid back the initial societal investment in his or her education—should be denied the right to die on his or her own terms. And anyone, at any age, with a terminal diagnosis or painful chronic illness should have the same right.

Yes, there should be rules. Certainly there should be counseling involved and a waiting period. It should never be easy to take one's life on a whim rather than taking arms against a sea of troubles. If it were, I and many others would probably not have made it through our teenage years. But we should not presume to leverage guilt and shame upon sane adults who wish to control the day of their final breath.

Nearly every day, and often multiple times in a day, someone tells me that they have no interest in living to 100, let alone many decades longer.

"If I get to a hundred, just shoot me," they say.

"I think that seventy-five healthy years sounds about right," they say.

"I just can't imagine having to live with my husband for even longer than I already have to," one rather distinguished scientist once told me.

That's fine.

Indeed, there seems to be little appetite for the idea of living in perpetuity. I recently gave a talk to a general audience of about a hundred people spread across ages 20 to 90, a good cross section of the local community. The main donor to the institute was late, so I had to fill in time. I grabbed the microphone and did a little experiment.

"How long do each of you want to live?" I asked.

By a show of hands, a third said they'd be happy with 80 years; I told that group that they should all apologize to all the audience members who were older than 80. That got a laugh.

Another third indicated that they'd like to see 120. "That's a good goal," I said, "and probably not an unrealistic one."

About a quarter wanted to make it to 150. "That's not a silly thing to dream about anymore," I said.

Only a few people wanted to live "forever."

The numbers were similar at a recent dinner at Harvard for scientists who study aging. Very few of the attendees said they were gunning for immortality.

I've talked to hundreds of people about this topic. Most people who want immortality are not afraid of death. They just love life. They love their family. They love their careers. They would love to see what the future holds.

I'm not a fan of death, either. It's not because I'm afraid of being dead. I can say this without reservation. On a plane, my wife, Sandra, clings to my arm at the first sign of turbulence, whereas my pulse doesn't change. I travel enough to have experienced mechanical troubles on planes more than a few times, so I know how I react when faced with possible death.

If the plane goes down, I die. Letting go of that fear was one of the best things I ever did.

Here's where things get really interesting: when I do this little survey and then tell the audience that they could retain their health no matter how many years they live, the numbers of those who say they'd like to live forever shoots way up. Almost everyone wants that.

It turns out that most people aren't afraid of losing their lives; they are afraid of losing their humanity.

And they should be. My wife's grandfather was sick for many years before he died in his early 70s. At that point, he'd been in a vegetative state for several years—a truly horrible fate—but he had a pacemaker, and so, whenever his body would try to die, he'd be zapped back to life.

Not back to health, mind you. Back to life. There's a big difference.

In my mind, there are few sins so egregious as extending life without health. This is important. It does not matter if we can extend lifespans if we cannot extend healthspans to an equal extent. And so if we're going to do the former, we have an absolute moral obligation to do the latter.

Like most people, I don't want unlimited years, just ones filled with less sickness and more love. And for most of those I know who are engaged in this work, the fight against aging isn't about ending death; it's about prolonging healthy life and giving more people the chance to meet death on far better terms—indeed, on their own terms. Quickly and painlessly. When they are ready.

Either by refusing the treatments and therapies that offer a prolonged healthy life or accepting those interventions and then deciding to leave whenever the time is right, no one who has returned what they have been given should have to stay on this planet if he or she does not wish to do so. And we need to begin the process of developing the cultural, ethical, and legal principles that will allow that to happen.

WE MUST ADDRESS CONSUMPTION WITH INNOVATION

The environmental writer and activist George Monbiot is among those who have observed that when it comes to the future health of our planet, people are overly preoccupied with the number of humans on Earth while ignoring the fact that consumption "bears twice as much responsibility for pressure on resources and ecosystems as population growth."[28] Monbiot, who is on the far left, isn't right about everything, but he's certainly right about that. The problem isn't population; it is consumption.

We know that humans *can* live healthily and quite happily while consuming far less than most do in the developed world. But we don't know *if* they will. It is for this reason that among scientists who subscribe to the idea that our planet has an absolute limit of people it can sustain, those who have offered a generous estimate of the Earth's carrying capacity are those that assume that our species will be capable of making more from less, even, perhaps, as we increase the standard of living of billions of people. The more pessimistic predictors, meanwhile, generally assume a global "tragedy of the commons" in which we greedily consume ourselves to death at an all-you-can-eat buffet of natural resources. Generally, people will be people, so which way we head will largely be determined by politics and by technology.

At least in one regard—the "stuff factor," so to speak—technology is already driving a tremendous and positive change, a global process of "dematerialization" that has replaced billions of tons of goods with digital products and human services. Thus it is that wall-to-wall shelves dedicated to records and compact discs have been replaced by streaming music services; people who once needed vehicles for once-in-a-while travel now open an app on their phones to request a ride share; and entire wings of hospitals once used for storing patients' records have been supplanted by handheld cloud-connected tablet computers.

As Steven Pinker has pointed out, a lot of the time, energy, and money we once spent making "stuff" is now "directed toward cleaner air, safer cars and drugs for 'orphan diseases.'"[29] Meanwhile, the "experiences, not things" movements and the like are transforming the ways in which we save and spend money—and leaving us with less crap in our basements. After a century of movement toward McMansions, the latter half of the 2010s saw a significant drop in the square footage of new homes and increasing demand for smaller apartments,[30] continuing a centuries-long migration from farm-based rural living to smaller, shared urban spaces. As the global success of WeWork proves, today's young adults are not only comfortable with much smaller working and living quarters, with shared community spaces such as offices, kitchens, gyms, laundries, and lounges, but increasingly are demanding them.[31]

The slow death of stuff is not the end of consumption, though. We're as addicted as ever to wasting food, water, and energy. As it stands, the United Nations has warned, we are polluting water far faster than nature can recycle and purify it. We literally throw away half of the world's edible food each year, more than a billion tons of it, even as millions of people are left hungry or malnourished.[32]

At the current pace of population growth and economic mobility, the United Nations estimates, by 2050 it will take the equivalent of nearly three of our planet's resources to sustain our lifestyles for one year. Yet the United Nations spends surprisingly little time debating consumption, let alone forging international agreements that would help build a world in which no society consumes more than its share of what the Earth can produce under contemporary technological conditions.

That last part is important: just as it is helping us reduce our "stuff" addiction, technology absolutely has to play a role in solving these other consumption problems—for there is no free nation in the world that can unilaterally force its citizens to consume less while others on the planet consume more. Laws can encourage businesses to conform, but we also have to make it attractive and easy for individuals to consume less.

Therefore, we must invest in research that allows us to grow more healthy food and transport it more effectively. And please make no mistake: that includes accepting genetically modified crops, those engineered to include a trait in the plant that doesn't occur in its wild form, such as resistance to insects, tolerance to drought, greater vitamin A production, or more efficient use of sunlight to convert CO_2 to sugar—as an absolutely necessary part of our food future. With more efficient plants, we could feed up to 200 million additional people, just from plants grown in the US Midwest. [33]

These crops have gotten a bad rap for being "unnatural," although many people who hold this view don't recognize that most of the food we think of as "natural" has already been subject to significant genetic manipulation. The ears corn you see at the grocery store look nothing like the wild plant from which modern corn came; over the course of nine thousand years, the spindly finger-length grass known as teosinte was cultivated to evolve larger cobs and more rows of plump, soft, sugary kernels, a process of modification that significantly altered the plant's genome.[34] The apples we've grown accustomed to eating have a bit more resemblance to their small, wild ancestors, but good luck finding one of those ancestors; they have been nearly wiped off the planet, and that's no great loss to our diet, since the biggest genetic contributor to modern apples, *Malus sylvestris*, is so tart it's darn near inedible.[35]

In 2016, the National Academy of Sciences, in a sweeping report on genetically engineered crops, noted that lab-modified plants could be vital for feeding the planet's growing human population if global warming threatens traditional farm products. And since numerous other reports over the past few decades had not been enough to assuage continuing public concern, the report's authors once again reaffirmed the academy's position that GMO crops are safe for both human consumption and the environment.

There is nothing wrong with skepticism, but after thousands of studies, the evidence is irrefutable: if you believe climate change is a threat, you can't say that GMOs are, because the evidence that GMOs are safe is stronger than the evidence that climate change is occurring.

The World Health Organization, the American Association for the Advancement of Science, and the American Medical Association have also affirmed that, as WHO puts it, "no effects on human health have been shown as a result of the consumption of such foods by the general population." Moreover, these foods could be vital to meeting the challenge of feeding the billions of people who are already going hungry in our world and the additional billions who will be joining us on this planet in coming years.

If we are to feed the world now and in the future, we need to embrace safe new technologies.

According to UNICEF, up to 2 million deaths each year could be prevented if poor families had access to more vitamin A in their diets in crops that are perfectly safe.[36] Vitamin A supplements aren't working as well as is needed. Between 2015 and 2016, vitamin A supplementation coverage dropped by more than half in the five countries with the highest child mortality rates.

An open letter signed by more than a hundred Nobel Prize winners called on governments to approve genetically modified organisms: "How many poor people in the world must die before we consider this a 'crime against humanity'?" they asked. We could feed a billion more people with more nutritional food. With climate change, we may have no choice.

To decrease the impact of humans, there's also a tremendous need to figure out how to satiate the global demand for protein without the tremendous environmental costs of farmed animal meat. Made with 99 percent less water, 93 percent less land, and 90 percent fewer greenhouse gases, innovations that are giving us damn-near-close-to-meat products—with plant "leghemoglobin" that "bleeds" and some good old-fashioned mad science—are booming and will need to continue to boom if we are to feed our appetite for tasty protein without further degrading our planet.

There's no question that one of the greatest technological advances in this century has been the discovery of precise, programmable "genome editing." As with most other breakthroughs, there were dozens of brilliant people involved in the lead-up to it,[37] but Emmanuelle Charpentier, then at the Laboratory for Molecular Infection Medicine in Sweden, and Jennifer Doudna at UC Berkeley have garnered the most fame for their remarkable discovery that the bacterial Cas9 protein is a DNA-cutting enzyme with an RNA-based "GPS" or "guide."[38] The next year, Feng Zhang at MIT and George Church at Harvard proved that the system could be used to edit human cells. They, too, garnered fame—and some very valuable patents.[39] News of the discovery spread quickly down the hall to my lab. It seemed too good to be true—except it was.

The technology is colloquially known as CRISPR, for "clustered regularly interspaced short palindromic repeats," which are the natural DNA targets of Cas9 cutting in bacteria. Cas9, and now dozens of other DNA-editing enzymes from other bacteria, can alter plant genes with accuracy, without using any foreign DNA. They can create exactly the same kind of alterations that occur naturally. Using CRISPR is far more "natural" than bombarding seeds with radiation, a treatment that is not banned.

That's why the decision by the Court of Justice of the European Union in 2018 was so unexpected and upsetting to the United States. The court ruled in favor of Confédération Paysanne, a French agricultural union that defends the interests of small-scale farming, and eight other groups, to ban CRISPR-made foods.[40]

The ruling defies science. It bans healthy foods that could relieve the environmental burden, increase the health of the poor, and allow Europe to cope better with global warming. The ruling also scared developing nations away from CRISPR-modified crops; there they could make a positive impact both on people's lives and on their land.

The text of the ruling makes it clear that it was not a decision to protect consumers from the dangers of GMO; it was part of a global trade war to prevent US-patented products from entering the European Union. The US secretary of agriculture, Sonny Perdue, made this abundantly clear in his response: "Government policies should encourage scientific innovation without creating unnecessary barriers or unjustifiably stigmatizing new technologies. Unfortunately, this week's ECJ ruling is a setback in this regard in that it narrowly considers newer genome editing methods to be within the scope of the European Union's regressive and outdated regulations governing genetically modified organisms."[41]

Of course nations should be able to help local farmers whose livelihoods are threatened, but there are other ways to do that. It's ultimately hurtful to everyone on the planet to use the cover of "dangerous science" to justify trade restrictions, especially to those who need the new technology most.

We also need to solve the shortage of fresh, drinkable water. Cities such as Las Vegas, a very thirsty town in the middle of the driest place in the United States, have demonstrated that by marrying conservation and innovation, efficient water recycling is not only possible but profitable; whereas metro Vegas grew by half a million people from 2000 to 2016, its total water use fell by a third.

We often adopt new technologies way too slowly, but when we finally do, they can solve some of our biggest problems. It was back in 1962 that scientist Nick Holonyak, Jr., created the first practical visible light–emitting diode. At General Electric they called it "the magic one." It took another half century to develop an LED house bulb, and even then, many US consumers revolted, preferring to slow the phase-out of incandescent bulbs even as other nations moved forward with the LED revolution. Eventually, a combination of tax incentives and laws that outlawed the Edison lightbulb forced the adoption of LED lighting. Today's LED lights use 75 percent less energy than incandescent lighting and last fifty times as long, which in a typical home is about two decades.

Widespread use of LEDs in the United States is set to save the equivalent of the annual output of forty-four large electric power plants, saving about $30 billion a year.[42] To put this into perspective, that money could double the budget of the National Institutes of Health and set forty thousand scientists to work on lifesaving medicines. Human ingenuity is not a zero-sum game.

Longer, healthier lives will do us little good if we consume ourselves into oblivion. The imperative is clear: whether or not we increase human longevity, our survival depends on consuming less, innovating more, and bringing balance to our relationship with the bounty of our natural world.

That might seem like a tall order. Indeed, it *is* a tall order. But I believe we can stand tall—and together—to meet it.

In many ways, we are already doing so.

At the 2018 Global Climate Action Summit, for instance, it was announced that twenty-seven cities had reached peak emission levels. A peak, not a plateau. All of those places were seeing steep emission declines. Among that group of cities was Los Angeles, which was once definable by its ubiquitous smog. It had cut its emissions by 11 percent. In one year.[43]

Yes, there are more people than ever in the cities of North America, South America, Europe, and Asia, but today the impact of each human in those regions is declining. We're rapidly moving from petroleum to natural gas, solar power, and electricity. When I first visited Bangkok, I experienced respiratory distress. Now, more days than not, there is blue sky. When I arrived in Boston in 1995, a splash of water from the harbor could land you in hospital—or in the grave. Now it is safe for swimming.[44] The same is true of Sydney Harbor, the Rhine River, and the Great Lakes.

Going backward or even staying put is not a viable solution to the current crisis. The only path forward is one in which we embrace human capital and ingenuity.

One of the best examples comes from a tiny town in South Australia. After the closure of the last coal-fired power station in the state in 2016, investors built Sundrop Farms on the barren coast, then hired 175 people who had recently become unemployed.[45] The farm uses free energy from the sun and seawater to make 180 Olympic-sized swimming pools' worth of freshwater per year, an effort that in the past would have burned a million gallons of diesel fuel. Today, 33 thousand pounds of fresh organic tomatoes are shipped each year from the port where coal used to come in.

Sundrop is an example of a Schumpeterian "gale of creative destruction," the type of technological paradigm shift we will need to usher in the age of longevity and prosperity. For this to happen, we need more visionary scientists, engineers, and investors. We need more smart legislation to speed, not impede, the adoption of Earth-saving technologies. This will free up money and human capital that are currently wasted. The freed-up money needs to be reinvested in people and technologies, not in meaningless "stuff," to ensure that humanity and the Earth endure—indeed thrive—together.

WE NEED TO RETHINK THE WAY WE WORK

The University of Pennsylvania was a wonderful school at which to study theology and the classics. It had recently launched a medical school, too. As a native Philadelphian, Joseph Wharton was proud of the local college. But the millionaire industrialist also believed the university was missing something essential.

"With industry now powered by steam and steel, we can no longer rely on apprenticeship alone to create future generations versed in business," he wrote to friends and associates on December 6, 1880, just months before officially opening the world's first business college, the Wharton School. "There needs to be institutions to instill a sense of the coming strife of the business life and of the immense swings upward or downward that await the incompetent soldier in this modern strife."[46]

But Wharton could scarcely have predicted the extent of the "strife" that was on the horizon: a nascent labor movement in Europe would soon go global, bringing with it revolutionary changes in workers' rights.

Among those changes was something that had never existed in the history of labor: the weekend. We tend to take the five-day workweek as a given, but it's an exceptionally recent innovation. It didn't exist as a concept—or even a phrase—until the late 1800s.[47] The same can be said of legal limits on daily working hours, the abolition of child labor, medical benefits, and health and safety regulations. All of this was a response to the needs and demands of labor—and, indeed, the best interests of business owners such as Wharton.

The global Schumpeterian transformation now at hand will reshape the world as profoundly as the Industrial Revolution. Every business school in the world should be preparing its students for what is coming—and labor advocates should be doing the same. The idea of connecting retirement to a person's chronological age will be an anachronism soon enough. And just like Social Security, the structures that support labor pensions will need to be reevaluated.

Skillbaticals, which might take the shape of a government-supported paid year off for every ten worked, might ultimately become cultural and even legal requisites, just as many of the labor innovations of the twentieth century have. In this way, those who are tired of "working harder" would be afforded every opportunity to "work smarter" by returning to school or a vocational training program paid for by employers or the government, a variation of the universal basic income that is being discussed in the United States and some countries in Europe.

Meanwhile, those who believe they are happy and secure in their careers can enjoy what has come to be known as "a miniretirement"—a year off to travel, learn a language or musical instrument, volunteer, or refresh and reconsider the ways in which they are spending their lives.

This is not a particularly crazy scheme; sabbatical leave is common in higher education. Yet an idea like this might seem ridiculous to those who only consider the way the world works today. Who would pay for such a benefit? How will companies retain workers in the long term without the promise of a "gold watch retirement" plan at the end of decades of service?

But whoever engages in this discussion now will have the upper hand when we decide how to redistribute the resources freed by the elimination of ever-skyrocketing insurance premiums and pyramid-scheme pensions. Yet few business professors are so much as thinking about this coming change and even fewer courses are being taught on the subject in places such as the Wharton School. Labor leaders, meanwhile, are locked in an understandable but ultimately futile fight for retirement and benefits for workers who in the past would have labored for forty or fifty years, retired for a short spell, and then rather promptly died. Almost no one is fighting over what the world of work will look like when age is truly nothing more than a number.

But that era is coming. And it is coming sooner than most people and institutions realize.

WE NEED TO GET READY TO MEET OUR GREAT-GREAT-GRANDKIDS

"I sure am glad I won't be around when that happens."

I hear this a lot—mostly, it seems, from people who are in, or soon approaching, retirement. These are folks who have already decided that their lives are going to end in the next couple of decades. They're certainly hoping to stay healthy during that time and maybe eke out a few extra good years if they can, but they don't think they're going to be around for much longer than that. To them, the middle of this century might as well be the next millennium. It's not on their radar.

And that's the world's biggest problem: the future is seen as someone else's concern.

In part, this stems from our relationship to the past. Very few of us had the opportunity to get to know our great-grandparents. Many of us don't even know their names. That relationship is an abstraction. And so most of us don't think about our great-grandkids as much more than a fuzzy, abstract idea.

Sure, we care about the world our children will live in because we love our children, but the conventional wisdom about aging and death tells us they'll be gone a few decades after we are. And yes, we care about our grandkids, but by the time *they* come around, we're often so close to the exit that it doesn't seem as though there's much we can do about their future anyway.

This is what I want to change—more than anything else in the world. I want everyone to expect that they will meet not only their grandchildren but their great-grandchildren and their great-great-grandchildren. Generations upon generations living together, working together, and making decisions together. We will be accountable—in *this* life—for the decisions we made in the past that will impact the future. We will have to look our family members, friends, and neighbors in the eye and account for the way we lived before they came along.

That, more than anything else, is how our understanding of aging and inevitable prolonged vitality is going to change the world. It will compel us to confront challenges that we currently push down the road. To invest in research that won't just benefit us now, but people 100 years from now. To worry about the planet's ecosystems and climate 200 years from now. To make the changes we need to make to ensure that the rich don't enjoy an increasingly lavish way of life while the middle class begins to tumble toward poverty. To ensure that new leaders have a fair and legitimate opportunity to displace old ones. To bring our consumption and waste into balance with what the world can sustain today and many centuries into the future.

This isn't going to be easy. The challenges are vast. We are not only going to have to "touch the third rail" of politics—Social Security—but to douse ourselves in water and lie down upon it, adjusting our expectations about work, retirement, and who deserves what and when. We're no longer going to be able to wait for prejudiced people to die; we're going to have to confront them and work to soften their hearts and change their minds. We can't just allow the Anthropocene extinction to continue—at a rate thousands of times higher than the natural rate—we need to slow it dramatically and, if we can, stop it altogether.

To build the next century, we're going to have to figure out where everyone is going to live, how they are going to live, under what rules they are going to live. We're going to have to ensure that the vast social and economic dividends we receive from prolonging people's lives are spent wisely.

We're going to have to be more empathetic, more compassionate, more forgiving, and more just.

My friends, we're going to have to be more human.

CONCLUSION

LET ME TAKE YOU ON A TOUR OF MY LAB AT HARVARD MEDICAL SCHOOL IN Boston, Massachusetts.

You'll find us in the Genetics Department in the New Research Building, arguably the best group of biologists in the world. That's the same place where Connie Cepko is working to grow mammalian eyes in a dish and studying the potential for gene therapy to restore lost vision. Down the hall in his clean room, David Reich, the author and scientist, is sequencing DNA from 20,000-year-old teeth and discovering that our ancestors liked to breed with other human subspecies. And a floor down, George Church is working on, among other wizardries, printing an entire human genome and reviving the woolly mammoth. Across the street, Jack Szostak has moved on from his Nobel Prize–winning work to uncover secrets about how life began four billion years ago; he comes by sometimes to visit.

Yeah, the elevator conversations are awesome.

My lab is on the ninth floor. The first person you'll see when you walk in the office is Susan DeStefano, who basically has kept our lab and my life under control for the past fourteen years. Susan is a devout Christian who believes in the literal version of Genesis. She figures that we are doing God's bidding by helping the sick and the needy; there's no reason why our views on God and science can't coincide. We both want to make the world a better place.

To the left of Susan's doorway you'll find lab manager Luis Rajman's office. Luis, who has a PhD in cellular and molecular biology, ran the transgenic mouse facility at the giant biotech company Biogen Idec, but when we first met he was managing a high-end framing company. He's worked on paintings worth more than my house—and probably more than all my neighbors' houses together, too—so he's the right sort of guy for a job that requires exceptional meticulousness. Sitting with her back to Luis is Karolina Chwalek, who has a PhD in regenerative medicine and is our chief of staff, a strict but fair manager who makes sure our team of thirty to forty scientists is funded and remains very much worth funding.

Daniel Vera sits next to Luis and is usually staring at at least one and often several screens. He's the lab's data guru, having established Florida State University's Center for Genomics. I'll never forget the day he showed me the whole-genome analysis of the epigenetic changes in the ICE mice that helped reinforce the Information Theory of Aging.

Down the hall, past framed copies of research papers we've published, there's a sign above a door that says "Operations Room," as a nod to Winston Churchill's central command. Inside you'll find the lab and an ever-rotating group of some of the best minds in the world. When I took a walk through the lab on one recent day, one of my favorite things to do, these were some of the characters who were there.

On my left were Israel Pichardo-Casas, a Mexican cell biologist, and Bogdan Budnik, a Ukrainian physicist, who have found more than five thousand new human genes within noncoding "junk DNA." These small genes make small proteins that course through our bloodstreams, any one of which could be a treatment to cure cancer, treat diabetes, or be the factor that allows young mice to rejuvenate the old. Then there were the Michaels: Bonkowski, Schultz, and Cooney. Bonkowski played a key role in our study on reversing vascular aging, making old mice run twice as far.[1] He holds the record for the creation of the longest-lived mouse in scientific history, at five years.

Schultz, his mentee, is studying the molecular events that cause age-related inflammation, looking for ways to suppress that reaction and thus remove a key driver of age-aggravated diseases. He and Bonkowski are using gene therapy to "infect" old mice with longevity genes, aiming to break their own mouse longevity record.

Cooney is working with NASA to introduce DNA repair genes—from the supertough, eight-legged micro creatures known as tardigrades—into human cells in an effort to provide astronauts with protection from cosmic radiation and, of course, to slow aging.

There was João Amorim from Portugal; he's studying resveratrol and a bunch of STACs in an effort to understand how they activate SIRT1 in the body. He's changed just one base pair in the mouse *SIRT1* gene that makes the enzyme resistant to resveratrol and other STACs. He's testing if that mutant mouse still receives the health and lifespan benefits of resveratrol. If resveratrol no longer works on the mutant, then it should resolve the debate about whether resveratrol is working by directly activating the SIRT1 enzyme, or via some other mechanism, like activating AMPK. So far, the results look promising for the SIRT1-activation hypothesis.

There was Jae-Hyun Yang from South Korea; he has spent the past six years tickling the chromosomes of cells and animals to understand how and why the ICE mice age prematurely. It was he and João who first showed that the epigenetic clock in the ICE mice ticks faster. Next to him was Yuancheng Lu, one of China's top students, who discovered the powerful epigenetic reprogramming system that can be delivered into aged animals via a modified virus.

Xiao Tian had just used that virus to protect human nerve cells from chemotherapy. The normal nerves had either died or shrunk into a ball. But the reprogrammed ones were completely healthy, with long, beautiful cellular projections extending out across the floor of the petri dish.

Some experiments aren't very conclusive; this was night and day. We plan to test our virus in patients with eye disorders within a couple of years.

Patrick Griffin, my most recent graduate student, wants to know if stimulating a response to DNA damage, without creating actual DNA damage, is sufficient to cause aging in mammals. To test this, he has designed a way to tether DNA damage signaling proteins to the genome using a non-cutting version of Cas9/CRISPR. If our theory is correct, he should still cause aging. Jaime Ross has engineered "NICE mice" to experience accelerated epigenomic noise just in neurons. She wants to know if the brain controls aging in the rest of the body, and if these mice operate more like eighty-year-old humans. If so, they could be used as better models for human brain aging and possibly Alzheimer's disease.

Joel Sohn has worked with some of the greatest biologists of the twentieth century, then spent thirty years as a fisherman catching and exporting marine life, and is now searching the seas for secrets of immortality. He's studying cnidarians, transparent ocean animals that can do amazing tricks with their bodies, such as regrow a new body part or spawn a baby from their feet. That day was a good day for Joel: his decapitated sea anemone was regrowing a head, and his immortal jellyfish were budding off baby clones. Perhaps these regenerative processes are the same as the ones that allow us to regenerate optic nerves. Perhaps these creatures have access to the biological equivalent of Shannon's observer, the one who stores youthful epigenetic information.

Abhirup Das, who spearheaded the old mouse marathon project, was studying the impact of precursors such as hydrogen sulfide and NMN on wound healing. Lindsay Wu, who also runs our labs in Sydney, Australia, at the University of New South Wales, was examining molecules that activate an enzyme called G6PD, which has been shown to extend the lifespans of multiple animals and, tragically, is mutated in 300 million people, the most common of all mutations. He has also restored fertility in old female mice by feeding them NMN and protecting their eggs from DNA damage.

There was our resident dental student, Roxanne Bavarian, who is working to identify the role of sirtuins on oral toxicities and cancer. And there was Kaisa Selesniemi, from Finland, who is one of the world's leading experts on culturing stem cells from ovaries and reversing female infertility.

Parvez Mohammed, from India, was creating new chemicals in the fume hood, and Conrad Rinaldi was testing if the latest batch worked to rejuvenate skin cells from aged people. Giuseppe Coppotelli, from Italy, was examining new human longevity genes we'd discovered, including one called Copine2 that is mutated in Parkinson's and Alzheimer's disease patients.

Alice Kane, an Australian, was examining some mice to develop a mouse frailty clock to predict how long a mouse will live and is helping all of us see and appreciate sex differences. Jun Li, our lab's senior biochemist, was investigating why our ability to repair DNA goes down with age and has found that NMN reverses the process.[2]

And those were just the people who were in the lab on that specific day. There are others—a lot of them—who are doing world-changing work.

These people are brilliant. They could be working on answering any question in the universe. But they've come to Harvard to work on aging. Some of them are introverts, as scientists often are. A few are cautious, conservative researchers, a trait I'm working to fix. Yet there isn't a single one who doesn't believe that prolonged human vitality is on its way.

And this is just one lab. There are three more labs in the Paul F. Glenn Center for the Biology of Aging at Harvard that are focused on helping people live longer and healthier lives. At Bruce Yankner's lab, they're exploring the impact of aging specifically on the human brain. Marcia Haigis's lab is investigating the role that mitochondria play in aging and disease and has uncovered the role of sirtuin mutations in cancer. Amy Wagers's lab was one of the first to show the blood of young mice rejuvenates the old, and vice versa, prompting people to infuse themselves with serum from young donors. Amy and I are collaborating to find the factors in the blood and develop new, advanced pharmaceuticals to treat age-related diseases without the creepiness.

At another Glenn Center, just across the river at MIT, Lenny Guarente, Angelika Amon, and Li-Huei Tsai are all working on fundamental questions related to slowing, stopping, and reversing aging. In other
cities in the United States, Thomas Rando, Anne Brunet, Tony Wyss-
Coray, Elizabeth Blackburn, Nir Barzilai, Rich Miller, and others are all
running large labs or centers aimed at changing the way we think about
aging. North of San Francisco there is an entire building, called the Buck
Institute for Research on Aging, completely devoted to understanding
and combating aging. The list goes on and on.

Those are just a few of the labs. Around the world, more than a dozen
independent research centers are working hard on these same questions,
and there is now at least one scientist in every major university in the
world who is working on aging. Most of these labs obtain their research
grants for other diseases but increasingly are turning their attention toward understanding aging, reasoning that fixing that problem will fix
whatever disease they're funded to fight. This is, after all, an environment
in which a huge source of research funding is off limits for those who are
fighting something that most people believe is inevitable and few people
recognize is a disease.

Meanwhile, private enterprise is leading the way in the development
of neural net-based initiatives for drug discovery and development, gene
analysis, biotracking, and disease detection to dramatically extend our
lives. And every day research into the simple things anyone can do to
prolong their lifespans and healthspans is mounting, too—offering better and better maps to good health and long lives.

A decade or two ago, when even the most optimistic scientists were
only just beginning to envision a world in which aging wasn't inevitable,
and when there were only a handful of researchers in the world specifically working to slow, stop, or reverse aging, I certainly could understand
when people listened politely about my work and then gave me a look as
though I were crazy. Today I have a hard time understanding how anyone
could look at this vast and brilliant army of researchers and not believe
that a tremendous change in human aging is coming—and soon.

I have some compassion for those who say "It can't be done." They are, in my view, the same kind of people who said vaccines couldn't work and humans couldn't fly. But given the benefits that longevity research can bring to the world, I have far less patience—indeed, really none at all—for those who say "It shouldn't be done."

BEYOND BELIEF

There are those who would have you believe that the people in my lab— and those like them in labs around the world—are engaged in an unnatural and even immoral campaign to change what it means to be human. That view is rooted in ideas about human nature that might charitably be described as subjective but that are probably more accurately called zealotry.

That, it seems to me, was the guiding force behind a 2003 report submitted to the White House by the President's Council on Bioethics entitled *Beyond Therapy: Biotechnology and the Pursuit of Happiness*, which ominously warned against aging research because it goes against "the human grain" and violates the purportedly orderly cycle of birth, marriage, and death.

Would people "be more or less inclined to swear lifelong fidelity 'until death do us part,' if their life expectancy at the time of marriage were eighty or a hundred more years, rather than, as today, fifty?"[3] the council wondered. What kind of unhappy marriages, I wondered in return, would drive people to even ask such a question? I would *love* to have an extra fifty years with my wife, Sandra.

Aging, the council proffered, "is a process that mediates our passage through life, and that gives shape to our sense of the passage of time"; without it, the council's members cautioned, we could become "unhinged from the life cycle."[4]

Our so-called natural life cycle, of course, is one in which the vast majority of our ancestors never got old enough to get gray hair or wrinkles and in which consumption by carnivore was a perfectly ordinary way to go. If you'd like to be hinged to that, be my guest.

"Might we be cheating ourselves," the council asked, "by departing from the contour and constraint of natural life (our frailty and finitude), which serve as a lens for a larger vision that might give all of life coherence and sustaining significance?"[5]

Oh, for goodness' sake, if we truly believed that frailty was a requisite for meaningful life, we'd never mend a broken bone, vaccinate against polio, or encourage women to stave off osteoporosis by maintaining adequate calcium levels and exercising.

I know I shouldn't get worked up about these sorts of things. It's a tale as old as science, after all—just ask Galileo what happens when you "disrupt the natural order of things."

But this was more than a trifling report from moralizing bureaucrats. The chair of the committee that wrote it, Leon Kass, is one of the most influential bioethicists of our time and came to be known, during the tenure of George W. Bush, as "the President's Philosopher." For years after the report was issued, aging research was framed not as a fight against a disease but as a fight against our humanity. That's hogwash, and, in my mind, it's rather deadly hogwash.

Yet once the framing has been set, the effort to shift ideas, understandings, and biases becomes a herculean one. The fight to help people see aging for what it is, rather than "just the way it goes," will be a long one.

More funding for the kind of research happening in my lab and others like it could bring about these advancements even sooner. But because of a lack of funding, people over sixty today may not live long enough to be helped. If you and your family members end up the last of humanity to live a life that ends all too early with decay and decrepitude, or our children never see the benefits of this research, you can thank those bioethicists.

After all of these arguments, if you still think that extending the healthy part of your life would not be for you—perhaps it would reduce your life's urgency or be against the natural course of things—consider your friends and family. Would you subject your loved ones to a decade or two of unnecessary hardship having to look after you physically, emotionally, and financially in your final years if you didn't have to?

Spend a day in a nursing home like my wife does every few days. Go feed people who can't chew. Wipe their bottoms. Bathe them with a sponge. Watch as they struggle to remember where they are and who they are. When you are done, I think you will agree that it would be negligent and cruel for you *not* to do what you can to combat your own age-related deterioration.

There are still a lot of people like Kass out there. But if they live long enough, they will also have to come to terms with reality. The momentum makes the future I have described, or one close to it, unstoppable. Prolonged healthspans are inevitable.

More and more people recognize this every day, and they want in.

Because no matter what people say or believe, whether they are optimists or scaremongers, scientists or bioethicists, there is change in the air.

On June 18, 2018, WHO released the eleventh edition of the *International Classification of Diseases*, known as *ICD-11*. It is a fairly unremarkable document, except that someone slipped in a new disease code. At first no one saw it. Here is the line, which you can find on the WHO website[6] if you type in code MG2A. It reads:

MG2A Old age

- old age without mention of psychosis
- senescence without mention of psychosis
- senile debility

Every country in the entire world is encouraged to start reporting using *ICD-11* on January 1, 2022. What this means is that it is now possible to be diagnosed with a condition called "old age." Countries will have to report back to the WHO with their statistics on who dies from aging as a condition.

Will this lead to changes at the regulatory level, directing billions of dollars in investment to develop the medicines we deserve? Will federal regulators and doctors finally accept that it is ethically okay to prescribe medicines to slow aging and all the diseases that aging causes? Will they recognize it is indeed within a patient's rights to receive them? Will insurance companies reimburse patients for the cost of antiaging treatments that will save money down the line?

We will see. I certainly hope the winds build. Until that time comes, though, there is plenty we can do.

WHAT I DO

Save for "Eat fewer calories," "Don't sweat the small stuff," and "Exercise," I don't give medical advice. I'm a researcher, not a medical doctor; it's not my place to tell anyone what to do, and I don't endorse supplements or other products.

I don't mind sharing what I do, though, albeit with some caveats:

- This isn't necessarily, or even likely, what *you* should do.
- I have no idea if this is even the right thing for *me* to be doing.
- While human trials are under way, there are *no* treatments or therapies for aging that have been through the sort of rigorous long-term clinical testing that would be needed to have a more complete understanding of the wide range of potential outcomes.

People often wonder, when I tell them things like this, why on earth I would subject myself to the potential for unexpected and adverse side effects or even the possibility—low though it seems to be—that I could expedite my own demise.

The answer is simple: I know exactly what is going to happen to me if I *don't* do anything at all—and it's not pretty. So what do I have to lose? And so, with all that on the table, what do I do?

- I take 1 gram (1,000 mg) of NMN every morning, along with 1 gram of resveratrol (shaken into my homemade yogurt) and 1 gram of met-formin.[7]
- I take a daily dose of vitamin D, vitamin K_2, and 83 mg of aspirin.
- I strive to keep my sugar, bread, and pasta intake as low as possible. I gave up desserts at age 40, though I do steal tastes.
- I try to skip one meal a day or at least make it really small. My busy schedule almost always means that I miss lunch most days of the week.
- Every few months, a phlebotomist comes to my home to draw my blood, which I have analyzed for dozens of biomarkers. When my levels of various markers are not optimal, I moderate them with food or exercise.
- I try to take a lot of steps each day and walk upstairs, and I go to the gym most weekends with my son, Ben; we lift weights, jog a bit, and hang out in the sauna before dunking in an ice-cold pool.
- I eat a lot of plants and try to avoid eating other mammals, even though they do taste good. If I work out, I will eat meat.
- I don't smoke. I try to avoid microwaved plastic, excessive UV expo-sure, X-rays, and CT scans.
- I try to stay on the cool side during the day and when I sleep at night.
- I aim to keep my body weight or BMI in the optimal range for health-span, which for me is 23 to 25.

About fifty times a day I'm asked about supplements. Before I an-swer, let me say that I never recommend supplements, I don't test or study products, nor do I endorse them; if you see a product implying that I do, it's certainly a scam. Supplements are far, far less regulated than medicines, so if I do take a supplement, I look for a large manufacturer with a good reputation, seek highly pure molecules (more than 98 percent is a good guide), and look for "GMP" on the label, which means the product was made under "good manufacturing practices." Nicotinamide riboside, or NR, is converted to NMN, so some people take NR instead of NMN because it is cheaper. Cheaper still are niacin and nicotinamide, but they don't seem to raise NAD levels as NMN and NR do.

Some people have suggested NAD boosters could be taken with a compound that provides cells with methyl groups, such as trimethylglycine, also known as betaine or methylfolate. Conceptually, this makes sense—the "N" in NR and NMN stands for nicotinamide, a version of vitamin B_3 that the body methylates and excretes in urine when it is in excess, potentially depleting cells of methyls—but this remains a theory.

My father follows almost the same regimen as I do, and I can't remember the last time he was sick. He claims he's speeding up. This summer, he left his busy social calendar behind in Australia, and, having helped us with home repairs in Boston for six weeks while working remotely in his second career at the University of Sydney, he then drove around the US East Coast for a few weeks with his lifelong friend on their annual pilgrimage to the Summer Theater Festival in Wooster, Ohio.

Dad flew home at the end of summer, only to come back a few weeks later to see me get "knighted," as he called it, in Washington, DC. Now that he's home in Sydney again, he's planning on driving six hundred miles north for a few days to "see a couple of friends." He is loving life, seemingly more than he ever has.

As I get older, I spend more and more time thinking how lucky I've been in life. As an Aussie, I was taught that "big boys don't cry." But, nowadays, when I have the time and the sense to pause for a few moments to ponder my life, it's easy to get a little teary.

I grew up in a free country, then moved to an even freer one. I have three amazing children and friends who treat my family like their own. I am very proud of Sandra, my wife, who was one of the top students in Germany. She aced her botany degree, then came to Boston to be with me, got into the PhD program at MIT, and worked in a lab that was cloning mice for the first time. To earn her PhD, she figured out how to cure mice of a lethal genetic disease called Rett syndrome that disrupts the epigenome and prevents brain development in infant girls. By strange coincidence, the gene she worked on, MECP2, binds to methylated DNA and might be a cellular observer that stores youthful age correction data.

Sandra has been teaching me a lot these past twenty-five years about how to be a better husband and parent, not to mention the names of all the plants, insects, and animals we see on our walks. When we were first married, we argued a lot. She had "ethical issues" with my research, which pained me. Now, having examined and discussed the wealth of biological and economic data over the years, we no longer argue as much, and in fact she's started taking NMN.

It's impossible to say if my regimen is working for us, but it doesn't seem to be hurting. I am now 50, and I feel the same as I did when I was 30. My heart looks 30, too, according to a video of my heart in 3D that one of my colleagues kindly made by inserting me into an experimental magnetic resonance imager. I don't have a gray hair, and I'm not superwrinkly—well, at least not yet.

A year ago, my younger brother, Nick, was going gray and losing his hair when he demanded to be put on the same regimen after accusing me, only half jokingly, of using him as a negative control. I insisted that I would never do that to my own brother, but I can't say the thought hadn't crossed my mind. He's now on my dad's regimen, too.

Living longer makes no sense if you don't have your friends and family around you. Even our three dogs—a small 10-year-old poodle cross named Charlie and two 3-year-old black Labradors, Caity and Melaleuca—have been on NMN for a couple of years. Charlie is a therapy dog whose job it is to calm people, but he becomes too hyperactive if Sandra gives him NMN the day he heads to work, so on those days he's off it. Caity suffers from a congenital kidney defect, and we hope NMN will allow her to make it past her predicted five-year expiry date. The results of tests in mice with kidney damage say it's possible.[8]

A lot of people think a regimen intended to promote prolonged vitality must be hard to stick to, but if it were, my family couldn't do it. We are just an average bunch trying to get through the day. I do live life as mindfully as possible, focus on feeling good, and check my blood markers occasionally. Over time, I've identified the diet, exercise, and supplement routines that work best for me. And I'm confident that my family and I will continue to fine-tune these practices in response to the evolving research as our lives go on.

And on.

And on.

Because, yes, I do hope to be here for a long time to come. There are plenty of X factors that could interfere with that goal. I could get hit by a bus tomorrow, after all. But it's getting easier and easier to imagine not being around—happy, healthy, and connected to friends, family members, and colleagues—past my 100th year.

How long past my 100th year?

Well, I think it would be nice to see the twenty-second century. That would mean making it to my 132nd year. To me, that is a remote chance but not beyond the laws of biology or way off our current trajectory. And if I do make it that far, perhaps I'll want to stick around even longer.

There's so much I want to do—and so many people I'd like to help. I'd love to keep nudging humanity down what I believe is a path to greater health, happiness, and prosperity, and to live long enough to know what path we take.

BUSHWALKING

I recently returned to the neighborhood where I grew up, in the northern suburbs of Sydney, on the edge of Garigal National Park. Dad and Sandra were both there, and so was my twelve-year-old son, Benjamin.

We'd come for a hike along the trail, the exact same one my grandmother Vera used to take my brother and me on when we were that age. She'd tell us stories about her difficult childhood, about how lucky we were to have grown up in a free country, and about the wisdom of A. A. Milne:

BUSHWALKING. If you head north from my childhood home, you'll move through hundreds of miles of consecutively larger national parks, a seemingly endless undulation of saltwater estuaries and craggy mountain ridges decorated by ancient rock carvings left by the original inhabitants, the Garigal clan. Dad is now 80, the age his mother, Vera, was when she lost the will to live—aging has that effect on people. Instead, my father hikes mountains, travels the world, and has started a new career, representing hope for all of us.

"What day is it?" asked Pooh.

"It's today," squeaked Piglet.

"My favorite day," said Pooh.

Dad was raring to go. Ben was, too. Tigers, those boys are. But as I stood at the beginning of the trail, on the edge of a high sandstone cliff overlooking a gully filled with ambrosial eucalyptus and with deafening cicadas overhead, I found myself frozen in awe by the way in which the city quickly gave way to the bush, how the present and the deep past came together, and how it felt to be on the edge of something so vast and so beautiful.

If you wind your way south, down the rocky trail leading off Melaleuca Drive, the street I lived on as a boy, you'll reach Middle Harbor, an estuary lined by a canopy of bloodwoods, angophoras, and scribbly gum trees that ends in Sydney Harbor. If instead you head north, you'll move through hundreds of miles of consecutively larger national parks: Garigal to Ku-ring-gai to Marramarra, Dharug to Yengo to Wollemi, a seemingly endless undulation of saltwater estuaries and craggy mountain ridges decorated by ancient rock carvings. You could walk for days, even for weeks, and not hear anyone, save for the distant echoes of the original inhabitants of this land.

On that day in Garigal Park we were planning on walking only a few hours, but I'd been looking forward to it for weeks.

There is, at least to me, a subtle but important difference between hiking and bushwalking. When people hike, they are most often looking for exercise, serenity, beauty, or time together with loved ones. When Australians go for a bushwalk, they are seeking all of those things but with the intention of finding wisdom, too.

I'm not sure how long I was standing on the cliff. A minute or two, perhaps. Five or ten, maybe. However long it was, my family didn't seem to mind. When the spell of nostalgia and wonderment released me, I found them a short way down the trail.

Ben was peeling the paperlike bark off a melaleuca tree, while Dad was trying to explain something to him about the cliffs being made of sand that had been deposited when mammals first appeared. Sandra was examining a banksia—the strange, prickly flower Sir Joseph Banks collected to show the Royal Society—which, she delighted in reminding us, for the umpteenth time, is a member of the Proteacea family.

As I write this, Ben is in seventh grade. He's a good boy. A smart boy. He wants to work in my lab someday and take over from me to "finish the job." I tell him he'll have a lot of competition and won't get any special treatment from me, and he says, "Well, if that's true, I can always work for Lenny Guarente."

Yes, he's a funny boy, too.

Our two older children are making their own paths: Natalie, I reckon, as a veterinarian; Alex, perhaps, as a diplomat or a politician.

Dad is now 80, the age his mother, Vera, was when the fire in her eyes had completely gone. She had lost the will to live and never ventured outside again. I cannot predict the future, but when I look at the full life Dad leads now, his world travel, his optimism, and the state of his health, I think he'll be around for a long time to come. I sure hope he will.

Not just because he represents hope for all of us, but because I'd like to come back to this place again and again with Dad and Sandra and all those I love. Searching for serenity. Hearing stories. Finding beauty. Making memories.

Sharing wisdom.

With Ben and Natalie and Alex, yes. But also with their children. And with their children's children.

Why not? Nothing is inevitable.

Acknowledgments

DAVID

I cannot begin to express my love and gratitude to my wife, Sandra Luikenhuis, who has stood by me for two decades and tolerated my writing and rewriting this book for one of them. To my children, Alex, Natalie, and Benjamin: I could not have asked for better offspring.

Writing a book requires considerable kinship among all those involved in the creative process. I'm indebted to Matthew LaPlante for his friendship, sense of humor, and polymath's ability to turn hundreds of discussions and dozens of whiteboard diagrams into a coherent narrative. Matt and I are both so lucky to have worked with Caity Delphia, this book's masterful and veracious illustrator, who bravely took on the challenge to turn our words and ideas into stunning works of art—and pulled it off. Every day I'm thankful for the camaraderie of Susan DeStefano, my assistant for the past fourteen years who keeps my life and our labs running smoothly; she deserves an entire page of thanks for her ability to handle all that's thrown at her.

I'm indebted to Luis Rajman and Karolina Chwalek, who help run the research lab in Boston, and Lindsay Wu, who runs our sister lab in Sydney. I'm fortunate to work with such a dedicated, intelligent, and practical team. Bruce Yankner, my co-director at Harvard's Glenn Center for Aging Research, has been a wonderful collaborator and colleague.

My deep and heartfelt thanks to Celeste Fine, John Maas, and Laurie Bernstein, my agents; to Sarah Pelz, our editor, for her careful and skillful editing; to Melanie Iglesias Pérez and Lisa Sciamba; to Lynn Anderson for copyediting; and all the staff at Simon & Schuster who believed in this book. Thanks to Laura Tucker for beginning this journey a decade ago; to the public relations team Carrie Cook, Sandi Mendelson, Rob Mohr, and Nicholas Platt. Matt and I are so thankful to everyone who read and made suggestions to improve the manuscript, especially Stephen Dark, who coedited the glossary and endnotes, Mark Jones, Sandra Luikenhuis, Mehmood Kahn, John Kempler, Lise Kempler, Tristan Edwards, Emil and Dariel Liathovetski (the RockCellos), Dave Deamer, Terri Sinclair, Andrew Sinclair, and Nick Sinclair. My appreciation to Brigitte Lacombe, master photographer, for the unkempt headshot (Instagram brigittelacombe).

Thanks to all the teams that work tirelessly to make this world a better place, including and in order of incorporation, CohBar, Vium, InsideTracker, MetroBiotech, Arc Bio, Liberty Biosecurity, Dovetail Genomics, Life Biosciences, Continuum Biosciences, Jumpstart Fertility, Senolytic Therapeutics, Animal Biosciences, Spotlight Therapeutics, Selphagy Therapeutics, and Iduna Therapeutics.

When I chose to be a scientist, I thought that the greatest reward would be discovering things, but it's actually the lifelong friends you make, the ones who stand up for you when times are tough. So I am grateful to have the friendship and receive the wise counsel of Nir Barzilai, Rafael de Cabo, Stephen Helfand, Edward Schulak, Jason Anderson, Todd Dickinson, Raj Apte, Anthony Sauve, David Livingston, Peter Elliott, Darren Higgins, Mark Boguski, Carlos Bustamante, Tristan Edwards, Lindsay Wu, Bruce Ksander, Meredith Gregory Ksander, Zhigang He, Michelle Berman, Pinchas "Hassy" Cohen, Mark Tatar, Alice Park, Sri Devi Narasimhan, Kyle Landry, James Watson, David Ewing Duncan, Joseph Maroon, John Henry, Duncan Purvis, Li-Huei

Tsai, Christoph Westphal, Rich Aldrich, Michelle Dipp, Bracken Darrell, Charles de Portes, Stuart Gibson, Adam Neumann, Adi Neumann, Ari Emanuel, Vonda Shannon, Joel and Cathy Sohn, Alejandro Quiroz Zarate, Mathilde Thomas, Bertrand Thomas, Joseph Vercauteren, Nicholas Wade, Karen Weintraub, Jay Mitchell, Marcia Haigis, Amy Wagers, Yang Shi, Raul Mostoslavsky, Tom Rando, Jennifer Cermak, Phil Lambert, Bruce Szczepankiewicz, Ekat Peheva, Matt Easterday, Rob Mohr, Kyle Meetze, Joanna Schulak, Ricardo Godinez, Pablo Costa, Andreas Pfenning, Fernando Fontove, Abraham Solis, Jaques Estaban, Carlos Sermeño and the entire C3 team, Peter Buchthal, Mark Tatar, Dean Ornish, Margaret Morris, Peter Smith, David Le Couteur, Thomas Watson, Kyle Landry, Meredith Carpenter, Margaret Morris, Steven Simpson, Mark Sumich, Adam Hanft, David Chin, Jim Cole, Ed Green, Phil Lambert, Shally Bhasin, Lawrence Gozlan, Daniel Kraft, Mark Hyman, Marc Hodosh, Felipe Sierra, Michael Sistenic, Bob Kain, David Coomber, Ken Rideout, Bob Bass, Tim Bass, John Monsky, Jose Morey, Michael Bonkowski, David Gold, Matt Westfall, Julia Dimon, Richard Hersey, Joe Hockey, Bjarke Ingels, Margo McInnes, Joe Rogan, Mhairi Anderson, Lon Augustenborg, Mike Harris, Sean Riley, Greg Keeley, Ari Patrinos, Andy, Henny, Ian, Josh, and all the other special people who have served and risked their lives to make the world a better place. To everyone I've worked with over the years: thank you for the encouragement and inspiration to keep working on this book. I am deeply indebted to those who took the time to be my mentors: my grandmother Vera; my father, Andrew; my mother, Diana; my uncle and aunt Barry and Anne Webb; my PhD mentors, Ian Dawes, Richard Dickinson, and Jeff Kornfeld; my postdoc mentor, Lenny Guarente; and my Harvard mentors, Peter Howley, George Church, and Cliff Tabin, and everyone who's stood up for and supported our research.

My lab and its research would not have been possible without the support of grants from the Helen Hay Whitney Foundation, an Australian graduate fellowship, the US National Institutes of Health, the National Health and Medical Research Council of Australia, Mark Collins,

Leonard Judson and Kevin Lee at the Glenn Foundation for Medical Research, the American Federation for Aging Research, Caudalie, Hood Foundation, Leukemia and Lymphoma Society, Lawrence Ellison Medical Foundation, Hank and Elenor Rasnow, Vincent Giampapa, and Edward Schulak. I'm so very grateful to the hundreds of donors, both large and small, who have contributed to our lab's research.

And finally, I cannot begin to express my gratitude for the vision, wisdom, and kindness of Paul Glenn, whose funding of aging research will change the world.

MATTHEW

While I truly value my professional partnership with David, I absolutely treasure our friendship, and I am so grateful to have him in my life. I am also deeply appreciative of Sandra, Alex, Natalie, and Ben Sinclair, who always treated me like family when I was in Boston, and of Susan DeStefano, who always greeted me with a hug at Harvard. I am in debt to the researchers in David's lab and to the leaders and employees of the companies in which David is involved for their kindness and patience during my visits. I would not have known any of these amazing people if not for my truly amazing agent, Trena Keating. Most of all, I am grateful to my wife, Heidi, and our daughter, Mia, who lifted me up through the simultaneous writing of two books.

Notes

INTRODUCTION: A GRANDMOTHER'S PRAYER

1. In a wide-ranging interview to promote his memoirs, Lanzmann said of his masterpiece film about the Holocaust, "I wanted to get as close as possible to death. No personal accounts are told in *Shoah*, no anecdotes. It's only about death. The film is not about the survivors." "'Shoah' Director Claude Lanzmann: 'Death Has Always Been a Scandal,'" *Spiegel*, September 10, 2010, http://www.spiegel.de /international/zeitgeist/shoah-director-claude-lanzmann-death-has-always-been-a -scandal-a-716722.html.

2. The study looked at three concepts about death that children come to understand before they are seven years old: irreversibility, nonfunctionality, and universality. M. W. Speece and S. B. Brent, "Children's Understanding of Death: A Review of Three Components of a Death Concept," *Child Development* 55, no. 5 (October 1984): 1671–86, https://www.ncbi.nlm.nih.gov/pubmed/6510050.

3. The author attended the birth of her daughter's first child along with her son-in-law. R. M. Henig, "The Ecstasy and the Agony of Being a Grandmother," *New York Times*, December 27, 2018, https://www.nytimes.com/2018/12/27/style/self-care /becoming-a-grandmother.html.

4. The film's exhortations to make the most of every day took on a darker hue after the suicide of its star, Robin Williams. P. Weir, director, *Dead Poets Society*, United States: Touchstone Pictures, 1999.

5. The author argues that rather than focusing on cancer and cardiovascular issues, medical research should be focusing on "reducing ageing and age-related morbidity, thereby increasing both our health and our wealth." G. C. Brown, "Living Too Long," *EMBO Reports* 16, no. 2 (February 2015): 137–41, https://www.ncbi.nlm .nih.gov/pmc/articles/PMC4328740/.

6. In a survey coconducted by the *Economist*, the majority of respondents from four countries expressed the wish to die at home, although only a small number thought that they would do so. With the exception of Brazilians, most felt that dying without pain was more important than extending life. "A Better Way to Care for the Dying," *Economist*, April 29, 2017, https://www.economist.com/international /2017/04/29/a-better-way-to-care-for-the-dying.

7. See my conflict disclosures at the end of this book and at https://genetics.med.har vard.edu/sinclair-test/people/sinclair-other.php.

8. My editor made me write self-centered things about myself to give me credibility. I hope she doesn't see this endnote and make me delete it.

9. In 2018, my family and I made a pilgrimage to London to see the original account of Captain James Cook's "voyage round the world" and the original Australian botanical specimens collected by Sir Joseph Banks. There were stop-offs to see Watson and Crick's original model of DNA, fossils of early life, a Moai statue from Rapa Nui, a cross-sectional cut through a 1,500-year-old sequoia tree trunk, a statue of Charles Darwin, the Broad Street pump, Winston Churchill's War Rooms, and the Royal Society, of course. Tracing the path of Cook along the lower east coast of Australia, or "New Holland," as it was called then, it is obvious that Banks already had a colony in mind, one that would never forget him. Not only was the original site named Botany Bay, the coast was named "Cape Banks." After exploring Botany Bay, the explorers' tall ship, the HMS *Endeavor*, sailed north, past the heads of a harbor they called Port Jackson, which, thanks to its much deeper waters and the presence of a stream to supply fresh water, ended up being a far superior site for Governor Phillip to start a penal colony eight years later.

10. "Phillip's Exploration of Middle Harbour Creek," Fellowship of the First Fleeters, Arthur Phillip Chapter, http://arthurphillipchapter.weebly.com/exploration-of -middle-harbour-creek.html.

11. The Spanish explorer and conquistador's search for the mystical spring known as the Fountain of Youth is apocryphal, but it makes for a good story. J. Greenspan, "The Myth of Ponce de León and the Fountain of Youth," "History Stories," April 2, 2013, A&E Television Networks, https://www.history.com/news/the-myth-of -ponce-de-leon-and-the-fountain-of-youth.

12. According to the Creation Wiki: the Encyclopedia of Creation Science (a website of the Northwest Creation Network, http://creationwiki.org/Human_longevity), in Genesis, most of us once got to 900 years, then we didn't. Then most of us got to 400, then we didn't. Then most of us got to 120, then we didn't. In more recent times, as Oeppen and Vaupel have written, "Mortality experts have repeatedly asserted that life expectancy is close to an ultimate ceiling; these experts have repeatedly been proven wrong. The apparent leveling off of life expectancy in various countries is an artifact of laggards catching up and leaders falling behind."

J. Oeppen and J. W. Vaupel, "Broken Limits to Life Expectancy," *Science* 296, no. 5570 (May 10, 2002): 1029–31.

13. There is some debate as to what constitutes verifiable age. There are humans who have claimed, and provided considerable evidence, of being of great age, but who don't have formal Western-style records of their year of birth. In any case, these people are one in a billion, if that. In November 2018, the Russian gerontologist Valery Novoselov and the mathematician Nikolay Zak claimed that after much research, they believe that Jeanne Calment's daughter, Yvonne, usurped Jeanne's identity in 1934, claiming that the daughter had died instead of the mother to avoid paying estate taxes. The debate continues. "French Scientists Dismiss Russian Claims over Age of World's Oldest Person," Reuters, January 3, 2019, https://www .reuters.com/article/us-france-oldest-woman-controversy/french-scientists-dismiss -russian-claims-over-age-of-worlds-oldest-person-idUSKCN1OX145.

14. Italian researchers found after studying 4,000 elderly people that if you make it to age 105, the risk of death effectively plateaus from one birthday to the next, the odds of dying in the next year becoming approximately fifty-fifty. E. Barbi, F. Lagona, M. Marsili, et al., "The Plateau of Human Mortality: Demography of Longevity Pioneers," *Science* 360, no. 396 (June 29, 2018): 1459–61, http://science .sciencemag.org/content/360/6396/1459.

15. "If people live on average to 80 or 90, like they do now, then the very long lived make it to 110 or 120," says Siegfried Hekimi, professor of genetics at McGill University in Canada. "So if the average lifespan keeps expanding, that would mean the long-lived would live even longer, beyond 115 years"; A. Park, "There's No Known Limit to How Long Humans Can Live, Scientists Say," *Time*, June 28, 2017, http:// time.com/4835763/how-long-can-humans-live/.

16. "Any sufficiently advanced technology is indistinguishable from magic." "Arthur C. Clarke," Wikiquote, https://en.wikiquote.org/wiki/Arthur_C._Clarke.

ONE. VIVA PRIMORDIUM

1. D. Damer and D. Deamer, "Coupled Phases and Combinatorial Selection in Fluctuating Hydrothermal Pools: A Scenario to Guide Experimental Approaches to the Origin of Cellular Life," *Life* 5, no. 1 (2015): 872–87, https://www.mdpi.com /2075-1729/5/1/872.

2. According to precise radiological and geological readings and recent discoveries about the early chemistry of life, this is an accurate picture of how the inanimate was animated and life took hold. M. J. Van Kranendonk, D. W. Deamer, and T. Djokic, "Life on Earth Came from a Hot Volcanic Pool, Not the Sea, New Evidence Suggests," *Scientific American*, August 2017, https://www.scientific american.com/article/life-on-earth-came-from-a-hot-volcanic-pool-not-the-sea -new-evidence-suggests/.

3. J. B. Iorgulescu, M. Harary, C. K. Zogg, et al., "Improved Risk-Adjusted Survival for Melanoma Brain Metastases in the Era of Checkpoint Blockade Immunotherapies: Results from a National Cohort," *Cancer Immunology Research*, 6, no. 9 (September 2018): 1039–45, http://cancerimmunolres.aacrjournals.org/content/6/9/1039.long; R. L. Siegel, K. D. Miller, and A. Jemal, "Cancer Statistics, 2019," *CA: A Cancer Journal for Clinicians* 69, no. 1 (January–February 2019): 7–34, https://onlinelibrary.wiley.com/doi/full/10.3322/caac.21551.

4. As far back as Aristotle, scientists and philosophers have struggled to resolve the enigma of aging, the authors wrote. D. Fabian and T. Flatt, "The Evolution of Aging," *Nature Education Knowledge* 3, no. 10 (2011): 9, https://www.nature.com/scitable/knowledge/library/the-evolution-of-aging-23651151.

5. A bat from Siberia set a world record when it reached 41 years of age. R. Locke, "The Oldest Bat: Longest-Lived Mammals Offer Clues to Better Aging in Humans," *BATS Magazine* 24, no. 2 (Summer 2006): 13–14, http://www.batcon.org/resources/media-education/bats-magazine/bat_article/152.

6. Small colonies of lizards on a series of Caribbean islands were likely to explore islands where there weren't predators, while less adventurous animals survived better when predators were present. O. Lapiedra, T. W. Schoener, M. Leal, et al., "Predator-Driven Natural Selection on Risk-Taking Behavior in Anole Lizards," *Science* 360, no. 3692 (June 1, 2018): 1017–20, http://science.sciencemag.org/content/360/6392/1017.

7. Richard Dawkins eloquently made this point in *River Out of Eden*, arguing that primitive societies don't have a place in science, using as an example their belief the moon is an old calabash tossed into the sky. R. Dawkins, *River Out of Eden* (New York: Basic Books, 1995).

8. See "The Scale of Things" at the end of this book.

9. Szilard spent his last years as a fellow of the Salk Institute for Biological Studies in La Jolla, California, as a resident fellow. He lived in a bungalow on the property of the Hotel del Charro and died on May 30, 1964.

10. R. Anderson, "Ionizing Radiation and Aging: Rejuvenating an Old Idea," *Aging* 1, no. 11 (November 17, 2009): 887–902, https://www.ncbi.nlm.nih.gov/pmc/articles/PMC2815743/.

11. L. E. Orgel, "The Maintenance of the Accuracy of Protein Synthesis and Its Relevance to Ageing," *Proceedings of the National Academy of Sciences of the United States of America* 49, no. 4 (April 1963): 517–21, https://www.ncbi.nlm.nih.gov/pmc/articles/PMC299893/.

12. Harman concluded that the diseases related to aging, as well as aging itself, stem fundamentally from "the deleterious side attacks of free-radicals on cell constituents and on the connective tissues." The source of the free radicals, he continued, was "molecular oxygen catalyzed in the cell by the oxidative enzymes" and metal traces. D. Harman, "Aging: A Theory Based on Free Radical and Radiation Chemistry,"

Journal of Gerontology 11, no. 3 (July 1, 1956): 298–300, https://academic.oup.com/geronj/article-abstract/11/3/298/616585?redirectedFrom=fulltext.

13. Nutraceuticals World predicts that a rising appetite for synthetic antioxidants at the same time as a fall in costs, combined with increasing demand for them by food and beverage companies, will power market growth for the next few years. "Global Antioxidants Market Expected to Reach $4.5 Billion by 2022," *Nutraceuticals World*, January 26, 2017, https://www.nutraceuticalsworld.com/contents/view_breaking-news/2017-01-26/global-antioxidants-market-expected-to-reach-45-billion-by-2022

14. The sharp growth in demand for drinks with a health benefit, a beverage industry website finds, goes hand in hand with consumers wanting ingredients they value. A. Del Buono, "Consumers' Understanding of Antioxidants Grows," *Beverage Industry*, January 16, 2018, https://www.bevindustry.com/articles/90832-consumers-understanding-of-antioxidants-grows?v=preview.

15. I. Martincorena, J. C. Fowler, A. Wabik, et al., "Somatic Mutant Clones Colonize the Human Esophagus with Age," *Science* 362, no. 6417 (November 23, 2018): 911–17, https://www.ncbi.nlm.nih.gov/pubmed/30337457.

16. The authors concluded that their data "calls into serious question the hypothesis that alterations in oxidative damage/stress play a role in the longevity of mice." V. I. Pérez, A. Bokov, H. Van Remmen, et al., "Is the Oxidative Stress Theory of Aging Dead?," *Biochimica et Biophysica Acta* 1790, no. 10 (October 2009): 1005–14, https://www.ncbi.nlm.nih.gov/pmc/articles/PMC2789432/.

17. A. P. Gomes, N. L. Price, A. J. Ling, et al., "Declining NAD(+) Induces a Pseudohypoxic State Disrupting Nuclear-Mitochondrial Communication During Aging," *Cell* 155, no. 7 (December 19, 2013): 1624–38, https://www.ncbi.nlm.nih.gov/pubmed/24360282.

18. W. Lanouette and B. Silard, *Genius in the Shadows: A Biography of Leo Szilard: The Man Behind the Bomb* (New York: Skyhorse Publishing, 1992).

19. According to the NIH fact sheet, "clones created from a cell taken from an adult might have chromosomes that are already shorter than normal, which may condemn the clones' cells to a shorter life span." "Cloning," National Human Genome Research Institute, March 21, 2017, https://www.genome.gov/25020028/cloning-fact-sheet/.

20. In the debates over Dolly the cloned sheep, the question that has proved to be challenging to answer is how old an animal is at birth when cloned from an adult's cell. The answer an author on the site The Conversation found was that other clones born from the same cell as Dolly lived normal lifespans. "The new Dollies are now telling us that if we take a cell from an animal of any age, and we introduce its nucleus into a nonfertilized mature egg, we can have an individual born with its lifespan fully restored." J. Cibell, "More Lessons from Dolly the Sheep: Is a Clone Really Born at Age Zero?," The Conversation, February 17, 2017, https://

theconversation.com/more-lessons-from-dolly-the-sheep-is-a-clone-really-born-at
-age-zero-73031.

21. Though some cloned animals match their species' rates of normal aging, it's a field that still needs further analysis to get beyond the largely anecdotal evidence so far collected. J. P. Burgstaller and G. Brem, "Aging of Cloned Animals: A Mini-Review," *Gerontology* 63, no. 5 (August 2017): 417–25, https://www.karger.com /Article/FullText/452444.

22. University of Bath researchers found in cloned mice that the telomeres protecting the ends of chromosomes were, surprisingly, slightly longer in successive generations and demonstrated no evidence of premature aging. T. Wakayama, Y. Shinkai, K. L. K. Tamashiro, et al., "Ageing: Cloning of Mice to Six Generations," *Nature* 407 (September 21, 2000): 318–19. "Despite the length of telomeres reported in different studies, most clones appear to be aging normally. In fact, the first cattle clones ever produced are alive, healthy, and are 10 years old as of January 2008"; "Myths About Cloning," U.S. Food & Drug Administration, August 29, 2018, https://www .fda.gov/animalveterinary/safetyhealth/animalcloning/ucm055512.htm.

23. The authors discovered mitochondrial DNA in a Neanderthal bone in Croatia that revealed older dates of survival than previously thought. T. Devièse, I. Karavanié, D. Comeskey, et al., "Direct Dating of Neanderthal Remains from the Site of Vindija Cave and Implications for the Middle to Upper Paleolithic Transition," *Proceedings of the National Academy of Sciences of the United States of America* 114, no. 40 (October 3, 2017): 10606–11, https://www.ncbi.nlm.nih.gov/pubmed /28874524.

24. A. S. Adikesevan, "A Newborn Baby Has About 26,000,000,000 Cells. An Adult Has About 1.9×10^3 Times as Many Cells as a Baby. About How Many Cells Does an Adult Have?," Socratic, January 26, 2017, https://socratic.org/questions/a-new born-baby-has-about-26-000-000-000-cells-an-adult-has-about-1-9-10-3-times-.

25. C. B. Brachmann, J. M. Sherman, S. E. Devine, et al., "The *SIR2* Gene Family, Conserved from Bacteria to Humans, Functions in Silencing, Cell Cycle Progression, and Chromosome Stability," *Genes & Development* 9, no. 23 (December 1, 1995): 2888–902, http://genesdev.cshlp.org/content/9/23/2888.long; X. Bi, Q. Yu, J. J. Sandmeier, and S. Elizondo, "Regulation of Transcriptional Silencing in Yeast by Growth Temperature," *Journal of Molecular Biology* 34, no. 4 (December 3, 2004): 893–905, https://www.ncbi.nlm.nih.gov/pubmed/15544800.

26. It is one of the most interesting and important papers I've ever read. C. E. Shannon, "A Mathematical Theory of Communication," *Bell System Technical Journal* 27, no. 3 (July 1948): 379–423, and 27, no. 4 (October 1948): 623–66, http:// math.harvard.edu/~ctm/home/text/others/shannon/entropy/entropy.pdf.

27. Research by the authors showed that mTORC1 signaling in cancer cells increases survival by "suppressing endogenous DNA damage, and may control cell fate through the regulation of CHK1." X. Zhou, W. Liu, X. Hu, et al., "Regulation of

CHK1 by mTOR Contributes to the Evasion of DNA Damage Barrier of Cancer Cells," *Nature Scientific Reports*, May 8, 2017, https://www.nature.com/articles
/s41598-017-01729-w; D. M. Sabatini, "Twenty-five Years of mTOR: Uncovering
the Link from Nutrients to Growth," *Proceedings of the National Academy of Sciences
of the United States of America* 114, no. 45 (November 7, 2017): 11818–25, https://
www.ncbi.nlm.nih.gov/pmc/articles/PMC5692607/.

28. E. J. Calabrese, "Hormesis: A Fundamental Concept in Biology," *Microbial Cell*
1, no. 5 (May 5, 2014): 145–49, https://www.ncbi.nlm.nih.gov/pmc/articles
/PMC5354598/.

TWO. THE DEMENTED PIANIST

1. Up to 69 percent of the human genome may be repetitive or derived from endogenous viral DNA repeats, compared to previous estimates of around half.
A. P. de Konig, W. Gu, T. A. Castoe, et al., "Repetitive Elements May Comprise
over Two-thirds of the Human Genome," *PLOS Genetics* 7, no. 12 (December 7,
2011), https://www.ncbi.nlm.nih.gov/pmc/articles/PMC3228813/.

2. Just what do we mean by the word *finished* when it comes to the sequencing of
the human genome? Turns out, more than we thought back in the early 2000s.
Regions of the genome previously thought of as nonfunctional are now emerging
as playing potential roles in cancer, autism, and aging. S. Begley, "Psst, the Human
Genome Was Never Completely Sequenced. Some Scientists Say It Should Be,"
STAT, June 20, 2017, https://www.statnews.com/2017/06/20/human-genome
-not-fully-sequenced/.

3. Dating back to the 1960s, every three or four years the center has published a catalog of its strains of *Saccharomyces cerevisiae*. R. K. Mortimer, "Yeast Genetic Stock
Center," Grantome, 1998, http://grantome.com/grant/NIH/P40-RR004231-10S1.

4. Yeast researchers have interesting names. John Johnston and my adviser Dick Dickinson are just two of them.

5. In 2016, Dr. Yoshinori Ohsumi won the Nobel Prize in Physiology or Medicine for
his work on autophagy in yeast. That's when cells stave off extinction during hard
times by digesting nonkey parts of themselves. B. Starr, "A Nobel Prize for Work in
Yeast. Again!," Stanford University, October 3, 2016, https://www.yeastgenome.org
/blog/a-nobel-prize-for-work-in-yeast-again.

6. Dawes's delightful tour of his experiences in the world of academe and cell biology
research is a refreshingly direct and personal account of a remarkable journey into
yeast research over four decades. I. Dawes, "Ian Dawes—the Third Pope—Lucky
to Be a Researcher," *Fems Yeast Research* 6, no. 4 (June 2016), https://academic.oup
.com/femsyr/article/16/4/fow040/2680350.

7. I also learned, the hard way, that I should not drink copious quantities of yeasty
beer.

8. For four years after that, I sent Professor Melton a bottle of red wine for New Year's,

just to say thanks for changing my life. He never acknowledged any of them or ever smiled at me, either because he didn't think that's what an awardee should do or because he's a very private person. At least he knew I was grateful. The selection of red wine turned out to be ironic, as that foodstuff helped propel my career a second time nine years later.

9. C. E. Yu, J. Oshima, Y. H. Fu, et al., "Positional Cloning of the Werner's Syndrome Gene," *Science* 27, no. 5259 (April 12, 1996): 258–62, https://www.ncbi.nlm.nih .gov/pubmed/8602509.

10. *SIR2* stands for "silent information regulator 2." When *SIR2* is written in capitals and italics, it refers to the gene; when it's written Sir2, it refers to the protein.

11. In a paper published in late 1997, I showed how ERCs—rDNA circles—cause aging and shorten the life of yeast cells. D. A. Sinclair and L. Guarente, "Extrachromosomal rDNA Circles—A Cause of Aging in Yeast," *Cell* 91, no. 7 (December 26, 1997): 1033–42, https://www.ncbi.nlm.nih.gov/pubmed/9428525.

12. One way to think of the epigenome is as a cell's software. In the same way digital files are stored in a phone's memory and the software uses the ones and zeros to turn a phone into a clock, calendar, or music player, a cell's information is stored as As, Ts, Gs, and Ts, and the epigenome uses those letters to direct a yeast cell to become male or a female and turn a mammalian cell into a nerve, a skin cell, or an egg.

13. I am not the first person to use this analogy. One of the earliest uses of the piano metaphor I can find came from a study guide intended to accompany a *Nova ScienceNOW* program on epigenetics in 2007. "Nova ScienceNOW: Epigenetics," PBS, http://www.pbs.org/wgbh/nova/education/viewing/3411_02_nsn.html.

14. C. A. Makarewich and E. N. Olson, "Mining for Micropeptides," *Trends in Cell Biology* 27, no. 9 (September 27, 2017): 685–96, https://www.ncbi.nlm.nih.gov /pubmed/28528987.

15. D. C. Dolinoy, "The Agouti Mouse Model: An Epigenetic Biosensor for Nutritional and Environmental Alterations on the Fetal Epigenome," *Nutrition Reviews* 66, suppl. 1 (August 2008): S7–11, https://www.ncbi.nlm.nih.gov/pmc/articles /PMC2822875/.

16. The more extroverted you are, the longer your lifespan, while, perhaps unsurprisingly, pessimists and psychotics see significant increases in the risk of death at an earlier age. That's according to a study of 3,752 twins 50 years or older that looked at the relationship between personality and lifespan through the prism of genetic influences. M. A. Mosing, S. E. Medland, A. McRae, et al., "Genetic Influences on Life Span and Its Relationship to Personality: A 16-Year Follow-up Study of a Sample of Aging Twins," *Psychosomatic Medicine* 74, no. 1 (January 2012): 16–22, https://www.ncbi.nlm.nih.gov/pubmed/22155943. The authors considered definitions of extreme longevity, using multiple European twin registries. A. Skytthe, N. L. Pedersen, J. Kaprio, et al., "Longevity Studies in GenomEUtwin," *Twin*

Research 6, no. 5 (October 2003): 448–54, https://www.ncbi.nlm.nih.gov/pubmed /14624729.

17. It was a eureka moment—discovering why yeast cells age. Supercoiled circles of ribosomal DNA pinch off the yeast chromosome and accumulate as the yeast divide, distracting the Sir2 enzyme from its main role of controlling genes for sex and reproduction. David A. Sinclair and Leonard Guarente, "Extrachromosomal rDNA Circles—A Cause of Aging in Yeast," *Cell* 91 (December 26, 1997): 1033–42.

18. D. A. Sinclair, K. Mills, and L. Guarente, "Accelerated Aging and Nucleolar Fragmentation in Yeast SGS1 Mutants," *Science* 277, no. 5330 (August 29, 1997): 1313–16, https://www.ncbi.nlm.nih.gov/pubmed/9271578.

19. Sinclair and Guarente, "Extrachromosomal rDNA Circles—A Cause of Aging in Yeast."

20. K. D. Mills, D. A. Sinclair, and L. Guarente, "MEC1-Dependent Redistribution of the Sir3 Silencing Protein from Telomeres to DNA Double-Strand Breaks," *Cell* 97, no. 5 (May 28, 1999): 609–20, https://www.ncbi.nlm.nih.gov/pubmed /10367890.

21. Sinclair, Mills, and Guarente, "Accelerated Aging and Nucleolar Fragmentation in Yeast SGS1 Mutants."

22. P. Oberdoerffer, S. Michan, M. McVay, et al., "SIRT1 Redistribution on Chromatin Promotes Genomic Stability but Alters Gene Expression During Aging," *Cell* 135, no. 5 (November 28, 2008): 907–18, https://www.cell.com/cell/fulltext /S0092-8674(08)01317-2; Z. Mao, C. Hine, X. Tian, et al., "SIRT6 Promotes DNA Repair Under Stress by Activating PARP1," *Science* 332, no. 6036 (June 2011): 1443–46, https://www.ncbi.nlm.nih.gov/pubmed/21680843.

23. A. Ianni, S. Hoelper, M. Krueger, et al., "Sirt7 Stabilizes rDNA Heterochromatin Through Recruitment of DNMT1 and Sirt1," *Biochemical and Biophysical Research Communications* 492, no. 3 (October 21, 2017): 434–40, https://www.ncbi.nlm .nih.gov/m/pubmed/28842251/.

24. The authors show how SIRT7, in protecting against the instability of rDNA, also guards against the death of human cells. S. Paredes, M. Angulo-Ibanez, L. Tasselli, et al., "The Epigenetic Regulator SIRT7 Guards Against Mammalian Cellular Senescence Induced by Ribosomal DNA Instability," *Journal of Biological Chemistry* 293 (July 13, 2018): 11242–50, http://www.jbc.org/content/293/28/11242.

25. Oberdoerffer et al., "SIRT1 Redistribution on Chromatin Promotes Genomic Stability but Alters Gene Expression During Aging."

26. M. W. McBurney, X. Yang, K. Jardine, et al., "The Mammalian SIR2alpha Protein Has a Role in Embryogenesis and Gametogenesis," *Molecular and Cellular Biology* 23, no. 1 (January 23, 2003): 38–54, https://mcb.asm.org/content/23 /1/38.long.

27. R.-H. Wang, K. Sengupta, L. Cuiling, et al., "Impaired DNA Damage Response, Genome Instability, and Tumorigenesis in SIRT1 Mutant Mice," *Cancer Cell*

14, no. 4 (October 7, 2008): 312–23, https://www.cell.com/cancer-cell/fulltext /S1535-6108(08)00294-8.

28. R. Mostoslavsky, K. F. Chua, D. B. Lombard, et al., "Genomic Instability and Aging-like Phenotype in the Absence of Mammalian SIRT6," *Cell* 124 (January 27, 2006): 315–29, https://doi.org/10.1016/j.cell.2005.11.044.

29. The treatments work better in male mice, for reasons that are not yet known, but my former postdoc Haim Cohen at Bar-Ilan University in Israel wins the award for the best-ever name given to a transgenic mouse strain: MOSES. A. Satoh, C. S. Brace, N. Rensing, et al., "Sirt1 Extends Life Span and Delays Aging in Mice Through the Regulation of Nk2 Homeobox 1 in the DMH and LH," *Cell Metabolism* 18, no. 3 (September 3, 2013): 416–30, https://www.ncbi.nlm.nih.gov/pmc /articles/PMC3794712.

30. When we write *SIR2* in capitals and italics, it refers to the gene; when we write Sir2, it refers to the protein the gene encodes.

31. It's possible that by not allowing mating-type genes to turn on, yeast with additional copies of *SIR2* have less efficient DNA repair by homologous recombination, which is what the expression of mating-type genes also does when switched on besides preventing mating. This needs to be tested. But at least under safe lab conditions, the cells grow perfectly fine.

32. M. G. L. Baillie, *A Slice Through Time: Dendrochronology and Precision Dating* (London: Routledge, 1995).

33. Along with bristlecones, Matthew LaPlante, my coauthor on *Lifespan*, looks at a wide variety of biology's outliers that define the very edges of our understanding of plants and animals, from ghost sharks and elephants to beetles and microbacteria. M. D. LaPlante, *Superlative: The Biology of Extremes* (Dallas: BenBella Books, 2019).

34. When researchers compared trees of a variety of ages to look for a steady incremental decline in annual shoot growth, they found "no statistically significant age-related differences." R. M. Lanner, and K. F. Connor, "Does Bristlecone Pine Senesce?," *Experimental Gerontology* 36, nos. 4–6 (April 2001): 675–85, https:// www.sciencedirect.com/science/article/pii/S0531556500002345?via%3Dihub.

35. Investigating mutations in the gene Daf-2, researchers made a remarkable find: the largest reported lifespan extension of any living thing, namely twice as long. This relied on the involvement of two genes, Daf-2 and Daf-16, opening the door to new horizons of ways to understand how to prolong life. C. Kenyon, J. Chang, E. Gensch, et al., "A *C. elegans* Mutant That Lives Twice as Long as Wild Type," *Nature* 366 (December 2, 1993): 461–64, https://www.nature.com/articles/36 6461a0; F. Wang, C.-H. Chan, K. Chen, et al., "Deacetylation of FOXO3 by SIRT1 or SIRT2 Leads to Skp2-Mediated FOXO3 Ubiquitination and Degradation," *Oncogene* 31, no. 12 (March 22, 2012): 1546–57, https://www.nature.com /articles/onc2011347.

36. Why do genes often have a variety of names? The language of genetics is just like any other language; its words contain the echoes of history. Knowing the entire genome of a yeast cell, a nematode worm, or a human was the stuff of dreams less than a quarter century ago. Now, of course, I can sequence my own genome in a day on a USB drive–sized sequencer. When I was a student, genes would be given a name based on the characteristics of mutants we would generate with mutagenic chemicals. Typically, all we knew about a gene when we named it was its rough location on a particular chromosome. Only later were its distant cousins identified.

37. A. Brunet, L. B. Sweeney, J. F. Sturgill, et al., "Stress-Dependent Regulation of FOXO Transcription Factors by the SIRT1 Deacetylase," *Science* 303, no. 5666 (March 24, 2004): 2011–15, https://www.ncbi.nlm.nih.gov/pubmed/1497 6264.

38. O. Medvedik, D. W. Lamming, K. D. Kim, and D. A. Sinclair, "*MSN2* and *MSN4* Link Calorie Restriction and TOR to Sirtuin-Mediated Lifespan Extension in *Saccharomyces cerevisiae*," *PLOS Biology*, October 2, 2007, http://journals.plos.org /plosbiology/article?id=10.1371/journal.pbio.0050261.

39. The authors found convincing evidence linking *FOXO3* and longevity in humans. L. Sun, C. Hu, C. Zheng, et al., "*FOXO3* Variants Are Beneficial for Longevity in Southern Chinese Living in the Red River Basin: A Case-Control Study and Meta-analysis," *Nature Scientific Reports*, April 27, 2015, https://www.nature.com/articles /srep09852.

40. H. Bae, A. Gurinovich, A. Malovini, et al., "Effects of *FOXO3* Polymorphisms on Survival to Extreme Longevity in Four Centenarian Studies," *Journals of Gerontology, Series A: Biological Sciences and Medical Sciences* 73, no. 11 (October 8, 2018): 1437–47, https://academic.oup.com/biomedgerontology/article/73/11/1439 /3872296.

41. If you're a dedicated exerciser in middle age or an athlete in her fifties, chances are your heart is going to resemble that of someone much younger, several studies have revealed. Not so for the office worker who doesn't exercise or someone who hits the gym or runs in the street on a sporadic basis. What isn't clear, though, is whether commencing an aggressive exercise program in your middle years can turn around the effects of a sedentary lifestyle on the heart's functioning and structure. G. Reynolds, "Exercise Makes the Aging Heart More Youthful," *New York Times*, July 25, 2018, https://www.nytimes.com/2018/07/25/well/exercise-makes-the -aging-heart-more-youthful.html.

42. "These findings have implications for improving blood flow to organs and tissues, increasing human performance, and reestablishing a virtuous cycle of mobility in the elderly."A. Das, G. X. Huang, M. S. Bonkowski, et al., "Impairment of an Endothelial NAD^+-H_2S Signaling Network Is a Reversible Cause of Vascular Aging," *Cell* 173, no. 1 (March 22, 2018): 74–89, https://www.cell.com/cell/pdf/S0092 -8674(18)30152-1.pdf.

THREE. THE BLIND EPIDEMIC

1. F. Bacon, *Of the Proficience and Advancement of Learning, Divine and Human* (Oxford, UK: Leon Lichfield, 1605). An original of this book sits on our mantelpiece at home, a gift from Sandra, my wife.

2. C. Kenyon, J. Chang, E. Gensch, et al., "A *C. elegans* Mutant That Lives Twice as Long as Wild Type," *Nature* 366, no. 6454 (December 2, 1993): 461–64, https://www.nature.com/articles/366461a0.

3. L. Partridge and P. H. Harvey, "Methuselah Among Nematodes," *Nature* 366, no. 6454 (December 2, 1993): 404–5, https://www.ncbi.nlm.nih.gov/pubmed/8247143.

4. "Decelerated aging," Gems wrote, "has an element of tragic inevitability: its benefits to health compel us to pursue it, despite the transformation of human society, and even human nature, that this could entail." D. Gems, "Tragedy and Delight: The Ethics of Decelerated Ageing," *Philosophical Transactions of the Royal Society B: Biological Sciences* 366 (January 12, 2011): 108–12, https://royalsocietypublishing.org/doi/pdf/10.1098/rstb.2010.0288.

5. "You know the cartoon where Bugs Bunny is driving an old car that suddenly falls apart, every bolt sprung, with the last hubcap rattling in a circle until it comes to rest?" *Washington Post* reporter David Brown wrote in 2010. "Some people die like that, too. The trouble is there's not a good name for it." D. Brown, "Is It Time to Bring Back 'Old Age' as a Cause of Death?" *Washington Post*, September 17, 2010, http://www.washingtonpost.com/wp-dyn/content/article/2010/09/17/AR2010091703823.html?sid=ST2010091705724.

6. "Really, people don't die of old age," Chris Weller wrote on Medical Daily. "Something else has to be going on." C. Weller, "Can People Really Die of Old Age?," "The Unexamined Life," Medical Daily, January 21, 2015, http://www.medicaldaily.com/can-people-really-die-old-age-318528.

7. B. Gompertz, "On the Nature of the Function Expressive of the Law of Human Mortality, and on a New Mode of Determining the Value of Life Contingencies," *Philosophical Transactions of the Royal Society* 115 (January 1, 1825): 513–85, https://royalsocietypublishing.org/doi/10.1098/rstl.1825.0026.

8. D. A. Sinclair and L. Guarente, "Extrachromosomal rDNA Circles—A Cause of Aging in Yeast," *Cell* 91, no. 7 (December 26, 1997): 1033–42, https://www.ncbi.nlm.nih.gov/pubmed/9428525.

9. Based on global population estimates and census reports, among other sources, the World Bank plotted out a fifty-six-year period ending in 2016 that showed life expectancy increasing from 52 to 72. "Life Expectancy at Birth, Total (Years)," The World Bank, https://data.worldbank.org/indicator/SP.DYN.LE00.IN.

10. I inherited the SERPINA1 mutation from my mother. Even though I have never smoked, I find it hard to breathe in some situations, such as when I am visiting a place with substantial pollution. Armed with this information, I avoid breathing

in dust and other contaminants when possible. I feel empowered knowing the genetic instructions within each of my cells, an experience that previous generations never had.

11. A. M. Binder, C. Corvalan, V. Mericq, et al., "Faster Ticking Rate of the Epigenetic Clock Is Associated with Faster Pubertal Development in Girls," *Epigenetics* 13, no. 1 (February 15, 2018): 85–94, https://www.tandfonline.com/doi/full/10.1080/15592294.2017.1414127.

12. Women over 65 are more prone to hip fractures, with sepsis being the main cause of death. Researchers have linked the sepsis to poor medical care, a lack of family support, and dementia. "Time wise, mortality was found to be higher within the first six months, with 10 deaths (50%), and within the first year, with six deaths (30%)." J. Negrete-Corona, J. C. Alvarano-Soriano, and L. A. Reyes-Santiago, "Hip Fracture as Risk Factor for Mortality in Patients over 65 Years of Age. Case-Control Study" (abstract translation from Spanish), *Acta Ortopédica Mexicana* 28, no. 6 (November–December 2014): 352–62, https://www.ncbi.nlm.nih.gov/pubmed/26016287, (Spanish) http://www.medigraphic.com/pdfs/ortope/or-2014/or146c.pdf.

13. Up to 74 percent of patients who have a foot amputated due to diabetes die within five years of surgery. The authors argue for more aggressive focus on the issue by doctors and patients alike. "New-onset diabetic foot ulcers should be considered a marker for significantly increased mortality and should be aggressively managed locally, systemically, and psychologically."J. M. Robbins, G. Strauss, D. Aron, et al., "Mortality Rates and Diabetic Foot Ulcers: Is It Time to Communicate Mortality Risk to Patients with Diabetic Foot Ulceration?," *Journal of the American Podiatric Medical Association* 98, no. 6 (November–December 2008): 489–93, https://www.ncbi.nlm.nih.gov/pubmed/19017860.

14. Have we made a deal with the medical devil that's backfired? Olshansky certainly thinks so, contrasting the quest for human longevity and health to the dark narrative of Faust's ultimately pyrrhic deal with Mephistopheles. "It's possible that humanity has squeezed about as much healthy life out of public health interventions as possible and that the human body is now running up against inherent limits that the genetically fixed attributes of our biology impose." S. J. Olshansky, "The Future of Health," *Journal of the American Geriatrics Society* 66, no. 1 (December 5, 2017): 195–97, https://onlinelibrary.wiley.com/doi/full/10.1111/jgs.15167.

15. The numbers are indeed staggering: close to 800,000 Americans die annually of cardiovascular-related diseases; medical costs related to cardiovascular issues are expected to be over $818 billion by 2030 and lost productivity costs above $275 billion. "Heart Disease and Stroke Cost America Nearly $1 Billion a Day in Medical Costs, Lost Productivity," CDC Foundation, April 29, 2015, https://www.cdcfoundation.org/pr/2015/heart-disease-and-stroke-cost-america-nearly-1-billion-day-medical-costs-lost-productivity.

16. As treatments for patients with disease have prolonged their lives, so the amount of

disease in society has augmented. This situation means that the only way to increase the human healthspan will be by "'delaying aging,' or delaying the physiological change that results in disease and disability," the author argues. Along with scientific breakthroughs, changes in socioeconomic inequalities, lifestyle, and behavior can contribute to improving both healthspan and lifespan. E. M. Crimmins, "Lifespan and Healthspan: Past, Present, and Promise," *Gerontologist* 55, no. 6 (December 2015): 901–11, https://www.ncbi.nlm.nih.gov/pmc/articles/PMC4861644/.

17. According to the World Health Organization, one DALY can be thought of as one lost year of "healthy" life. The sum of these DALYs across the population, or the burden of disease, can be thought of as a measurement of the gap between current health status and an ideal health situation in which the entire population lives to an advanced age, free of disease and disability. "Metrics: Disability-Adjusted Life Year (DALY)," World Health Organization, https://www.who.int/healthinfo/global _burden_disease/metrics_daly/en/.

18. And almost everyone at that age spends a considerable part of his or her life visiting the doctor. According to the study, published in 2009 by the *British Medical Journal*, 94 percent of 85-year-olds had had contact with a doctor in the past year, and one in ten was in institutional care. J. Collerton, K. Davies, C. Jagger, et al., "Health and Disease in 85 Year Olds: Baseline Findings from the Newcastle 85+ Cohort Study," *British Medical Journal*, December 23, 2009, https://www.bmj.com /content/339/bmj.b4904.

19. The possibility that both genetic and epigenetic aging are needed for a tumor to develop we've termed "geroncogenesis," and it explains why tumors don't occur in young people even after extreme sun exposure, why it often takes decades for DNA damage to lead to a tumor even if you avoid the sun later in life, and why cancers often have an unusual metabolism (named after the physicist Otto Warburg), one that directly consumes glucose, has decreased mitochondrial activity, and uses less oxygen to make energy, similar to the metabolism of old cells.

20. According to the World Health Organization, "The State of Global Tobacco Control," 2008, http://www.who.int/tobacco/mpower/mpower_report_global_control _2008.pdf.

21. R. A. Miller, "Extending Life: Scientific Prospects and Political Obstacles," *Milbank Quarterly* 80, no. 1 (March 2002): 155–74, https://www.ncbi.nlm.nih .gov/pmc/articles/PMC2690099/; graph redrawn from D. L. Hoyert, K. D. Kochanek, and S. L. Murphy, "Deaths: Final Data for 1997," *National Vital Statistics Report* 47, no. 19 (June 30, 1999):1–104, https://www.ncbi.nlm.nih.gov /pubmed/10410536.

22. Using a survey of 593 people that was then repeated four years later, the authors explored the role of "subjective age" (meaning how old an individual feels in contrast to his or her biological age) in shaping the process of aging. A. E. Kornadt, T. M. Hess, P. Voss, and K. Rothermund, "Subjective Age Across the Life Span: A

Differentiated, Longitudinal Approach," *Journals of Gerontology: Psychological Sciences* 73, no. 5 (June 1, 2018): 767–77, http://europepmc.org/abstract/med/273 34638.

23. "David A. Sinclair's Past and Present Advisory Roles, Board Positions, Funding Sources, Licensed Inventions, Investments, Funding, and Invited Talks," Sinclair Lab, Harvard Medical School, November 15, 2018, https://genetics.med.harvard .edu/sinclair-test/people/sinclair-other.php.

FOUR. LONGEVITY NOW

1. It seems likely that he had sex at least once again, as he had one daughter, Clara, with his wife, Veronica. L. Cornaro, *Sure and Certain Methods of Attaining a Long and Healthful Life: With Means of Correcting a Bad Constitution, &c.*, https://babel .hathitrust.org/cgi/pt?id=dul1.ark:/13960/t0dv2fm86;view=1up;seq=1.

2. There are other translations. This comes from the edition published in Milwaukee by William F. Butler in 1903.

3. A 3-year-old rat measured in terms of human lifespan would be akin to a 90-year-old human, according to a researcher quoted by the authors. One of their rats, raised on an experimental diet from 6 weeks of age, lived to 40 months, while of those rats raised on a normal diet, the oldest reached 34 months, with "less than a third of the rats in our colony . . . expected to live to be more than two years old." T. B. Osborne, L. B. Mendel, and E. L. Ferry, "The Effect of Retardation of Growth upon the Breeding Period and Duration of Life of Rats," *Science* 45, no. 1160 (March 23, 1917): 294–95, http://science.sciencemag.org/content/45/1160/294.

4. I. Bjedov, J. M. Toivonen, F. Kerr, et al., "Mechanisms of Life Span Extension by Rapamycin in the Fruit Fly *Drosophila melanogaster*," *Cell Metabolism* 11, no. 1 (January 6, 2010): 35–46, https://www.ncbi.nlm.nih.gov/pmc/articles/PMC2824086/.

5. Among Kagawa's findings on the impact of Western diets on the Japanese were significant increases in colon and lung cancer and decreases in stomach and uterine cancers, although the subjects' food consumption was still much smaller than that of Americans or Europeans. When he looked at the residents of Okinawa, they had "the lowest total energy, sugar and salt, and the smallest physique, but had healthy longevity and the highest centenarian rate." Y. Kagawa, "Impact of Westernization on the Nutrition of Japanese: Changes in Physique, Cancer, Longevity and Centenarians," *Preventive Medicine* 7, no. 2 (June 1978): 205–17, https://www.sciencedi rect.com/science/article/pii/0091743578902463.

6. Two of the authors of the report were themselves part of the crew who elected to be locked up inside the Biosphere for two years and live on a low-calorie diet, with just 12 percent protein and 11 percent fat in terms of calorie consumption. Despite this calorie restriction and a 17±5 percent weight loss, all eight crew members were healthy and highly active during the two-year period. R. L. Walford, D. Mock, R. Verdery, and T. MacCallum, "Calorie Restriction in Biosphere 2: Alterations

in Physiologic, Hematologic, Hormonal, and Biochemical Parameters in Humans Restricted for a 2-Year Period," *Journals of Gerontology, Series A: Biological Sciences and Medical Sciences* 57, no. 6 (June 2002): 211–24, https://www.ncbi.nlm.nih.gov /pubmed/12023257.

7. L. K. Heilbronn, and E. Ravussin, "Calorie Restriction and Aging: Review of the Literature and Implications for Studies in Humans," *American Journal of Clinical Nutrition* 3, no. 178 (September 2003): 361–69, https://academic.oup.com/ajcn /article/78/3/361/4689958.

8. The authors used the results of a publicly accessible, 24-month trial run by the National Institute on Aging of calorie restriction in nonobese youth. D. W. Belsky, K. M. Huffman, C. F. Pieper, et al., "Change in the Rate of Biological Aging in Response to Caloric Restriction: CALERIE Biobank Analysis," *Journals of Gerontology, Series A: Biological Sciences and Medical Sciences* 73, no. 1 (January 2018): 4–10, https://academic.oup.com/biomedgerontology/article/73/1/4/3834057.

9. McGlothin wrote in an article, "I am delighted that a 70-year-old can have biomarkers that are like those of a healthy school-age child." P. McGlothin, "Growing Older and Healthier the CR Way®," *Life Extension Magazine*, September 2018, https://www.lifeextension.com/Magazine/2018/9/Calorie-Restriction-Update /Page-01.

10. The authors are in no doubt of the potential benefits calorie restriction offers humans in terms of addressing diseases and aging. "A clear understanding of the biology of ageing, as opposed to the biology of individual age-related diseases, could be the critical turning point for novel approaches in preventative strategies to facilitate healthy human ageing," they wrote. "Caloric restriction (CR) offers a powerful paradigm to uncover the cellular and molecular basis for the age-related increase in overall disease vulnerability that is shared by all mammalian species." J. A. Mattison, R. J. Colman, T. M. Beasley, et al., "Caloric Restriction Improves Health and Survival of Rhesus Monkeys," *Nature Communications*, January 17, 2017, https:// www.nature.com/articles/ncomms14063.

11. Y. Zhang, A. Bokov, J. Gelfond, et al., "Rapamycin Extends Life and Health in C57BL/6 Mice," *Journals of Gerontology, Series A: Biological Sciences and Medical Sciences* 69, no. 2 (February 2014): 119–30, https://www.ncbi.nlm.nih.gov /pubmed/23682161.

12. "We really study this as a paradigm to understand aging," she told *Scientific American* in 2017. "We're not recommending people do it." R. Conniff, "The Hunger Gains: Extreme Calorie-Restriction Diet Shows Anti-aging Results," *Scientific American*, February 16, 2017, https://www.scientificamerican.com/article/the -hunger-gains-extreme-calorie-restriction-diet-shows-anti-aging-results/.

13. "The optimum amount of fasting appeared to be fasting 1 day in 3 and this increased the life span of littermate males about 20% and littermate females about 15%." A. J. Carlson and F. Hoelzel, "Apparent Prolongation of the Life Span of

Rats by Intermittent Fasting: One Figure," *Journal of Nutrition* 31, no. 3 (March 1, 1946): 363–75, https://academic.oup.com/jn/article-abstract/31/3/363/4725632 ?redirectedFrom=fulltext.

14. H. M. Shelton, "The Science and Fine Art of Fasting," in *The Hygienic System*, vol. III, *Fasting and Sunbathing* (San Antonio, Texas: Dr. Shelton's Health School, 1934).

15. C. Tazearslan, J. Huang, N. Barzilai, and Y. Suh, "Impaired IGF1R Signaling in Cells Expressing Longevity-Associated Human IGF1R Alleles," *Aging Cell* 10, no. 3 (June 2011): 551–54, https://onlinelibrary.wiley.com/doi/full/10.1111/j.1474 -9726.2011.00697.

16. One in three Ikarians reaches the age of 90, and most do so free of dementia and many other chronic diseases of aging. "Ikaria, Greece. The Island where People For- get to Die," Blue Zones, https://www.bluezones.com/exploration/ikaria-greece/.

17. The fasting extends to 180 days in the year and requires abstinence from primarily dairy products and red-blooded animals and fish, which means that octopus and squid can still be eaten. In the run-up to Holy Communion, fasting encompasses all food. N. Gaifyllia, "Greek Orthodox 2018 Calendar of Holidays and Fasts," The Spruce Eats, October 6, 2018, https://www.thespruceeats.com/greek-orthodox-cal endar-1706215.

18. Bapan has been widely ignored by Western researchers, mostly because people in that part of southern China—a region long reputed to have large populations of very healthy centenarians—don't have formal birth records. The cardiologist John Day and his colleagues, however, have argued that there is good reason to believe their claims. J. D. Day, J. A. Day, and M. LaPlante, *The Longevity Plan: Seven Life- Transforming Lessons from Ancient China* (New York: HarperCollins, 2017).

19. Avoiding animal protein is not easy. One of the main reasons is that protein con- sumption produces satiety. No one has done more to understand why eating carbo- hydrates doesn't stave off hunger than Stephen Simpson, the director of the Charles Perkins Centre in Sydney, Australia. Simpson started his career trying to understand why locusts swarm. If he could figure that out, he felt, perhaps he could prevent the global loss of millions of tons of crops each year. What he discovered was that locusts seek protein. They crave it. They march along consuming anything edible in their path, but if there's not enough protein in their diet, they transform into raven- ous, hungry creatures that seek protein from any possible source. And the closest source of protein is the locust in front of it. Under these conditions, the best way to stay alive is to keep moving forward, occasionally pausing to snack on a slower relative. Simpson's latest work is fascinating: it shows that this same trigger exists in the mammalian brain. When we lack protein, we also turn ravenous, and although we don't normally try to eat our neighbors, in the throes of extreme hunger, who hasn't considered it? What this all tells us is that it's best not to eat a lot of animal protein, but it's hard to avoid it altogether. F. P. Zanotto, D. Raubenheimer, and

S. J. Simpson, "Selective Egestion of Lysine by Locusts Fed Nutritionally Unbalanced Foods," *Journal of Insect Physiology* 40, no. 3 (March 1994): 259–65, https://www.sciencedirect.com/science/article/pii/0022191094900493.

20. Though it seems the occasional hot dog or hamburger is acceptable, a review of 800 studies by twenty-two experts found that a daily diet that included 50 grams of processed meat appeared to increase subjects' chances of developing colorectal cancer by 18 percent. S. Simon, "World Health Organization Says Processed Meat Causes Cancer," American Cancer Society, October 26, 2015, https://www.cancer.org/latest-news/world-health-organization-says-processed-meat-causes-cancer.html.

21. With processed, calorie-rich food largely absent from their diet and a lifestyle dominated by physical activity, there's little in the way of obesity or cardiovascular disease to be found in hunter-gatherer communities. H. Pontzer, B. M. Wood, and D. A. Raichlen, "Hunter-Gatherers as Models in Public Health," *Obesity Reviews* 19, suppl. 1 (December 2018): 24–35, https://onlinelibrary.wiley.com/doi/full/10.1111/obr.12785.

22. M. Song, T. T. Fung, F. B. Hu, et al., "Association of Animal and Plant Protein Intake with All-Cause and Cause-Specific Mortality," *JAMA Internal Medicine* 176, no. 10 (October 1, 2016): 1453–63, https://jamanetwork.com/journals/jamainternalmedicine/fullarticle/2540540.

23. A 2011 study identified a new signaling pathway used by amino acids to activate mTOR. I. Tato, R. Bartrons, F. Ventura, and J. L. Rosa, "Amino Acids Activate Mammalian Target of Rapamycin Complex 2 (mTORC2) via PI3K/Akt Signaling," *Journal of Biological Chemistry* 286, no. 8 (February 25, 2011): 6128–42, http://www.jbc.org/content/286/8/6128.full.

24. C. Hine, C. Mitchell, and J. R. Mitchell, "Calorie Restriction and Methionine Restriction in Control of Endogenous Hydrogen Sulfide Production by the Transsulfuration Pathway," *Experimental Gerontology* 68 (August 2015): 26–32, https://www.ncbi.nlm.nih.gov/pubmed/25523462.

25. Rather than caloric restriction, the researchers at the Lamming Lab devised a short-term methionine deprivation regimen that reduced fat mass, restored normal body weight, and reinstituted glycemic control to male and female mice alike. D. Yu, S. E. Yang, B. R. Miller, et al., "Short-Term Methionine Deprivation Improves Metabolic Health via Sexually Dimorphic, mTORC1-Independent Mechanisms," *FASEB Journal* 32, no. 6 (June 2018): 3471–82, https://www.ncbi.nlm.nih.gov/pubmed/29401631.

26. The eternal quest for a well-balanced diet, the answer, the authors suggest, may have to do with how "longevity can be extended in ad libitum–fed animals by manipulating the ratio of macronutrients to inhibit mTOR activation." S. M. Solon-Biet, A. C. McMahon, J. W. Ballard, et al., "The Ratio of Macronutrients, Not Caloric Intake, Dictates Cardiometabolic Health, Aging, and Longevity in Ad

Libitum–Fed Mice," *Cell Metabolism* 3, no. 19 (March 4, 2014): 418–30, https://www.ncbi.nlm.nih.gov/pmc/articles/PMC5087279/.

27. In other words, the specific amino acid composition of a person's diet may be more important than limiting all aminos. The easiest way to do this, though, is still to reduce meat intake. L. Fontana, N. E. Cummings, S. I. Arriola Apelo, et al., "Decreased Consumption of Branched Chain Amino Acids Improves Metabolic Health," *Cell Reports* 16, no. 2 (July 12, 2016): 520–30, https://www.ncbi.nlm.nih.gov/pmc/articles/PMC4947548/.

28. Some have suggested that a better understanding of this connection could help researchers develop mTOR-targeted therapies to prevent muscle wasting. M.-S. Yoon, "mTOR as a Key Regulator in Maintaining Skeletal Muscle Mass," *Frontiers in Physiology* 8 (2017): (October 17, 2017): 788, https://www.ncbi.nlm.nih.gov/pmc/articles/PMC5650960/.

29. Just cutting one's consumption of branched-chain amino acids for one day rapidly improves insulin sensitivity. F. Xiao, J. Yu, Y. Guo, et al., "Effects of Individual Branched-Chain Amino Acids Ceprivation on Insulin Sensitivity and Glucose Metabolism in Mice," *Metabolism* 63, no. 6 (June 2014): 841–50, https://www.ncbi.nlm.nih.gov/pubmed/24684822/.

30. There are certainly other lifestyle factors at play. But a meta-analysis of seven studies including nearly 125,000 participants, published in 2012 in *Annals of Nutrition and Metabolism*, is compelling evidence. Among vegetarians, the researchers who conducted the study observed a 16 percent lower mortality from circulatory diseases and a 12 percent lower mortality from cerebrovascular disease. T. Huang, B. Yang, J. Zheng, et al., "Cardiovascular Disease Mortality and Cancer Incidence in Vegetarians: A Meta-analysis and Systematic Review," *Annals of Nutrition & Metabolism* 4, no. 60 (June 1, 2012): 233–40, https://www.karger.com/Article/FullText/337301.

31. The study looked at nearly 6,000 men and women enrolled in the National Health and Nutrition Examination Survey. If you want a reminder of how little a sedentary life does for prolonging existence, the following jumps out of the report: "Adults with High activity were estimated to have a biologic aging advantage of 9 years (140 base pairs ÷ 15.6) over Sedentary adults. The difference in cell aging between those with High and Low activity was also significant, 8.8 years, as was the difference between those with High and Moderate PA (7.1 years)." L. A. Tucker, "Physical Activity and Telomere Length in U.S. Men and Women: An NHANES Investigation," *Preventive Medicine* 100 (July 2017): 145–51, https://www.sciencedirect.com/science/article/pii/S0091743517301470.

32. Intrigued by the potential insights into aging afforded by the health and physicality of middle-aged, regular bike riders, British scientists, among whose ranks were recreational athletes, looked at how exercise might influence longevity. They recruited older male and female cyclists between 55 and 79 for their study and contrasted

them to older and younger sedentary people. "The cyclists proved to have reflexes, memories, balance and metabolic profiles that more closely resembled those of 30-year-olds than of the sedentary older group." G. Reynolds, "How Exercise Can Keep Aging Muscles and Immune Systems 'Young,'" *New York Times*, March 14, 2018, https://www.nytimes.com/2018/03/14/well/move/how-exercise-can-keep -aging-muscles-and-immune-systems-young.html.

33. D. Lee, R. R. Pate, C. J. Lavie, et al., "Leisure-Time Running Reduces All-Cause and Cardiovascular Mortality Risk," *Journal of the American College of Cardiology* 54, no. 5 (August 2014): 472–81, http://www.onlinejacc.org/content/64/5/472.

34. The authors show how cardiorespiratory fitness algorithms can identify those at risk of cardiovascular disease and also potentially help to develop appropriate exercise regimes depending on an individual's initial fitness level. E. G. Artero, A. S. Jackson, X. Sui, et al., "Longitudinal Algorithms to Estimate Cardiorespiratory Fitness: Associations with Nonfatal Cardiovascular Disease and Disease-Specific Mortality," *Journal of the American College of Cardiology* 63, no. 21 (June 3, 2014): 2289–96, https://www.sciencedirect.com/science/article/pii/S0735109714016301 ?via%3Dihub.

35. T. S. Church, C. P. Earnest, J. S. Skinner, and S. N. Blair, "Effects of Different Doses of Physical Activity on Cardiorespiratory Fitness Among Sedentary, Overweight or Obese Postmenopausal Women with Elevated Blood Pressure: A Randomized Controlled Trial," *Journal of the American Medical Association* 297, no. 19 (May 16, 2007): 2081–91, https://jamanetwork.com/journals/jama/fullarticle/110 8370.

36. M. M. Robinson, S. Dasari, A. R. Konopka, et al., "Enhanced Protein Translation Underlies Improved Metabolic and Physical Adaptations to Different Exercise Training Modes in Young and Old Humans," *Cell Metabolism* 25, no. 3 (March 7, 2017): 581–92, https://www.cell.com/cell-metabolism/comments/S1550-4131 (17)30099-2.

37. Sage recommendations from the Mayo Clinic include dedicating 150 minutes a week to activities such as swimming or mowing the lawn or doing 75 minutes of more demanding exercise, such as spinning or running. "Be realistic and don't push yourself too hard, too fast," the clinic staff wrote. "Fitness is a lifetime commitment, not a sprint to a finish line," "Exercise Intensity: How to Measure It," Mayo Clinic, June 12, 2018, https://www.mayoclinic.org/healthy-lifestyle/fitness /in-depth/exercise-intensity/art-20046887.

38. Investigating how the hypothalamus potentially controls aspects of aging, the authors found that "immune inhibition or GnRH restoration in the hypothalamus /brain" offers two possible directions for extending lifespan and fighting health issues that come with aging. G. Zhang, J. Li, S. Purkayasatha, et al., "Hypothalamic Programming of Systemic Ageing Involving IKK-β, NF-κB and GnRH," *Nature* 497, no. 7448 (May 9, 2013): 211–16, https://www.nature.com/articles/nature12143.

39. The team couldn't say why this happened, only that it happened. Back then they theorized that lowering the mice's body temperature might slow down metabolism and thus reduce the notorious free radicals. We've learned a lot since then. B. Conti, M. Sanchez-Alvarez, R. Winskey-Sommerer, et al., "Transgenic Mice with a Reduced Core Body Temperature Have an Increased Life Span," *Science* 314, no. 5800 (November 3, 2006): 825–28, https://www.ncbi.nlm.nih.gov/pubmed/17082459.

40. The mice suffered from increased rates of obesity, beta cell dysfunction, and type 2 diabetes. C.- Y. Zhang, G. Baffy, P. Perret, et al., "Uncoupling Protein-2 Negatively Regulates Insulin Secretion and Is a Major Link Between Obesity, β Cell Dysfunction, and Type 2 Diabetes," *Cell* 105, no. 6 (June 15, 2001): 745–55, https://www.sciencedirect.com/science/article/pii/S0092867401003786.

41. The researchers also believed this occurred because of a reduction in oxidative damage. Y.-W. C. Fridell, A. Sánchez-Blanco, B. A. Silvia, et al., "Targeted Expression of the Human Uncoupling Protein 2 (hUCP2) to Adult Neurons Extends Life Span in the Fly," *Cell Metabolism* 1, no. 2 (February 2005): 145–52, https://www.sciencedirect.com/science/article/pii/S155041310500032X.

42. The researchers concluded that UCP2 regulates brown adipose tissue thermogenesis through nonesterified fatty acids. A. Caron, S. M. Labbé, S. Carter, et al., "Loss of UCP2 Impairs Cold-Induced Non-shivering Thermogenesis by Promoting a Shift Toward Glucose Utilization in Brown Adipose Tissue," *Biochimie* 134 (March 2007): 118–26, https://www.sciencedirect.com/science/article/pii/S030090841630270X?via%3Dihub.

43. The researchers, led by Justin Darcy at the University of Alabama, demonstrated enhanced brown adipose tissue function in animals that lived 40 to 60 percent longer than their littermates. J. Darcy, M. McFadden, Y. Fang, et al., "Brown Adipose Tissue Function Is Enhanced in Long-Lived, Male Ames Dwarf Mice," *Endocrinology* 157, no. 12 (December 1, 2016): 4744–53, https://academic.oup.com/endo/article/157/12/4744/2758430.

44. "How brown fat is regulated in humans and how it relates to metabolism, though, remain unclear," the authors of a study wrote in 2014. Since then the mechanism has become clearer. Endocrine Society, "Cold Exposure Stimulates Beneficial Brown Fat Growth," *Science Daily*, June 23, 2014, https://www.sciencedaily.com/releases/2014/06/140623091949.htm.

45. T. Shi, F. Wang, E. Stieren, and Q. Tong, "SIRT3, a Mitochondrial Sirtuin Deacetylase, Regulates Mitochondrial Function and Thermogenesis in Brown Adipocytes," *Journal of Biological Chemistry* 280, no. 14 (April 8, 2005): 13560-67, http://www.jbc.org/content/280/14/13560.long.

46. A. S. Warthin, "A Fatal Case of Toxic Jaundice Caused by Dinitrophenol," *Bulletin of the International Association of Medical Museums* 7 (1918): 123–26.

47. W. C. Cutting, H. G. Mertrens, and M. L. Tainter, "Actions and Uses of Dinitrophenol: Promising Metabolic Applications," *Journal of the American Medical*

Association 101, no. 3 (July 15, 1933): 193–95, https://jamanetwork.com/journals/jama/article-abstract/244026.

48. The authors calculated that with 1.2 million capsules supplied by the Stanford Clinics in 1934, that corresponded to 4,500 patients taking the drug over a three-month period. Overall, they estimated that in the United States, at least 100,000 people had been treated with the drug. M. L. Tainter, W. C. Cutting, and A. B. Stockton, "Use of Dinitrophenol in Nutritional Disorders: A Critical Survey of Clinical Results," *American Journal of Public Health* 24, no. 10 (1935): 1045–53, https://ajph.aphapublications.org/doi/pdf/10.2105/AJPH.24.10.1045.

49. Dinitrophenol has a variety of names on the internet. The authors list, along with DNP, "'Dinosan,' 'Dnoc,' 'Solfo Black,' 'Nitrophen,' 'Alidfen,' and 'Chemox.'" In the 2000s, there was a spike in DNP-related deaths as it was marketed online to bodybuilders and the weight conscious. J. Grundlingh, P. I. Dargan, M. El-Zanfaly, and D. M. Wood, "2,4-Dinitrophenol (DNP): A Weight Loss Agent with Significant Acute Toxicity and Risk of Death," *Journal of Medical Toxicology* 7, no. 3 (September 2011): 205–12, https://www.ncbi.nlm.nih.gov/pmc/articles/PMC3550200/.

50. T. L. Kurt, R. Anderson, C. Petty, et al., "Dinitrophenol in Weight Loss: The Poison Center and Public Health Safety," *Veterinary and Human Toxicology* 28, no. 6 (December 1986): 574–75, https://www.ncbi.nlm.nih.gov/pubmed/3788046.

51. A horrifying death from an overdose of DNP is described in a story on Vice; see G. Haynes, "The Killer Weight Loss Drug DNP Is Still Claiming Young Lives," Vice, August 6, 2018, https://www.vice.com/en_uk/article/bjbyw5/the-killer-weight-loss-drug-dnp-is-still-claiming-young-lives; see also Grundlingh et al., "2,4-Dinitrophenol (DNP)."

52. This happens differently from species to species, but the general trend is clear: cold and exercise together build brown fat. F. J. May, L. A. Baer, A. C. Lehnig, et al., "Lipidomic Adaptations in White and Brown Adipose Tissue in Response to Exercise Demonstrates Molecular Species-Specific Remodeling," *Cell Reports* 18, no. 6 (February 7, 2017): 1558–72, https://www.ncbi.nlm.nih.gov/pmc/articles/PMC5558157/.

53. "Until further research is available," an international team of researchers concluded in 2014, "athletes should remain cognizant that less expensive modes of cryotherapy, such as local ice-pack application or cold-water immersion, offer comparable physiological and clinical effects." C. M. Bleakley, F. Bieuzen, G. W. Davison, and J. T. Costello, "Whole-Body Cryotherapy: Empirical Evidence and Theoretical Perspectives," *Open Access Journal of Sports Medicine* 5 (March 10, 2014): 25–36, https://www.ncbi.nlm.nih.gov/pmc/articles/PMC3956737/.

54. The average time spent in the sauna was 15 minutes at 80°C. T. E. Strandberg, A. Strandberg, K. Pitkälä, and A. Benetos, "Sauna Bathing, Health, and Quality of Life Among Octogenarian Men: The Helsinki Businessmen Study," *Aging*

Clinical and Experimental Research 30, no. 9 (September 2018): 1053–57, https://www.ncbi.nlm.nih.gov/pubmed/29188579.

55. T. Laukkanen, H. Khan, F. Zaccardi, and J. A. Laukkanen, "Association Between Sauna Bathing and Fatal Cardiovascular and All-Cause Mortality Events," *JAMA Internal Medicine* 175, no. 4 (April 2015): 542–48, https://www.ncbi.nlm.nih.gov /pubmed/25705824.

56. H. Yang, T. Yang, J. A. Baur, et al., "Nutrient-Sensitive Mitochondrial NAD+ Levels Dictate Cell Survival," *Cell* 130, no. 6 (September 21, 2007): 1095–107, https://www.ncbi.nlm.nih.gov/pmc/articles/PMC3366687/.

57. R. Madabhushi, F. Gao, A. R. Pfenning, et al., "Activity-Induced DNA Breaks Govern the Expression of Neuronal Early-Response Genes," *Cell* 161, no. 7 (June 18, 2015): 1592–605, https://www.ncbi.nlm.nih.gov/pmc/articles/PMC4886855/.

58. H. Katoka, "Quantitation of Amino Acids and Amines by Chromatography," *Journal of Chromatography Library* 70 (2005): 364–404, https://www.sciencedirect.com /topics/chemistry/aromatic-amine.

59. Another highly prevalent chemical used in plastic bottles and food and drink cans is bisphenol A, or BPA. It's so ubiquitous that it can be found in the urine of nearly every American; in high quantities it has been linked to "cardiovascular disease and diabetes and may be associated with an increased risk for miscarriages with abnormal embryonic karyotype." P. Allard and M. P. Colaiácovo, "Bisphenol A Impairs the Double-Strand Break Repair Machinery in the Germline and Causes Chromosome Abnormalities," *Proceedings of the National Academy of Sciences of the United States of America* 107, no. 47 (November 23, 2010): 20405–10, http://www.pnas .org/content/107/47/20405.

60. "Our findings suggest that this colorant could cause harmful effects to humans if it is metabolized or absorbed through the skin." F. M. Chequer, V. de Paula Venâncio, et al., "The Cosmetic Dye Quinoline Yellow Causes DNA Damage in Vitro," *Mutation Research/Genetic Toxicology and Environmental Mutagenesis* 777 (January 1, 2015): 54–61, https://www.ncbi.nlm.nih.gov/pubmed/25726175.

61. Beer drinkers take note: "Beer is one source of NDMA, in which as much as 70 micrograms l(-1) has been reported in some types of German beer, although usual levels are much lower (10 or 5 micrograms l(-1)); this could mean a considerable intake for a heavy beer drinker of several liters per day." The good news, the writer adds, is that in recent decades there's been not only a reduction in the level of nitrates in food but also "greater control of exposure of malt to nitrogen oxides in beer making." W. Lijinsky, "N-Nitroso Compounds in the Diet," *Mutation Research* 443, nos. 1–2 (July 15, 1999): 129–38, https://www.ncbi.nlm.nih.gov/pubmed/10415436.

62. L. Robbiano, E. Mereto, C. Corbu, and G. Brambilla, "DNA Damage Induced by Seven N-nitroso Compounds in Primary Cultures of Human and Rat Kidney Cells," *Mutation Research* 368, no. 1 (May 1996): 41–47, https://www.ncbi.nlm .nih.gov/pubmed/8637509.

63. The state of Massachusetts did a study in 1988 to get to grips with the prevalence of radon by county. It found that one in four houses apparently had levels over the EPA-identified level of 4pCi/L, which requires additional investigation. "Public Health Fact Sheet on Radon," Health and Human Services, Commonwealth of Massachusetts, 2011, http://web.archive.org/web/20111121032816/http://www .mass.gov/eohhs/consumer/community-health/environmental-health/exposure -topics/radiation/radon/public-health-fact-sheet-on-radon.html.

64. "Most of the mercury that contaminates fish comes from household and industrial waste that is incinerated or released during the burning of coal and other fossil fuels. Products containing mercury that are improperly thrown in the garbage or washed down drains end up in landfills, incinerators, or sewage treatment facilities." "Contaminants in Fish," Washington State Department of Health, https:// www.doh.wa.gov/CommunityandEnvironment/Food/Fish/ContaminantsinFish.

65. S. Horvath, "DNA Methylation Age of Human Tissues and Cell Types," *Genome Biology* 14, no. 10 (2013): R115, https://www.ncbi.nlm.nih.gov/pubmed /24138928.

FIVE. A BETTER PILL TO SWALLOW

1. If Schrödinger couldn't exactly answer the question of what life is, his book arguably did everything else but that. It's credited with being a key influencer of the development of scientific thought in the twentieth century and helped lay the groundwork for the emergence of molecular biology and the discovery of DNA. E. Schrödinger, *What Is Life? The Physical Aspect of the Living Cell* (Cambridge, UK: Cambridge University Press, 1944).

2. V. L. Schramm and S. D. Schwartz, "Promoting Vibrations and the Function of Enzymes. Emerging Theoretical and Experimental Convergence," *Biochemistry* 57, no. 24 (June 19, 2018): 3299–308, https://www.ncbi.nlm.nih.gov/pubmed /29608286.

3. "Cell Size and Scale," Genetic Science Learning Center, University of Utah, http:// learn.genetics.utah.edu/content/cells/scale/.

4. The macromolecular biological catalysts whose names end in -*ase* are enzymes.

5. Out of so many eminent quotes, this is one that scientists hold as wisdom for the ages: "The first principle is that you must not fool yourself—and you are the easiest person to fool." R. P. Feynman, *The Quotable Feynman*, ed. Michelle Feynman (Princeton, NJ: Princeton University Press, 2015), 127.

6. After Sehgal's employer was bought by the international health care company Wyeth, he resumed his work on rapamycin. "In 1999, the U.S. Food and Drug Administration approved rapamycin as a drug for transplant patients. Sehgal died a few years after the FDA approval, too soon to see his brainchild save the lives of thousands of transplant patients and go on to make Wyeth hundreds of millions of dollars." B. Gifford, "Does a Real Anti-aging Pill Already Exist?," Bloomberg,

February 12, 2015, https://www.bloomberg.com/news/features/2015-02-12/does
-a-real-anti-aging-pill-already-exist-.

7. The authors concluded that "up-regulation of a highly conserved response to
starvation-induced stress is important for life span extension by decreased TOR
signaling in yeast and higher eukaryotes." R. W. Powers III, M. Kaeberlein,
S. D. Caldwell, et al., "Extension of Chronological Life Span in Yeast by Decreased
TOR Pathway Signaling," *Genes & Development* 20, no. 2 (January 15, 2006):
174–84, https://www.ncbi.nlm.nih.gov/pmc/articles/PMC1356109/.

8. I. Bjedov, J. M. Toivonen, F. Kerr, et al., "Mechanisms of Life Span Extension by Rapamycin in the Fruit Fly *Drosophilia melanogaster*," *Cell Metabolism* 11, no. 1 (January 6, 2010): 35–46, https://www.ncbi.nlm.nih.gov/pmc/articles/PMC2824086/.

9. The authors noted that these were the first results to show that mTOR could play
a role in extending life: "Rapamycin may extend lifespan by postponing death from
cancer, by retarding mechanisms of ageing, or both." D. E. Harrison, R. Strong,
Z. D. Sharp, et al., "Rapamycin Fed Late in Life Extends Lifespan in Genetically
Heterogeneous Mice," *Nature* 460 (July 8, 2009): 392–95, https://www.nature
.com/articles/nature08221.

10. K. Xie, D. P. Ryan, B. L. Pearson, et al., "Epigenetic Alterations in Longevity Regulators, Reduced Life Span, and Exacerbated Aging-Related Pathology in Old Father
Offspring Mice," *Proceedings of the National Academy of Sciences of the United States
of America* 115, no. 10 (March 6, 2018): E2348–57, https://www.pnas.org/content
/115/10/E2348.

11. How do they pick so many winners? According to a press release, a Thomson
Reuters' executive explained it thus: "Highly-cited papers turn out to be one of
the most reliable indicators of world-class research, and provide a glimpse at what
research stands the best chance at being recognized with a Nobel Prize." Thomson Reuters, "Web of Science Predicts 2016 Nobel Prize Winners," PR Newswire,
September 21, 2016, https://www.prnewswire.com/news-releases/web-of-science
-predicts-2016-nobel-prize-winners-300331557.html.

12. In this case, the authors showed that 3 months of rapamycin increased middle-aged
mice's life expectancy by 60 percent as well as improving their healthspan. A. Bitto,
K. I. Takashi, V. V. Pineda, et al., "Transient Rapamycin Treatment Can Increase
Lifespan and Healthspan in Middle-Aged Mice," *eLife* 5 (August 23, 2016): 5,
https://www.ncbi.nlm.nih.gov/pmc/articles/PMC4996648/.

13. Low-level doses of a drug called everolimus were given to people over 65. Their
response to flu vaccines improved by around 20 percent. A. Regalado, "Is This
the Anti-aging Pill We've All Been Waiting For?," *MIT Technology Review*, March
28, 2017, https://www.technologyreview.com/s/603997/is-this-the-anti-aging-pill
-weve-all-been-waiting-for/.

14. Metformin, given to patients with diabetes, was particularly promising, two researchers noted. "While there are caveats with any study of this nature, the findings

suggest that metformin may be affecting basic aging processes that underlie multiple chronic diseases and not just type II diabetes." B. K. Kennedy, and J. K. Pennypacker, "Aging Interventions Get Human," *Oncotarget* 6, no. 2 (January 2015): 590–91, https://www.ncbi.nlm.nih.gov/pmc/articles/PMC4359240/.

15. C. J. Bailey, "Metformin: Historical Overview," *Diabetologia* 60 (2017): 1566–76, https://link.springer.com/content/pdf/10.1007%2Fs00125-017-4318-z.pdf.

16. Patients who took metformin displayed lower rates of mortality not only compared to diabetics but also compared to nondiabetics, the researchers found. Other results included less cancer and less cardiovascular disease in those being treated with metformin. J. M. Campbell, S. M. Bellman, M. D. Stephenson, and K. Lisy, "Metformin Reduces All-Cause Mortality and Diseases of Ageing Independent of Its Effect on Diabetes Control: A Systematic Review and Meta-analysis," *Ageing Research Reviews* 40 (November 2017): 31–44, https://www.sciencedirect.com/science/article/pii/S1568163717301472.

17. R. A. DeFronzo, N. Barzilai, and D. C. Simonson, "Mechanism of Metformin Action in Obese and Lean Noninsulin-Dependent Diabetic Subjects," *Journal of Clinical Endocrinology & Metabolism* 73, no. 6 (December 1991): 1294–301, https://www.ncbi.nlm.nih.gov/pubmed/1955512.

18. A. Martin-Montalvo, E. M. Mercken, S. J. Mitchell, et al., "Metformin Improves Healthspan and Lifespan in Mice," *Nature Communications* 4 (2013): 2192, https://www.ncbi.nlm.nih.gov/pmc/articles/PMC3736576/.

19. V. N. Anisimov, "Metformin for Aging and Cancer Prevention," *Aging* 2, no. 11 (November 2010): 760–74.

20. S. Andrzejewski, S.-P. Gravel, M. Pollak, and J. St-Pierre, "Metformin Directly Acts on Mitochondria to Alter Cellular Bioenergetics," *Cancer & Metabolism* 2 (August 28, 2014): 12, https://www.ncbi.nlm.nih.gov/pmc/articles/PMC4147388/.

21. N. Barzilai, J. P. Crandall, S. P. Kritchevsky, and M. A. Espeland, "Metformin as a Tool to Target Aging," *Cell Metabolism* 23 (June 14, 2016): 1060–65, https://www.cell.com/cell-metabolism/pdf/S1550-4131(16)30229-7.pdf.

22. C.-P. Wang, C. Lorenzo, S. L. Habib, et al. "Differential Effects of Metformin on Age Related Comorbidities in Older Men with Type 2 Diabetes," *Journal of Diabetes and Its Complications* 31, no. 4 (2017): 679–86, https://www.ncbi.nlm.nih.gov/pmc/articles/PMC5654524/.

23. J. M. Campbell, S. M. Bellman, M. D. Stephenson, and K. Lisy, "Metformin Reduces All-Cause Mortality and Diseases of Ageing Independent of Its Effect on Diabetes Control: A Systematic Review and Meta-analysis," *Ageing Research Reviews* 40 (November 2017): 31–44, https://www.ncbi.nlm.nih.gov/pubmed/28802803.

24. N. Howlader, A. M. Noone, M. Krapcho, et al., "SEER Cancer Statistics Review, 1975–2009," National Cancer Institute, August 20, 2012, https://seer.cancer.gov/archive/csr/1975_2009_pops09/.

25. By the time you hit 90, the authors found, there's a threefold decrease in the probability of developing cancer. If you make it to 100, from there the probability is minimal, 0 to 4 percent. N. Pavlidis, G. Stanta, and R. A. Audisio, "Cancer Prevalence and Mortality in Centenarians: A Systematic Review," *Critical Reviews in Oncology/Hematology* 83, no. 1 (July 2012): 145–52, https://www.ncbi.nlm.nih.gov /pubmed/22024388.

26. I. Elbere, I. Silamikelis, M. Ustinova, et al., "Significantly Altered Peripheral Blood Cell DNA Methylation Profile as a Result of Immediate Effect of Metformin Use in Healthy Individuals," *Clinical Epigenetics* 10, no. 1 (2018), https://doi.org/10.1186 /s13148-018-0593-x.

27. B. K. Kennedy, M. Gotta, D. A. Sinclair, et al., "Redistribution of Silencing Proteins from Telomeres to the Nucleolus Is Associated with Extension of Lifespan in *S. cerevisiae,*" *Cell* 89, no. 3 (May 2, 1997): 381–91, https://www.ncbi.nlm.nih .gov/pubmed/?term=SIR4-42+sinclair+gotta; D. A. Sinclair and L. Guarente, "Extrachromosomal rDNA Circles—A Cause of Aging in Yeast," *Cell* 91, no. 7 (December 26, 1997): 1033–42, https://www.ncbi.nlm.nih.gov/pubmed/9428525; D. Sinclair, K. Mills, and L. Guarente, "Accelerated Aging and Nucleolar Fragmentation in Yeast *SGS1* Mutants," *Science* 277, no. 5330 (August 29, 1997): 1313–16, https://www.ncbi.nlm.nih.gov/pubmed/9271578.

28. The research into resveratrol suggests that it is promising for both cancer and cardiovascular disease prevention. Resveratrol's ability to act on tumor growth points to other possibilities. "Since tumor promoting agents alter the expression of genes whose products are associated with inflammation, chemoprevention of cardiovascular diseases and cancer may share the same common mechanisms." E. Ignatowicz and W. Baer-Dubowska, "Resveratrol, a Natural Chemopreventive Agent Against Degenerative Diseases, "*Polish Journal of Pharmacology* 53, no. 6 (November 2001): 557–69, https://www.ncbi.nlm.nih.gov/pubmed/11985329.

29. The title of our paper is a combination of two Greek words: "*xenos,* the Greek word for stranger, and *hormesis,* the term for health benefits provided by mild biological stress, such as cellular damage or a lack of nutrition." K. T. Howitz and D. A. Sinclair, "Xenohormesis: Sensing the Chemical Cues of Other Species," *Cell* 133, no. 3 (May 2, 2008): 387–91, https://www.ncbi.nlm.nih.gov/pmc/articles /PMC2504011/.

30. An average glass of red wine contains about 1 to 3 mg of resveratrol. There is no resveratrol in white wine because resveratrol is produced largely by the skins of the grape, which are not used in white wine production. For more information on and dietary sources of resveratrol, see J. A. Baur and D. A. Sinclair, "Therapeutic Potential of Resveratrol: The *in Vivo* Evidence," *Nature Reviews Drug Discovery* 5, no. 6 (June 2006): 493–506, https://www.ncbi.nlm.nih.gov/pubmed/16732220.

31. Continuing on from our work, the researchers proposed "a novel pathway by which products of the plant stress response confer stress tolerance and extend longevity

in animals." They also highlighted how xenohormesis may boost the health-giving and medicinal properties of plants, while also tackling issues around adapting in a world that's forever changing. P. L. Hooper, P. L. Hooper, M. Tytell, and L. Vigh, "Xenohormesis: Health Benefits from an Eon of Plant Stress Response Evolution," *Cell Stress & Chaperones* 15, no. 6 (November 2010): 761–70, https://www.ncbi .nlm.nih.gov/pmc/articles/PMC3024065/.

32. The implications for overweight humans were clear, we found. "This study shows that an orally available small molecule at doses achievable in humans can safely reduce many of the negative consequences of excess caloric intake, with an overall improvement in health and survival." J. A. Baur, K. J. Pearson, N. L. Price, et al., "Resveratrol Improves Health and Survival of Mice on a High-Calorie Diet," *Nature* 444, no. 7117 (November 1, 2006): https://www.ncbi.nlm.nih.gov/pmc/articles /PMC4990206/.

33. J. A. Baur and D. A. Sinclair, "Therapeutic Potential of Resveratrol: The *In Vivo* Evidence," *Nature Reviews Drug Discovery* 5, (2006): 493–506, https://www.nature .com/articles/nrd2060.

34. K. J. Pearson, J. A. Baur, K. N. Lewis, et al., "Resveratrol Delays Age-Related Deterioration and Mimics Transcriptional Aspects of Dietary Restriction Without Extending Life Span," *Cell Metabolism* 8, no. 2 (August 6, 2008): 157–68, https:// www.cell.com/cell-metabolism/abstract/S1550-4131%2808%2900182-4.

35. Our findings inevitably stirred up excitement in the media that drinking red wine may increase longevity, as well as admittedly a more sedate example, such as the article "Life-Extending Chemical Is Found in Some Red Wines" in the *New York Times*. K. T. Howitz, K. J. Bitterman, H. Y. Cohen, et al., "Small Molecule Activators of Sirtuins Extend *Saccharomyces cerevisiae* Lifespan," *Nature* 425, no. 6954 (September 11, 2003): 191–96, https://www.ncbi.nlm.nih.gov/pubmed/12939617.

36. To combat aging in mice, we fed them the equivalent of about 100 glasses of red wine a day, not "1,000," neither of which I recommend.

37. Martin-Montalvo et al., "Metformin Improves Healthspan and Lifespan in Mice."

38. Forty patients with varying degrees of psoriasis took part in the study, of whom just over a third had "good to excellent" improvement, according to skin biopsies. J. G. Kreuger, M. Suárez-Fariñas, I. Cueto, et al., "A Randomized, Placebo-Controlled Study of SRT2104, a SIRT1 Activator, in Patients with Moderate to Severe Psoriasis," *PLOS One*, November 10, 2015, https://journals.plos.org /plosone/article?id=10.1371/journal.pone.0142081.

39. Hydrogen is used for hundreds of so-called redox reactions in the cell. NAD is a "hydrogen carrier." The plus sign on "NAD+" indicates the form of NAD that doesn't have a hydrogen atom attached. When it has a hydrogen atom attached, it is called "NADH."

40. As NAD levels decline with age, so the body becomes more susceptible to disease, as two collaborators and I noted: "Restoration of NAD$^+$ levels in old or diseased

animals can promote health and extend lifespan, prompting a search for safe and efficacious NAD-boosting molecules that hold the promise of increasing the body's resilience, not just to one disease, but to many, thereby extending healthy human lifespan." L. Rajman, K. Chwalek, and D. A. Sinclair, "Therapeutic Potential of NAD-Boosting Molecules: The *in Vivo* Evidence," *Cell Metabolism* 27, no. 3 (March 6, 2018): 529–47, https://www.ncbi.nlm.nih.gov/pubmed/29514064.

41. Y. A. R. White, D. C. Woods, Y. Takai, et al., "Oocyte Formation by Mitotically Active Germ Cells Purified from Ovaries of Reproductive Age Women," *Nature Medicine* 18 (February 26, 2012): 413–21, https://www.nature.com/articles /nm.2669.

42. J. L. Tilly and D. A. Sinclair, "Germline Energetics, Aging, and Female Infertility," *Cell Metabolism* 17, no. 6 (June 2013): 838–50, https://www.sciencedirect.com /science/article/pii/S1550413113001976.

43. Our paper in which we showed that SIRT2 is a key player in regulating lifespan in a living organism came out in 2014. B. J. North, M. A. Rosenberg, K. B. Jeganathan, et al., "SIRT2 Induces the Checkpoint Kinase BubR1 to Increase Lifespan," *EMBO Journal* 33, no. 13 (July 1, 2014): 1438–53, https://www.ncbi.nlm.nih.gov/pmc /articles/PMC4194088/.

44. The researchers frame their results within the epidemic of obesity in developing countries and its link to reproductive health issues, including not only polycystic fibrosis but also gestational diabetes mellitus and endometrial cancer. They conclude that "Metformin may be a valuable alternative to, or adjunct for, modifying the toxic effects of obesity in these populations." V. N. Sivalingam, J. Myers, S. Nicholas, et al., "Metformin in Reproductive Health, Pregnancy and Gynaecological Cancer: Established and Emerging Indications," *Human Reproduction* 20, no. 6 (November 2014): 853–68, https://academic.oup.com/humupd/article/20/6/853 /2952671.

45. "Chemotherapy-treated animals had significantly fewer offspring compared with all other treatment groups, whereas cotreatment with mTOR inhibitors preserved normal fertility." K. N. Goldman, D. Chenette, R. Arju, et al., "mTORC1/2 Inhibition Preserves Ovarian Function and Fertility During Genotoxic Chemotherapy," *Proceedings of the National Academy of Sciences of the United States of America* 114, no. 2 (March 21, 2017): 3196–91, http://www.pnas.org/content/114/12 /3186.full.

46. Mice deficient in mTORC1, the authors found, "present spermatozoa with decreased motility, suggesting that mTORC1, besides controlling glandular size and seminal vesicle fluid composition, also regulates sperm physiology during passage through the epididymis." P. F. Oliveira, C. Y. Cheng, and M. G. Alves, "Emerging Role for Mammalian Target of Rapamycin in Male Fertility," *Trends in Endocrinology and Metabolism* 28, no. 3 (March 2017): 165–67, https://www.ncbi.nlm.nih .gov/pmc/articles/PMC5499664/.

47. The term "aging in place" refers to a recently evolved philosophy in Western countries of encouraging the elderly to grow old in places that meet their needs and circumstances. Australia, like so many other countries, is facing a demographic explosion in the number of its elderly, which has significant budgetary and societal implications. Australia's 65- to 84-year-old population is expected to double or more by 2050. H. Bartlett and M. Carroll, "Aging in Place Down Under," *Global Ageing: Issues & Action* 7, no. 2 (2011): 25–34, https://www.ifa-fiv.org/wp-content /uploads/global-ageing/7.2/7.2.bartlett.carroll.pdf.

SIX. BIG STEPS AHEAD

1. In a wide-ranging survey of interventions, the authors covered the health and life-prolonging benefits of various small molecules, exercising, and fasting regimes. "The current epidemics of obesity, diabetes, and related disorders constitute major impediments for healthy aging," they wrote. "It is only by extending the healthy human lifespan that we will truly meet the premise of the Roman poet Cicero: 'No one is so old as to think that he may not live a year.'" R. de Cabo, D. Carmona-Guttierez, M. Bernier, et al., "The Search for Antiaging Interventions: From Elixirs to Fasting Regimens," *Cell* 157, no. 7 (June 19, 2014): 1515–26, https://www.cell .com/fulltext/S0092-8674(14)00679-5.

2. J. Yost and J. E. Gudjonsson, "The Role of TNF Inhibitors in Psoriasis Therapy: New Implications for Associated Comorbidities," *F1000 Medicine Reports* 1, no. 30 (May 8, 2009), https://www.ncbi.nlm.nih.gov/pmc/articles/PMC2924720/.

3. Killing off senescent cells in mice led to their having healthier lives, the author wrote in a story for *Nature* on Baker and van Deursen's work. Their kidney function improved, and their hearts were more resilient to stress, they tended to explore their cages more, and developed cancers at a later age. E. Callaway, "Destroying Worn-out Cells Makes Mice Live Longer," *Nature*, February 3, 2016, https://www.nature .com/news/destroying-worn-out-cells-makes-mice-live-longer-1.19287.

4. The impact of injected senescent cells on young mice was also remarkable in their destructiveness. "As early as two weeks after transplantation, the SEN mice showed impaired physical function as determined by maximum walking speed, muscle strength, physical endurance, daily activity, food intake, and body weight," according to the NIH press release. "In addition, the researchers saw increased numbers of senescent cells, beyond what was injected, suggesting a propagation of the senescence effect into neighboring cells." "Senolytic Drugs Reverse Damage Caused by Senescent Cells in Mice," National Institutes of Health, July 9, 2018, https:// www.nih.gov/news-events/news-releases/senolytic-drugs-reverse-damage-caused -senescent-cells-mice.

5. R.-M. Laberge, Y. Sun, A. V. Orjalo, et al., "MTOR Regulates the Pro-tumorigenic Senescence-Associated Secretory Phenotype by Promoting IL1A Translation,"

Nature Cell Biology 17, no. 8 (July 6, 2015): 1049–61, https://www.ncbi.nlm.nih .gov/pmc/articles/PMC4691706/.

6. P. Oberdoerffer, S. Michan, M. McVay, et al., "DNA Damage–Induced Alterations in Chromatin Contribute to Genomic Integrity and Age-Related Changes in Gene Expression," *Cell* 135, no. 5 (November 28, 2008): 907–18, https://www.ncbi.nlm .nih.gov/pmc/articles/PMC2853975/.

7. M. De Cecco, S. W. Criscione, E. J. Peckham, et al., "Genomes of Replicatively Senescent Cells Undergo Global Epigenetic Changes Leading to Gene Silencing and Activation of Transposable Elements," *Aging Cell* 12, no. 2 (April 2013): 247–56, https://www.ncbi.nlm.nih.gov/pmc/articles/PMC3618682/.

8. "Adoptive transfer of T cells isolated from vaccine-treated tumor-bearing mice inhibited tumor growth in unvaccinated recipients, indicating that the iPSC vaccine promotes an antigen-specific anti-tumor T cell response," the researchers found. N. G. Kooreman, K. Youngkyun, P. E. de Almeida, et al., "Autologous iPSC-Based Vaccines Elicit Anti-tumor Responses *in Vivo*," *Cell Stem Cell* 22, no. 4 (April 5, 2018), http://www.cell.com/cell-stem-cell/fulltext/S1934-5909(18)30016-X.

9. Cells were taken from inside Streisand's dog's cheek and belly skin and sent to a lab in Texas. The cloning process resulted in four puppies, although one died shortly after birth. Streisand wrote that the dogs' resembling her beloved Samantha physically was enough. "You can clone the look of a dog, but you can't clone the soul. Still, every time I look at their faces, I think of my Samantha . . . and smile." B. Streisand, "Barbara Streisand Explains: Why I Cloned My Dog," *New York Times*, March 2, 2018, https://www.nytimes.com/2018/03/02/style/barbra -streisand-cloned-her-dog.html.

10. It is one of the most interesting and important papers I've ever read. C. E. Shannon, "A Mathematical Theory of Communication," *Bell System Technical Journal* 27, no. 3 (July 1948): 379–423 and no. 4 (October 1948): 623–66, http://math .harvard.edu/~ctm/home/text/others/shannon/entropy/entropy.pdf.

11. The results of their experiments were extremely promising when it came to slowing down aging by putting a halt to the molecular changes that cause it. "Molecular alterations induced by in vivo reprogramming may potentially lead to a better maintenance of tissue homeostasis and lifespan extension," they wrote. A. Ocampo, P. Reddy, P. Martinez-Redondo, et al., "In Vivo Amelioration of Age-Associated Hallmarks by Partial Reprogramming," *Cell* 167, no. 7 (December 15, 2016): 1719–33, https://www.cell.com/cell/pdf/S0092-8674(16)31664-6.pdf.

12. "I feel a strong responsibility that it's not just to make a first, but also make it an example," he told the Associated Press. "Society will decide what to do next" as to whether such experiments should continue or be banned. M. Marchione, "Chinese Researcher Claims First Gene-Edited Babies," Associated Press, November 26, 2018, https://www.apnews.com/4997bb7aa36c45449b488e19ac83e86d.

SEVEN. THE AGE OF INNOVATION

1. H. Singh, , A.N.D. Meyer, and E. J. Thomas, "The Frequency of Diagnostic Errors in Outpatient Care: Estimations from Three Large Observational Studies Involving US Adult Populations," *BMJ Quality & Safety* 23, no. 9 (August 12, 2014), https:// qualitysafety.bmj.com/content/23/9/727.

2. M. Jain, S. Koren, K. H. Miga, et al., "Nanopore Sequencing and Assembly of a Human Genome with Ultra-long Reads," *Nature Biotechnology* 36, no. 4 (2018): 338–45, https://www.nature.com/articles/nbt.4060.

3. The evolution of such technology is tied by its inventors to benefiting the community, rather than corporations. That said, this particular company was also promoting the idea of a "coin," or digital currency, not for investment or as a security, according to the writer, but to incentivize individuals to share their genomic data with scientists. "The underlying idea is to incentivize users to make their personal genomic data available for biomedical and health-related research for the greater good of medical discovery." B. V. Bigelow, "Luna DNA Uses Blockchain to Share Genomic Data as a 'Public Benefit,'" *Exome*, January 22, 2018, https://xconomy .com/san-diego/2018/01/22/luna-dna-uses-blockchain-to-share-genomic-data-as -a-public-benefit/.

4. S. W. H. Lee, N. Chaiyakunapruk, and N. M. Lai, "What G6PD-Deficient Individuals Should Really Avoid," *British Journal of Clinical Pharmacology* 83, no. 1 (January 2017): 211–12, https://www.ncbi.nlm.nih.gov/pmc/articles/PMC5338146/; "Glucose-6-Phosphate Dehydrogenase Deficiency," MedlinePlus, https://medline plus.gov/ency/article/000528.htm.

5. J. A. Sparano, R. J. Gray, D. F. Makower, et al., "Adjuvant Chemotherapy Guided by a 21-Gene Expression Assay in Breast Cancer," *New England Journal of Medicine* 379 (July 12, 2018): 111–21, https://www.nejm.org/doi/full/10.1056/NEJ Moa1804710.

6. K. A. Liu and N. A. D. Mager, "Women's Involvement in Clinical Trials: Historical Perspective and Future Implications," *Pharmacy Practice* 14, no. 1 (January–March 2016): 708–17, https://www.pharmacypractice.org/journal/index.php/pp/article /view/708/424.

7. Female mice that had received mTOR treatment lived 20 percent longer than the untreated mice in the control group. Leibniz Institute on Aging, Fritz Lipmann Institute, "Less Is More? Gene Switch for Healthy Aging Found," Medical Xpress, May 25, 2018, https://medicalxpress.com/news/2018-05-gene-healthy-aging.html.

8. Swedish records showed that in every single year since 1800, women had lived longer than men. "This remarkably consistent survival advantage of women compared with men in early life, in late life, and in total life is not confined to Sweden but is seen in every country in every year for which reliable birth and death records exist. There may be no more robust pattern in human biology," the authors noted. S. N. Austad and A. Bartke, "Sex Differences in Longevity and in Responses to

Anti-aging Interventions: A Mini-review," *Gerontology* 62, no. 2 (2015): 40–46, https://www.karger.com/Article/FullText/381472.

9. E. J. Davis, I. Lobach, and D. B. Dubal, "Female XX Sex Chromosomes Increase Survival and Extend Lifespan in Aging Mice," *Aging Cell* 18, no. 1 (February 2019), e12871, https://www.ncbi.nlm.nih.gov/pmc/articles/PMC6351820/.

10. One example where pharmacogenomics information is already used to inform drug prescription is in the treatment of HIV. Patients with HIV are tested for a specific genetic variant to see if they may have a bad reaction to an antiviral drug called abacavir, according to a fact sheet on the National Human Genome Research Institute's website; see "Frequently Asked Questions About Pharmacogenomics," National Human Genome Research Institute, May 2, 2016, https://www.genome.gov /27530645/.

11. The autopsy of a mummified corpse of a fourteenth-century Italian warlord lent credence to centuries-old rumors that days after his triumphant conquest of Treviso, 38-year-old Cangrande I della Scala had been poisoned with digitalis. H. Thompson, "Poison Hath Been This Italian Mummy's Untimely End," Smithsonian.com, January 14, 2015, https://www.smithsonianmag.com/science-nature/poison-hath -been-italian-mummys-untimely-end-digitalis-foxglove-180953822/.

12. M. Vamos, J. W. Erath, and S. H. Hohnloser, "Digoxin-Associated Mortality: A Systematic Review and Meta-analysis of the Literature," *European Heart Journal* 36, no. 28 (July 21, 2015): 1831–38, https://academic.oup.com/eurheartj/article/36 /28/1831/2398087.

13. M. N. Miemeijer, M. E. van den Berg, J. W. Deckers, et al., "*ABCB1* Gene Variants, Digoxin and Risk of Sudden Cardiac Death in a General Population," *BMJ Heart* 101, no. 24 (December 2015), https://heart.bmj.com/content/101/24/1973 ?heartjnl-2014-307419v1=; A. Oni-Orisan and D. Lanfear, "Pharmacogenomics in Heart Failure: Where Are We Now and How Can We Reach Clinical Application?," *Cardiology in Review* 22, no. 5 (September 1, 2015): 193–98, https://www.ncbi .nlm.nih.gov/pmc/articles/PMC4329642/.

14. Back in 2015, Johnson felt it would be only another ten years until our genome is defined and stored for use while we're alive. "When that happens, using genetic information to inform decisions about the right drug and the right dose will likely involve computerized approaches that marry the genetic data with knowledge about drugs and genes, to lead to a personalized treatment recommendation," she wrote. J. A. Johnson, "How Your Genes Influence What Medicines Are Right for You," *Conversation*, November 20, 2015, https://theconversation.com/how-your-genes -influence-what-medicines-are-right-for-you-46904.

15. That seems to be changing, though, according to the authors, as more of their colleagues publish papers on this area, ensuring "that the gut microbiota are moving out of the shadows and are moving towards centre stage in drug safety studies and personalized health care." I. D. Wilson and J. K. Nicholson, "Gut

Microbiome Interactions with Drug Metabolism, Efficacy and Toxicity," *Translational Research: The Journal of Laboratory and Clinical Medicine* 179 (January 2017): 204–22, https://www.ncbi.nlm.nih.gov/pmc/articles/PMC5718288/; see also B. Das, T. S. Ghosh, S. Kedia, et al., "Analysis of the Gut Microbiome of Rural and Urban Healthy Indians Living in Sea Level and High-Altitude Areas," *Nature Scientific Reports* 8 (July 4, 2018), https://www.nature.com/articles /s41598-018-28550-3.

16. P. Lehouritis, J. Cummins, M. Stanton, et al., "Local Bacteria Affect the Efficacy of Chemotherapeutic Drugs," *Nature Scientific Reports* 5 (September 29, 2015), https://www.nature.com/articles/srep14554.

17. The wait increased from 18.5 days in 2014 to 24 days in 2017, according to a study by MerrittHawkins. B. Japsen, "Doctor Wait Times Soar 30% in Major U.S. Cities," *Forbes*, March 19, 2017, https://www.forbes.com/sites/brucejapsen/2017/03 /19/doctor-wait-times-soar-amid-trumpcare-debate/#7ac0753b2e74.

18. The website of myDNAge offers some encouragement: "You can't change your genes, but you can change how your genes behave through epigenetics," runs the tagline. All you have to do is send in your bodily fluids (blood or urine) and they'll determine your biological age by measuring the epigenetic modifications on your DNA. "Reveal Your Biological Age Through Epigenetics," myDNAge, 2017, https://www.mydnage.com/. TeloYears offers to track your cellular age based on your telomeres, which, it informs its website readers, are "the caps on your DNA that, unlike your ancestry, you can actually change." TeloYears, 2018, https://www .teloyears.com/home/.

19. M. W. Snyder, M. Kircher, A. J. Hill, et al., "Cell-free DNA Comprises an *in Vivo* Nucleosome Footprint That Informs Its Tissues-of-Origin," *Cell* 164, nos. 1–2 (January 14, 2016): 57–68, https://www.ncbi.nlm.nih.gov/pmc/articles/PM C4715266/.

20. "Global Automotive Level Sensor Market Analysis, Trends, Drivers, Challenges & Forecasts 2018–2022, with the Market Set to Grow at a CAGR of 4.13%— ResearchAndMarkets.com," Business Wire, May 2, 2018, https://www.business wire.com/news/home/20180502005988/en/Global-Automotive-Level-Sensor -Market-Analysis-Trends.

21. University of Cincinnati senior scientist Jason Heikenfeld and his team worked with a US Air Force Research Laboratory in Ohio on a simple way to keep track of how airmen respond to everything from diet, stress, and injury to medication and disease. They came up with patches that both stimulate and monitor sweat and then send data to a smartphone. J. Heikenfeld, "Sweat Sensors Will Change How Wearables Track Your Health," *IEEE Spectrum*, October 22, 2014, https://spec trum.ieee.org/biomedical/diagnostics/sweat-sensors-will-change-how-wearables -track-your-health.

22. Owlstone has already started lung cancer clinical trials in the United Kingdom,

testing hundreds of patients for early signs. In the United Kingdom, it notes on its website, "only 14.5 percent of people are diagnosed with early stage, treatable lung cancer. If we are able to increase this to 25% we'd save 10,000 lives in the U.K. alone." D. Sfera, "Breath Test Detects Cancer Markers," Medium, August 2, 2018, https://medium.com/@TheRealDanSfera/breath-test-detects-cancer-markers-c57dcc86a583. With advancements in drug treatments, early detection, the company points out, is a more powerful tool to save lives than the development of new drugs. "A Breathalyzer for Disease," Owlstone Medical, https://www.owlstone medical.com/.

23. Two examples are Öura Ring (https://ouraring.com/) and Motiv Ring (https://my motiv.com/).

24. "A growing body of evidence suggests that an array of mental and physical conditions can make you slur your words, elongate sounds, or speak in a more nasal tone." R. Robbins, "The Sound of Your Voice May Diagnose Disease," *Scientific American*, June 30, 2016, https://www.scientificamerican.com/article/the-sound -of-your-voice-may-diagnose-disease/.

25. Researchers used the amount of time subjects took to press and release a key on the computer and converted it to a Parkinson's disease motor index. L. Giancardo, A. Sánchez-Ferro, T. Arroyo-Gallego, et al., "Computer Keyboard Interaction as an Indicator of Early Parkinson's Disease," *Nature Scientific Reports* 6 (October 5, 2016): 34468, https://www.nature.com/articles/srep34468.

26. For a more detailed account of what's just around the corner, this book is well worth the read: E. Topol, *The Creative Destruction of Medicine: How the Digital Revolution Will Create Better Health Care*, Kindle edition (New York: Basic Books, 2011).

27. I am an investor in and former member of the board of InsideTracker, a Segterra company based in Massachusetts, http://www.insidetracker.com/. I have invested in and advise the company, and I am an inventor on a patent application filed to calculate biological age based on markers that are known to change with age.

28. The app is called Clue. E. Avey, "'The Clue App Saved My Life': Early Detection Through Cycle Tracking," Clued In, September 24, 2017, https://medium .com/clued-in/the-clue-app-saved-my-life-early-detection-through-cycle-tracking -91732dd29d25.

29. Over the past three decades, a new infectious disease has appeared every single year in some part of the world. In total, researchers put the number of unknown viruses in birds and mammals that could infect humans between 631,000 and 827,000. Though there are ongoing efforts to identify all these viruses, "we likely won't ever be able to predict which will spill over next; even long-known viruses like Zika, which was discovered in 1947, can suddenly develop into unforeseen epidemics." E. Yong, "The Next Plague Is Coming. Is America Ready?," *The Atlantic*, July–August 2018, https://www.theatlantic.com/magazine/archive/2018/07/when-the-next-plague -hits/561734/.

30. L. M. Mobula, M. MacDermott, C. Hoggart, et al., "Clinical Manifestations and Modes of Death Among Patients with Ebola Virus Disease, Monrovia, Liberia, 2014," *American Journal of Tropical Medicine and Hygiene* 98, no. 4 (April 2018): 1186–93, https://www.ncbi.nlm.nih.gov/pmc/articles/PMC5928808/.

31. The measures to prepare for a future pandemic that Gates argues in an editorial should be put into play include building up public health systems in countries vulnerable to epidemics and mimicking how the military preps for war with "germ games and other preparedness exercises so we can better understand how diseases will spread, how people will respond in a panic, and how to deal with things like overloaded highways and communications systems." B. Gates, "Bill Gates: A New Kind of Terrorism Could Wipe Out 30 Million People in Less than a Year—and We Are Not Prepared," Business Insider, February 18, 2017, http://www.business insider.com/bill-gates-op-ed-bio-terrorism-epidemic-world-threat-2017-2.

32. It was only after a 2009 law was passed that companies had to inform the public and the government of any breaches. Since then, the volume of breaches at health care providers has consistently climbed, from 150 in 2010 to 250 seven years later. Consumer Reports, "Hackers Want Your Medical Records. Here's How to Keep Your Info from Them," *Washington Post*, December 17, 2018, https://www .washingtonpost.com/national/health-science/hackers-want-your-medical-records -heres-how-to-keep-your-info-from-them/2018/12/14/4a9c9ab4-fc9c-11e8-ad40 -cdfd0e0dd65a_story.html?utm_term=.ea4e14662e4a.

33. A. Sulleyman, "NHS Cyber Attack: Why Stolen Medical Information Is So Much More Valuable than Financial Data," *Independent*, May 12, 2017, https://www.in dependent.co.uk/life-style/gadgets-and-tech/news/nhs-cyber-attack-medical-data -records-stolen-why-so-valuable-to-sell-financial-a7733171.html.

34. S. S. Dominy, C. Lynch, F. Ermini, et al., "*Porphyromonas gingivalis* in Alzheimer's Disease Brains: Evidence for Disease Causation and Treatment with Small-Molecule Inhibitors," *Science Advances* 5, no. 1 (January 23, 2019), http://advances .sciencemag.org/content/advances/5/1/eaau3333.full.pdf.

35. That rate of decline continued over the next few years, thanks to fewer older adults requiring hospitalization for pneumonia. "By 2009, more than half the nationwide decline in pneumonia hospitalizations could be attributed to older adults, with some 70,000 fewer annual hospitalizations for those age 85 and older." "Infant Vaccine for Pneumonia Helps Protect Elderly," VUMC Reporter, July 11, 2013, http:// news.vumc.org/2013/07/11/infant-vaccine-for-pneumonia-helps-protect-elderly/.

36. M. R. Moore, R. Link-Gelles, W. Schaffner, et al., "Impact of 13-Valent Pneumococcal Conjugate Vaccine Used in Children on Invasive Pneumococcal Disease in Children and Adults in the United States: Analysis of Multisite, Population-Based Surveillance," *Lancet Infectious Diseases* 15, no. 3 (March 2015): 301–09, https:// www.ncbi.nlm.nih.gov/pmc/articles/PMC4876855/.

37. If you have a companion animal, it can still get the Lyme disease vaccine.

38. "'The [research and development] model is broken,' said Kate Elder, vaccines pol-icy adviser at Médecins sans Frontières. 'Priorities are chosen based on where the money is . . . diseases predominantly in the developed world,' she said." H. Collis, "Vaccines Need a New Business Model," *Politico*, April 27, 2016, https://www.po litico.eu/article/special-report-vaccines-need-a-new-business-model/.

39. "The analysis was conducted by Ronald Evens, adjunct research professor at Tufts CSDD and Tufts University School of Medicine and adjunct professor in the Thomas J. Long School of Pharmacy and Health Sciences at University of the Pa-cific, using data from company reports, periodic biotechnology reports from the Pharmaceutical Research and Manufacturers of America, IMS sales data, and FDA and Tufts CSDD databases." M. Powers, "Tufts: The Vaccine Pipeline Is Soaring and Global Sales Could Hit $40B by 2020," BioWorld, April 21, 2016, http://www.bioworld.com/content/tufts-vaccine-pipeline-soaring-and-global-sales-could -hit-40b-2020.

40. Africa has borne the brunt of more than 90 percent of the globe's malaria cases and deaths. "Malaria," World Health Organization, November 19, 2018, https://www .who.int/news-room/fact-sheets/detail/malaria.

41. "Ghana, Kenya and Malawi to Take Part in WHO Malaria Vaccine Pilot Pro-gramme," World Health Organization, Regional Office for Africa, April 24, 2017, http://www.afro.who.int/news/ghana-kenya-and-malawi-take-part-who-malaria -vaccine-pilot-programme.

42. Crises such as an Ebola outbreak highlight a fundamental flaw in medical research and drug development, researchers told a *Boston Globe* reporter. Unless there's public concern, researchers and pharmaceutical companies have "scant incentive to quickly develop vaccines and drugs for little-seen diseases." Y. Abutaleb, "Speeding Up the Fight Against Ebola, Other Diseases," *Boston Globe*, August 22, 2014, https:// www.bostonglobe.com/metro/2014/08/21/faster-development-vaccines-and-drugs -targeting-diseases-such-ebola-horizon/yrkrN56VgehrSzCtETPzzH/story.html.

43. An equally sobering statistic is that each day twenty people die waiting for a trans-plant, while just one organ donor can save eight lives. "Transplant Trends," United Network for Organ Sharing, https://unos.org/data/.

44. That said, Crouch notes, in Cruise's most recent *Mission Impossible* epic, "*Mission: Impossible—Fallout*," the 56-year-old Cruise's character, Ethan Hunt, appears to ac-knowledge that with advancing years come limitations, as in, for example, needing a younger associate to help him defeat a bad guy in a lengthy brawl or having an eye for ever-younger girlfriends. I. Crouch, "The Wilford Brimley Meme That Helps Measure Tom Cruise's Agelessness," "Rabbit Holes," *New Yorker*, August 11, 2018, https://www.newyorker.com/culture/rabbit-holes/the-wilford-brimley-meme-that -helps-measure-tom-cruises-agelessness.

EIGHT. THE SHAPE OF THINGS TO COME

1. A. Jenkins, "Which 19th century physicist famously said that all that remained to be done in physics was compute effects to another decimal place?," Quora, June 26, 2016, https://www.quora.com/Which-19th-century-physicist-famously-said-that -all-that-remained-to-be-done-in-physics-was-compute-effects-to-another-decimal -place.

2. "*The Road Ahead* (Bill Gates book)," Wikipedia, https://en.wikipedia.org/wiki/The _Road_Ahead_(Bill_Gates_book)#cite_note-Weiss06-3.

3. Kelly added a key point to that excellent mantra: "It's by use [that] we figure out what things are good for. Which is perhaps another way of saying "Go with the flow and see where it takes you." J. Altucher, "One Rule for Predicting What You Never Saw Coming . . . ," The Mission, July 15, 2016, https://medium.com/the-mission /kevin-kelly-one-rule-for-predicting-what-you-never-saw-coming-1e9e4eeae1da.

4. L. Gratton and A. Scott, *The 100 Year Life: Living and Working in an Age of Longevity* (London and New York: Bloomsbury Publishing, 2018).

5. A phrase originating with the theologian Theodore Parker but made famous by Dr. Martin Luther King, Jr., and used several times by President Barack Obama.

6. It was a time of sufficient population density that people began to take an interest in their appearance, notably changing how they looked with beads and pigments. E. Trinkaus, "Late Pleistocene Adult Mortality Patterns and Modern Human Establishment," *Proceedings of the National Academy of Sciences of the United States of America* 108, no. 4 (January 25, 2011): 12267–71, https://www.ncbi.nlm.nih.gov /pubmed/21220336.

7. Up until 4,000 years ago, human numbers were minimal, according to a writer at the Global Environmental Alert Service. Since then, growth has climbed ever faster, the rate peaking in the 1960s. In 2012, the United Nations estimated that by the end of the century world population will be 10.1 billion. "One Planet, How Many People? A Review of Earth's Carrying Capacity," UNEP Global Environmental Alert Service, June 2012, https://na.unep.net/geas/archive/pdfs/geas_jun_12_car rying_capacity.pdf.

8. Similar sentiments are held by the American public, according to a Pew Research Center survey, which found that 59 percent took a "pessimistic view about the effect of population growth saying it will be a major problem because there will not be enough food and resources to go around." "Attitudes and Beliefs on Science and Technology Topics," Pew Research Center, Science & Society, January 29, 2015, http:// www.pewinternet.org/2015/01/29/chapter-3-attitudes-and-beliefs-on-science -and-technology-topics/#population-growth-and-natural-resources-23-point-gap.

9. M. Blythe, "Professor Frank Fenner, Microbiologist and Virologist," Australian Academy of Science, 1992 and 1993, https://www.science.org.au/learning/general -audience/history/interviews-australian-scientists/professor-frank-fenner.

10. Fenner contrasted the fate of humanity with that of the residents of Easter Island,

who were decimated in the 1600s by their reliance on the forests they themselves had cut down. Dwindling food sources, followed by civil war and the arrival of foreign sailors who brought violence and disease, made its population plunge to 111 individuals by 1872. Though the numbers have since rebounded, Fenner's views on humanity's future did not hold up such a generous possibility, he told a reporter from the *Australian*. "As the population keeps growing to seven, eight or nine billion, there will be a lot more wars over food," he said. "The grandchildren of today's generations will face a much more difficult world." C. Jones, "Frank Fenner Sees No Hope for Humans," *Australian*, June 16, 2010, https://www.theaustralian.com.au/higher-education/frank-fenner-sees-no-hope-for-humans/news-story/8d77f0806a8a3591d47013f7d75699b9?nk=099645834c69c221f8ecf836d72b8e4b-1520269044.

11. "There are 925 million people who go hungry every day, despite the amazing economic prosperity we've enjoyed over the past 60 years," wrote Michael Schuman in a piece on Malthus's predictions for *Time*. "And twice in the past three years we've suffered through destabilizing spikes in the cost of food that have trapped tens of millions in poverty. Today, prices are nearly at historic highs." M. Schuman, "Was Malthus Right?," *Time*, July 15, 2011, http://business.time.com/2011/07/15/was-malthus-right/.

12. P. R. Ehrlich, *The Population Bomb* (New York: Ballantine Books, 1968), 1.

13. Ibid., 3.

14. Some of the statistics are simply mind-boggling. Not only is our global population growing by 74 million a year, but "we have consumed more resources in the last 50 years than the whole of humanity before us." S. Dovers, "Population and Environment: A Global Challenge," Australian Academy of Science, August 7, 2015, https://www.science.org.au/curious/earth-environment/population-environment.

15. "Municipal Solid Waste," Environmental Protection Agency, March 29, 2016, https://archive.epa.gov/epawaste/nonhaz/municipal/web/html/.

16. If you run your dryer two hundred times a year, according to a *Guardian* column on the carbon footprint of everyday items, that would generate approximately half a ton of CO_2. M. Berners-Lee and D. Clark, "What's the Carbon Footprint of . . . a Load of Laundry?," *Guardian*, November 25, 2010, https://www.theguardian.com/environment/green-living-blog/2010/nov/25/carbon-footprint-load-laundry.

17. MIT students estimated that "Whether you live in a cardboard box or a luxurious mansion, whether you subsist on homegrown vegetables or wolf down imported steaks, whether you're a jet-setter or a sedentary retiree, anyone who lives in the U.S. contributes more than twice as much greenhouse gas to the atmosphere as the global average." Massachusetts Institute of Technology, "Carbon Footprint of Best Conserving Americans Is Still Double Global Average," Science Daily, https://www.sciencedaily.com/releases/2008/04/080428120658.htm.

18. Residents of Luxembourg, Qatar, Australia, and Canada have greater average levels

of consumption and waste, according to the Global Footprint Network, https://www.footprintnetwork.org/.

19. "Country Overshoot Days," Earth Overshoot Day, https://www.overshootday.org/about-earth-overshoot-day/country-overshoot-days/.

20. The Yale economist William D. Nordhaus has argued that although 2° Celsius is not obtainable, 2.5° might be possible, although it would take extreme global policy measures to get there. W. D. Nordhaus, "Protections and Uncertainties about Climate Change in an Era of Minimal Climate Policies," Cowles Foundation for Research in Economics, Yale University, December 2016, https://cowles.yale.edu/sites/default/files/files/pub/d20/d2057.pdf.

21. Pennsylvania State University professor David Titley conjures up a powerful metaphor for gradations in temperature limits over 2°C. Think of 2° as the 30-mph signposted speed for a truck going down a hill. Then each fraction or whole degree beyond 2 increases the speed of the descending truck and consequently the ever-shortening odds of disaster. D. Titley, "Why Is Climate Change's 2 Degrees Celsius of Warming Limit So Important?," The Conversation, August 23, 2017, https://theconversation.com/why-is-climate-changes-2-degrees-celsius-of-warming-limit-so-important-82058.

22. The Great Barrier Reef isn't only one of the most stunning and unique ecosystems in the world, it's also a huge part of Australia's tourism industry. It rakes in $4.5 billion annually in revenue from tourists and provides jobs for 70,000 people. B. Kahn, "Bleaching Hits 93 Percent of the Great Barrier Reef," *Scientific American*, April 20, 2016, https://www.scientificamerican.com/article/bleaching-hits-93-percent-of-the-great-barrier-reef/.

23. Unless the global temperature increase can be held down to 1.5°C, the reef, whose area is equivalent to the size of Italy, will not survive, coral scientists have concluded. N. Hasham, "Australian Governments Concede Great Barrier Reef Headed for 'Collapse,'" *Sydney Morning Herald*, July 20, 2018, https://www.smh.com.au/politics/federal/australian-governments-concede-great-barrier-reef-headed-for-collapse-20180720-p4zsof.html.

24. By the end of the century, scientists predict, the sea level could rise by between .5 and 1.4 meters. A rise of 5 meters could swamp 3.2 million square kilometers of coastlines, impacting 670 million people. As warming waters impact Greenland and Antarctic ice, they will speed up the pace of the sea level's rising worldwide. "Study Says 1 Billion Threatened by Sea Level Rise," Worldwatch Institute, January 27, 2019, http://www.worldwatch.org/node/5056.

25. The WHO breaks down its estimate of 250,000 extra deaths due to climate change per year between 2030 and 2050 into these categories: heat exposure killing the elderly (38,000), diarrhea (48,000), malaria (60,000), and childhood undernutrition (95,000). "Climate Change and Health," World Health Organization, February 1, 2018, http://www.who.int/mediacentre/factsheets/fs266/en/.

26. Max Planck's *Wissenschaftliche Selbstbiographie* was translated from German by Frank Gaynor and published as *A Scientific Autobiography* in 1949 by Greenwood Press Publishers, Westport, Connecticut.

27. Brexit is a good example of this, Onder noted. Whereas only a quarter of youths voted to leave the European Union, six out of ten people 65 or older voted to leave. H. Onder, "The Age Factor and Rising Nationalism," Brookings, July 18, 2016, https://www.brookings.edu/blog/future-development/2016/07/18/the-age-factor -and-rising-nationalism/.

28. Those who are 80 or older—the "oldest-old," according to the United Nations— are increasing in number faster than are older people (those over 60) overall. In 2015, there were 125 million 80-plus-year-olds; by 2050 there are expected to be close to 450 million. Department of Economic and Social Affairs, Population Division, *World Population Ageing 2015* (New York: United Nations, 2015), http:// www.un.org/en/development/desa/population/publications/pdf/ageing/WPA2015 _Report.pdf.

29. "Strom Thurmond's Voting Records," Vote Smart, https://votesmart.org/candidate /key-votes/53344/strom-thurmond.

30. In an incisive piece in the *Nation*, UCLA and Columbia Law 'School professor Kimberlé Williams Crenshaw highlighted some of the abhorrent double standards surrounding Thurmond. "For most critics of sexual racism, this is simply a text-book case of a white man getting away with sexual behavior that would have sent an African-American man to his death," she wrote. Indeed, in 1942, then Judge Thurmond sent a black man to the electric chair "based on an alleged rape victim's cross-race identification, testimony now known to be extremely unreliable." K. W. Crenshaw, "Was Strom a Rapist?," *Nation*, February 26, 2004, https://www .thenation.com/article/was-strom-rapist/.

31. The elderly poor's only alternatives were family, friends, or the poorhouse. B. Veghte, "Social Security, Past, Present and Future," National Academy of Social Insurance, August 13, 2015, https://www.nasi.org/discuss/2015/08/social-security %E2%80%99s-past-present-future.

32. Men who reached age 65 in 1940 lived on average for another 12.7 years. By 1990, that average had climbed to 15.3 years. Women's average life expectancy for the same period (assuming they, too, survived to 65), increased by almost 5 years, to 19.6 years. "Life Expectancy for Social Security," Social Security, https://www.ssa .gov/history/lifeexpect.html.

33. By 2015, around 8 percent of the elderly were below the poverty line. "Per Capita Social Security Expenditures and the Elderly Poverty Rate, 1959–2015," The State of Working America, September 26, 2014, http://www.stateofworkingamerica.org /chart/swa-poverty-figure-7r-capita-social-security/.

34. "Actuarial Life Table," Social Security, 2015, https://www.ssa.gov/oact/STATS/table 4c6.html.

35. William Safire tracked down the source of this quote for the *New York Times* in 2007: it was Kirk O'Donnell, a top aide to Tip O'Neill. W. Safire, "Third Rail," *New York Times*, February 18, 2007, http://www.nytimes.com/2007/02/18/maga zine/18wwlnsafire.t.html.

36. "Social Security Beneficiary Statistics," Social Security, https://www.ssa.gov/oact /STATS/OASDIbenies.html.

37. "Quick Facts: United States," United States Census Bureau, https://www.census .gov/quickfacts/fact/table/US/PST045217.

38. Older voters have notably more impact when it comes to the primaries, according to Harvard professor of government Stephen Ansolabehere. "Older people tend to vote more often in primaries," he says. "And since the primary turnout tends to be lower, that means that bloc can be even more important." D. Bunis, "The Immense Power of the Older Voter," *AARP Bulletin*, April 30, 2018, https://www.aarp.org /politics-society/government-elections/info-2018/power-role-older-voters.

39. The days of long summer vacations (leaving continental Europe's capitals all but empty), early retirement, and blanket medical insurance seem to be becoming a thing of the past in Europe, wrote the *Washington Post*'s Edward Cody. "In the new reality, workers have been forced to accept salary freezes, decreased hours, postponed re-tirements and health-care reductions." E. Cody, "Europeans Shift Long-Held View That Social Benefits Are Untouchable," *Washington Post*, April 24, 2011, https:// www.washingtonpost.com/world/europeans-shift-long-held-view-that-social -benefits-are-untouchable/2011/02/09/AFLdYzdE_story.html?utm_term=.bcf29d 628eea.

40. Part of the reason there's such a stark difference in life expectancy, according to public health researchers, is the disappearance of smoking from the lifestyles of the rich and educated. S. Tavernise, "Disparity in Life Spans of the Rich and the Poor Is Growing," *New York Times*, February 12, 2016, https://www.nytimes.com/2016 /02/13/health/disparity-in-life-spans-of-the-rich-and-the-poor-is-growing.html.

41. Joint Committee on Taxation, U.S. Congress, "History, Present Law, and Analysis of the Federal Wealth Transfer Tax System," JCX-52-15, March 16, 2015, https:// www.jct.gov/publications.html?func=startdown&id=4744.

42. "SOI Tax Stats—Historical Table 17," IRS, August 21, 2018, https://www.irs.gov /statistics/soi-tax-stats-historical-table-17.

43. Horses pulling carriages littered the streets with dung. Corpses rotted in overflow-ing graveyards. Garbage piled up in the streets. L. Jackson, *Dirty Old London: The Victorian Fight Against Filth* (New Haven, CT: Yale University Press, 2015).

44. W. Luckin, "The Final Catastrophe—Cholera in London, 1886," *Medical History* 21, no. 1 (January 1977): 32–42, https://www.ncbi.nlm.nih.gov/pmc/articles/PMC 1081893/?page=5.

45. H. G. Wells highlighted the possibility of world destruction stemming from the splitting of the atom, wrote *Smithsonian* writer Brian Handwerk, as well as the

future threat of portable devices capable of mass destruction. "Wells also clearly saw the dangers of nuclear proliferation, and the doomsday scenarios that might arise both when nations were capable of 'mutually assured destruction' and when non-state actors or terrorists got into the fray." B. Handwerk, "The Many Futuristic Predictions of H. G. Wells That Came True," Smithsonian.com, September 21, 2016, https://www.smithsonianmag.com/arts-culture/many-futuristic-predictions -hg-wells-came-true-180960546/.

46. According to the author and film historian Mark Clark, Wells's sci-fi classic *Things to Come* and the subsequent 1936 film, which Clark claims was made under the author's creative control, was his attempt "to save the world. Literally." It's a story of a world racked by war, only to be offered salvation by the Airmen. "They are a society of scientists and engineers who, hidden away from the rest of the world, have made great scientific advances and are now prepared to lead humanity to a brighter future—so long as it submits to their benevolent rule." M. Clark, "Common Thread: Wells and Roddenberry," Onstage and Backstage, July 29, 2013, https://onstageandbackstage.wordpress.com/tag/gene-roddenberry/.

47. Roddenberry's work echoed Wells's utopian-themed writing, Clark has pointed out. Ibid.

48. As Wells noted time and again in his writings, these are the only two options for humanity.

49. A. van Leeuwenhoek, "Letters 43–69," Digitale Bibliotheek oor de Nederlandse, April 25, 1679, http://www.dbnl.org/tekst/leeu027alle03_01/leeu027alle03_01 _0002.php#b0043.

50. We have long been losing the battle to balance technological developments in goods and services with the environmental impact of population growth. "One Planet, How Many People?'"

51. Edward O. Wilson, *The Future of Life* (2002; repr., New York: Vintage Books), 33. In a review in the *New Yorker* (March 4, 2002): "Wilson, an eminent evolutionary biologist, shows the extent to which human prosperity, even in the information age, rests on the foundation of a diverse natural world, since the more species any ecosystem has, the more stable and productive it will be."

52. Nowhere has the debate about a land's carrying capacity been greater than in Australia. The Dutch may have been the first Europeans to discover the great southern land Terra Australis, but it was the British who permanently colonized the habitable southeastern coastal strip in 1788. One hundred years after the prisoners took their first steps on the beaches of Sydney and most of the original inhabitants had been pushed out or wiped out by guns and smallpox, the British brimmed with optimism about the country's future. That made sense: the American colonies were thriving, albeit a little too well for British tastes, and the Australian continent was just as large as America. In 1888, a story in the *Spectator* sounded like a proud mother talking about her child's future, with only a hint of racism, sexism, and

disdain for Americans: "There is every reasonable probability that in 1988 Australia will be a Federal Republic, peopled by 50 millions of English speaking men, who, sprung from the same races as the Americans of the Union, will have developed a separate and recognisable type. . . . The Australians, we conceive, with more genial and altogether warmer climate, without Puritan traditions, with wealth among them from the first . . . will be a softer, though not a weaker people, fonder of luxury, and better fitted to enjoy art . . . The discontent which permeates the whole American character will be absent and, if not exactly happier, they will be more at ease. The typical Australian will be a sunnier man." "Topics of the Day: The Next Centenary of Australia," *Spectator* 61 (January 28, 1888): 112–13.

Though the prediction about Australian male sunniness and relative paucity of Puritanism was spot on, it was off on its math. After 1888, Australia's population grew less than half as fast as predicted, in large part due to the absence of arable land. In 2018, the country had a population of only 25 million. But most Australians disagree with Sheridan, even more so after a few beers. They believe that the country is already overcrowded and the land is nearing its carrying capacity. Calls to limit immigration have dominated conversations, talk shows, and politics for three decades, long before it was fashionable in the United States. Many are deeply angry about the ever-increasing housing costs and commuting times. Some are simply racist. Others are professional scaremongers. Ted Trainer, Australia's answer to Paul Ehrlich, has been arguing his whole career that levels of human consumption and resource use are already unsustainable. I know because I took his university course in 1988. According to Trainer, gasoline would run out before the 2000s and we were all supposed to be starving by now. Trainer's version of utopia is his alternative-lifestyle, unkempt educational farm an hour's drive south of Sydney, called Pigface Point. I spent a day there, learning by example, for example, that to save the world we needed to start living on three-acre farms, use solar-powered ovens to cook homegrown eggs, and commute an hour each way in a rusty, smoke-belching car to give lectures on green living. Yes, we have major issues to solve—climate change being the most threatening. But contrary to Trainer's teachings, technology is not the enemy. In the arc of human history, technology has ultimately come to our rescue. For most of us, our daily life is improving, and it will continue to improve, just as London in 1840 and New York in 1900 did. There are more people than ever in the cities of North America, Europe, and Australasia, but today the impact of each human is rapidly declining and, contrary to what I was taught in the 1980s about the future, cities are getting cleaner. We're moving from petroleum to natural gas to solar and electricity. A visit to Bangkok used to provoke respiratory distress. Now there's blue sky. When I arrived in Boston in 1995, a splash of water from the harbor could land you in the hospital or the grave. Now it is safe for swimming.

53. E. C. Ellis, "Overpopulation Is Not the Problem," *New York Times*, September 13,

2013, https://www.nytimes.com/2013/09/14/opinion/overpopulation-is-not-the
-problem.html.

54. "World Population Projections," Worldometers, http://www.worldometers.info
/world-population/world-population-projections/.

55. Ibid. and Population Division, Department of Economic and Social Affairs, United
Nations Secretariat, "2017 Revision of World Population Prospects," https://popu
lation.un.org/wpp/.

56. Gates's argument is simple enough: when you improve children's health so they
don't die at an early age, families choose to have fewer children. B. Gates, "Does
Saving More Lives Lead to Overpopulation?," YouTube, February 13, 2018, https://
www.youtube.com/watch?v=obRG-2jurz0.

57. The others were Denmark, Finland, Norway, Great Britain, Germany, and France.
M. Roser, "Share of the Population Who Think the World Is Getting Better," Our
World in Data, https://ourworldindata.org/wp-content/uploads/2016/12/Opti
mistic-about-the-future-2.png.

58. *The Guardian* asked, "To what extent, if at all, do you feel that today's youth will
have had a better or worse life than their parent's [*sic*] generation?" Though the
Chinese polled were optimistic for the future of their youths, only 20 percent in
the United Kingdom thought things will be better for future young people, with 54
percent expecting them to get worse. The poll was taken in the face of rising rents,
house prices, and university fees in Great Britain, together with a steep drop in
wages, which had in turn driven austerity policies. S. Malik, "Adults in Developing
Nations More Optimistic than Those in Rich Countries," *Guardian*, April 14, 2014,
https://www.theguardian.com/politics/2014/apr/14/developing-nations-more
-optimistic-richer-countries-survey.

59. In developing countries, which may well still have high child mortality rates, they
nevertheless are also seeing declines in numbers. Our World in Data's Max Roser
pointed out that in sub-Saharan Africa child mortality has dropped consistently
over the past fifty years; whereas it was one in four in the 1960s, it's now one in
ten. M. Roser, "Child Mortality," Our World in Data, https://ourworldindata.org
/child-mortality.

60. Steven Pinker, *Enlightenment Now: The Case for Reason, Science, Humanism, and
Progress* (New York: Viking, 2018), 51.

61. Among her many charms, gifts, and abilities, there was also a dry, self-deprecating wit.
At a luncheon for women executives shortly before her death, Thompson said about
what was her last marathon, "I didn't get much attention, even though I was coming in first—I was the only one in my age group." R. Sandomir, "Harriette Thompson, Marathon Runner into Her 90s, Dies at 94," *New York Times*, October 19,
2017, https://www.nytimes.com/2017/10/19/obituaries/harriette-thompson-dead
-ran-marathons-in-her-90s.html.

62. "Old Age: Personal Crisis, U.S. Problem," *Life*, July 13, 1959, pp. 14–25.

63. The price that older unemployed workers pay for this discrimination is harsh. AARP writer Nathaniel Reade laid out a few of the statistics: "Forty-four percent of jobless workers 55 or older had been unemployed for over a year in 2012, a Pew study reported. And while older workers have a lower unemployment rate overall, the ones who lose their jobs can find the long hunt for work unbearable." Many are forced to tap into their Social Security, which puts not only their benefits but a financially secure retirement at risk. N. Reade, "The Surprising Truth About Older Workers," *AARP The Magazine*, September 2015, https://www.aarp.org/work/job -hunting/info-07-2013/older-workers-more-valuable.html.

64. That's according to research by Fabrizio Carmignani, a professor of business at Griffith University in Australia. F. Carmignani, "Does Government Spending on Education Promote Economic Growth?," The Conversation, June 2, 2016, https://the conversation.com/does-government-spending-on-education-promote-economic -growth-60229.

65. M. Avendano, M. M. Glymour, J. Banks, and J. P. Mackenbach, "Health Disadvantage in US Adults Aged 50 to 74 Years: A Comparison of the Health of Rich and Poor Americans with That of Europeans," *American Journal of Public Health* 99, no. 3 (March 2009): 540–48, https://www.ncbi.nlm.nih.gov/pubmed/19150903.

66. Of all the European countries, the United Kingdom will have the oldest working population, having set the retirement age to rise to 69 by 2046. "Retirement in Europe," Wikipedia, https://en.wikipedia.org/wiki/Retirement_in_Europe.

67. "Impact of Automation," *Life*, July 19, 1963, 68–88.

68. A. Swift, "Most U.S. Employed Adults Plan to Work Past Retirement Age," Gallup, May 8, 2017, http://news.gallup.com/poll/210044/employed-adults-plan-work -past-retirement-age.aspx?g_source=Economy&g_medium=lead&g_campaign =tiles.

69. Only 25 percent said they'd stop working altogether at retirement age, according to Gallup. Those who planned to work part-time after retirement age comprised 63 percent of those polled. Ibid.

70. In 2014, Massachusetts ranked fifth nationwide for the number of patents issued, having had an 81.3 percent increase over the prior ten years in patents issued to state inventors. E. Jensen-Roberts, "When It Comes to Patents, Massachusetts Is a Big Player," *Boston Globe*, August 9, 2015, https://www.bostonglobe.com/magazine /2015/08/08/when-comes-patents-massachusetts-big-player/3AmNfmSE8xW zzNbUnDzvPK/story.html.

71. D. Goldman, "The Economic Promise of Delayed Aging," *Cold Spring Harbor Perspectives in Medicine* 6, no. 2 (December 18, 2015): a025072, http://perspectivesin medicine.cshlp.org/content/6/2/a025072.full.

72. The authors contend that "the social, economic, and health benefits that would result from such advances may," in the same way a "peace dividend" enables countries

to rise out of poverty, "be thought of as 'longevity dividends,' and that they should be aggressively pursued as the new approach to health promotion and disease prevention in the 21st century." S. J. Olshansky, D. Perry, R. A. Miller, and R. N. Butler, "Pursuing the Longevity Dividend: Scientific Goals for an Aging World," *Annals of the New York Academy of Sciences* 114 (October 2017): 11–13, https://www.ncbi .nlm.nih.gov/pubmed/17986572.

73. Though 0.1 percent might not sound like much of the global population, that's still 7.8 million full-time researchers. "Facts and Figures: Human Resources," UNESCO, https://en.unesco.org/node/252277.

74. If the experiment sounds vaguely familiar, perhaps it's because its inspiration came from the 1964 murder of Kitty Genovese in Queens, New York. Her screams for help reached the ears of thirty-eight neighbors, but none tried to help her. I. Shenker, "Test of Samaritan Parable: Who Helps the Helpless?," *New York Times*, April 10, 1971, https://www.nytimes.com/1971/04/10/archives/test-of-samaritan-parable -who-helps-the-helpless.html.

75. Seneca, the philosopher who lived from c. 5 BC to AD 65, wrote about the brevity of life, the art of living, and the importance of morality and reason. Seneca, *On the Shortness of Life: Life Is Long if You Know How to Use It*, trans. G.D.N. Costa, Penguin Books Great Ideas (New York: Penguin Books, 2004).

NINE. A PATH FORWARD

1. J. M. Spaight, *Aircraft in War* (London: Macmillan, 1914), 3.

2. This was one of what has become known as three "laws" penned by Clarke, each famous in its own right. The other two were "The only way of discovering the limits of the possible is to venture a little way past them into the impossible" and "Any sufficiently advanced technology is indistinguishable from magic." A. C. Clarke, "Hazards of Prophecy: The Failure of Imagination," in *Profiles of the Future: An Inquiry into the Limits of the Possible* (New York: Orion, 1962), 14, 21, 36

3. L. Gratton and A. Scott, *The 100 Year Life: Living and Working in an Age of Longevity* (London and New York: Bloomsbury Publishing, 2018).

4. "And the days of Isaac were an hundred and fourscore years. And Isaac gave up the ghost, and died, and was gathered unto his people, being old and full of days: and his sons Esau and Jacob buried him." Genesis 35:28, King James Version.

5. It was initially funded by a clerk from the Treasury Department garnishing twenty cents a month from each merchant seaman's wages to pay for a number of contract hospitals. "A Short History of the National Institutes of Health," Office of NIH History, https://history.nih.gov/exhibits/history/index.html.

6. That's according to the Buck Institute for Research on Aging, which also noted that "if we move outside academic research to money spent on commercialized research applications by private companies, the pie changes quite a bit. In aggregate, drug companies outspend the NIH on R&D every year by over $20 billion." "Who

funds basic aging research in the US?," Fight Aging!, March 25, 2015, https://www
.fightaging.org/archives/2015/03/who-funds-basic-aging-research-in-the-us/.

7. The authors highlight the looming crisis the globe faces in terms of the increasing
number of aged. Come 2050, they estimate, the number of those over 60 will be
just over 2 billion, five times the number they were a century before. And 1.5 billion will be from developing countries. L. Fontana, B. K. Kennedy, V. D. Longo,
et al., "Medical Research: Treat Ageing," *Nature* 511, no. 750 (July 23, 2014):
405–7, July 24, 2014, https://www.nature.com/news/medical- research-treat-age
ing-1.15585.

8. "Estimates of Funding for Various Research, Condition, and Disease Categories
(RCDC)," National Institutes of Health, May 18, 2018, https://report.nih.gov
/categorical_spending.aspx.

9. R. Brookmeyer, D. A. Evans, L. Hebert, et al., "National Estimates of the Prevalence
of Alzheimer's Disease in the United States," *Alzheimer's & Dementia* 7, no. 1 (January 2011): 61–73, https://www.ncbi.nlm.nih.gov/pmc/articles/PMC3052294/.

10. The average American spends $1,100 on coffee every year. "2017 Money matters re-
port," Acorns, 2017, https://sqy7rm.media.zestyio.com/Acorns2017_MoneyMatters
Report.pdf.

11. "Actuarial Life Table," Social Security, 2015, https://www.ssa.gov/OACT/STATS
/table4c6.html.

12. Looking back over his life in a wide-ranging interview with Nautilus's Jordana Cepelewicz, Hayflick noted that what money was invested in aging research did not
go to where he felt it should. Most studies of aging focus on longevity determinants
or diseases linked to age, he said. "Less than 3 percent of the budget of the National Institute on Aging in the past decade or more has been spent on research on
the fundamental biology of aging." J. Cepelewicz, "Ingenious: Leonard Hayflick,"
Nautilus, November 24, 2016, http://nautil.us/issue/42/fakes/ingenious-leonard
-hayflick.

13. The film *Gattaca* is about a future society driven by eugenics in which children are
genetically selected to ensure that they possess the best hereditary traits. A father
asks the geneticist, "We were wondering if we should leave some things to chance."
The geneticist responds, "You want to give your child the best possible start. Believe
me, we have enough imperfection built in already. Your child doesn't need any additional burdens." A. Nicols, director, *Gattaca*, 1997.

14. The authors calculated that "from 1970 to 2000, gains in life expectancy added
about $3.2 trillion *per year* to national wealth, with half of these gains due to prog-
ress against heart disease alone." A cure for cancer "would be worth about $50
trillion." K. M. Murphy and R. H. Topel, "The Value of Health and Longevity,"
Journal of Political Economy 114, no. 5 (October 2006): 871–904, https://ucema
.edu.ar/u/je49/capital_humano/Murphy_Topel_JPE.pdf.

15. D. Goldman, B. Shang, J. Bhattacharya, and A. M. Garber, "Consequences of

Health Trends and Medical Innovation for the Future Elderly," *Health Affairs* 24, suppl. 2 (February 2005): W5R5–17, https://www.researchgate.net/publication /7578563_Consequences_Of_Health_Trends_And_Medical_Innovation_For _The_Future_Elderly.

16. For a good read, I recommend Bill Bryson's books on this topic. *Notes from a Big Country* (UK)/*I'm a Stranger Here Myself* (USA), 1999; and *Down Under*, 2000, are my favorites.

17. A phrase often used by US politicians, from the Parable of Salt and Light in Jesus's Sermon on the Mount. In Matthew 5:14, Jesus tells his listeners, "You are the light of the world. A city that is set on a hill cannot be hidden."

18. It's clearly impacting the workforce. The trend to live longer "is one factor contributing to a steady rise in the workforce participation rate of older Australians, especially women," wrote Matt Wade in the *Sydney Morning Herald*. "About one in five Australian workers are now aged over 55 years, compared with less than one in 10 in the 1980s and 1990s." M. Wade, "Trend for Australians to Live Longer Reshapes Economy," *Sydney Morning Herald*, August 12, 2018, https://www.smh .com.au/business/the-economy/trend-for-australians-to-live-longer-reshapes-econ omy-20180810-p4zwuv.html?btis.

19. What does the phrase "Medicare for all" mean? According to a CNBC article, Reuters defines it as "a publicly financed, privately delivered system with all Americans enrolled and all medically necessary services covered." Meanwhile, the cost of health care for US citizens keeps on climbing. "The average annual deductible for employer-sponsored health care plans, which make up most of the plans in the U.S., was $1,505 in 2017, compared to $303 in 2006," according to the Kaiser Family Foundation, wrote Yoni Blumberg for CNBC Make It. Y. Blumberg, "70% of Americans Now Support Medicare-for-All—Here's How Single-Payer Could Affect You," CNBC Make It, August 28, 2018, https://www.cnbc.com/2018/08/28 /most-americans-now-support-medicare-for-all-and-free-college-tuition.html.

20. "Australians Living Longer but Life Expectancy Dips in US and UK," *Guardian*, August 16, 2018, https://www.theguardian.com/society/2018/aug/16/australians -living-longer-but-life-expectancy-dips-in-us-and-uk.

21. Indeed, Americans who live in the top-income counties enjoy on average twenty years more of life than those living in the poorest counties, wrote Senator Bernie Sanders, and that is at least partly due to what he termed "grossly unequal access to quality healthcare." B. Sanders, "Most Americans Want Universal Healthcare. What Are We Waiting For?," *Guardian*, August 14, 2017, https://www.theguard ian.com/commentisfree/2017/aug/14/healthcare-a-human-right-bernie-sanders -single-payer-system.

22. In fact, the Patient Factor lists at the top of the world's health care systems (courtesy of the World Health Organization) the following countries: (1) France, (2) Italy, (3) San Marino, (4) Andorra, (5) Malta. "World Health Organization's Ranking

of the World's Health Systems," The Patient Factor, http://thepatientfactor.com
/canadian-health-care-information/world-health-organizations-ranking-of-the
-worlds-health-systems/.

23. "My father says that America has the best healthcare system in the world. What can
I say to prove him wrong?," Quora, https://www.quora.com/My-father-says-that
-America-has-the-best-healthcare-system-in-the-world-What-can-I-say-to-prove
-him-wrong.

24. N. Hanauer, "The Pitchforks Are Coming . . . For Us Plutocrats," *Politico*, July
/August 2014, https://www.politico.com/magazine/story/2014/06/the-pitchforks
-are-coming-for-us-plutocrats-108014.

25. See *International Journal of Astrobiology*, https://www.cambridge.org/core/journals
/international-journal-of-astrobiology.

26. P. Dayal, C. Cockell, K. Rice, and A. Mazumdar, "The Quest for Cradles of Life:
Using the Fundamental Metallicity Relation to Hunt for the Most Habitable
Type of Galaxy," *Astrophysical Journal Letters*, July 15, 2015, https://arxiv.org/abs
/1507.04346.

27. "List of Nearest Terrestrial Exoplanet Candidates," Wikipedia, https://en.wikipedia
.org/wiki/List_of_nearest_terrestrial_exoplanet_candidates.

28. George Monbiot, "Cutting Consumption Is More Important Than Limiting Population," "George Monbiot's Blog," *Guardian*, February 25, 2009, https://www.the
guardian.com/environment/georgemonbiot/2009/feb/25/population-emissions
-monbiot.

29. S. Pinker, *Enlightenment Now: The Case for Reason, Science, Humanism, and Progress*
(New York: Penguin Random House, 2018), 333.

30. One home builder told CNBC reporter Diana Olick that young adults are reluctant to leave rental apartments decked out by owners to be "resortlike," given
that they can't afford to buy an apartment with similar amenities. Olick found
that younger Americans "seem to be drawn to smaller, simpler living," citing the
tiny-house trend, one underpinned by technology equipping small spaces with
"big amenities." D. Olick, "Why Houses in America Are Getting Smaller," CNBC,
August 23, 2016, https://www.cnbc.com/2016/08/23/why-houses-in-america-are
-getting-smaller.html.

31. A $20 billion New York start-up called WeWork focuses on offering shared work-
ing environments rich with immediately accessible amenities. The eight-year-old
business, at the time of David Gelles's *New York Times* business profile, had "built a
network of 212 shared working spaces around the globe" and was in the process of
putting up the fifteen-story Dock 72 on the East River. Along with a vast cowork-
ing space, "there will be a juice bar, a real bar, a gym with a boxing studio, an out-
door basketball court and panoramic vistas of Manhattan. There will be restaurants
and maybe even dry cleaning services and a barbershop." D. Gelles, "The WeWork

Manifesto: First, Office Space. Next, the World," *New York Times*, February 17, 2018, https://www.nytimes.com/2018/02/17/business/the-wework-manifesto-first-office-space-next-the-world.html.

32. And let us not forget the water we expend on producing crops and meat that we never eat. 2013 estimates put the amount of water being used for food production by 2050 at 10 to 13 trillion cubic meters a year, which is 3.5 times as much as the amount of fresh water currently consumed by the planet's population. J. von Radowitz, "Half of the World's Food 'Is Just Thrown Away,'" *Independent*, January 10, 2013, https://www.independent.co.uk/environment/green-living/half-of-the-worlds-food-is-just-thrown-away-8445261.html.

33. Carl R. Woese Institute for Genomic Biology, University of Illinois at Urbana-Champaign, "Scientists Engineer Shortcut for Photosynthetic Glitch, Boost Crop Growth 40%," Science Daily, January 3, 2019, https://www.sciencedaily.com/releases/2019/01/190103142306.htm.

34. P. Mirocha, and A. Mirocha, "What the Ancestors Ate," Edible Baja Arizona, September/October 2015, http://ediblebajaarizona.com/what-the-ancestors-ate.

35. J. Wenz, "The Mother of All Apples Is Disappearing," *Discover*, June 8, 2017, http://blogs.discovermagazine.com/crux/2017/06/08/original-wild-apple-going-extinct/#.W_3i8ZNKjOQ.

36. "Vitamin A deficiency is the leading cause of preventable childhood blindness and increases the risk of death from common childhood illnesses such as diarrhoea," according to a late-2017 UNICEF report. It stated that vitamin A had been shown to "reduce all-cause mortality by 12 to 24 percent, and is therefore an important programme in support of efforts to reduce child mortality." "Vitamin A Deficiency," UNICEF, February 2019, https://data.unicef.org/topic/nutrition/vitamin-a-deficiency/.

37. Luciano Marraffini and Erik Sontheimer at Northwestern University in Evanston, Illinois, were the first to show how CRISPR protects bacteria from foreign DNA: the interference machinery targets DNA directly. "From a practical standpoint, the ability to direct the specific, addressable destruction of DNA . . . could have considerable functional utility, especially if the system can function outside of its native bacterial or archaeal context," they wrote. L. A. Marraffini and E. J. Sontheimer, "CRISPR Interference Limits Horizontal Gene Transfer in Staphylococci by Targeting DNA," *Science* 322, no. 5909 (December 19, 2008): 1843–45, https://www.ncbi.nlm.nih.gov/pmc/articles/PMC2695655/; see also J. Cohen, "How the Battle Lines over CRISPR Were Drawn," *Science*, February 17, 2017, https://www.sciencemag.org/news/2017/02/how-battle-lines-over-crispr-were-drawn.

38. M. R. O'Connell, B. L. Oakes, S. H. Sternberg, et al., "Programmable RNA Recognition and Cleavage by CRISPR/Cas9," *Nature* 516, no. 7530 (December 11, 2014): 263–66, https://www.ncbi.nlm.nih.gov/pubmed/25274302.

39. L. Cong, F. A. Ran, D. Cox, et al., "Multiplex Genome Engineering Using CRISPR /Cas Systems," *Science* 339, no. 6121 (February 15, 2013): 819–23, https://www .ncbi.nlm.nih.gov/pubmed/23287718.

40. Court of Justice of the European Union, "Organisms Obtained by Mutagenesis Are GMOs and Are, in Principle, Subject to the Obligations Laid Down by the GMO Directive," July 25, 2018, https://curia.europa.eu/jcms/upload/docs/application /pdf/2018-07/cp180111en.pdf.

41. "Secretary Perdue Statement on ECJ Ruling on Genome Editing," U.S. Department of Agriculture, July 27, 2018, https://www.usda.gov/media/press-releases /2018/07/27/secretary-perdue-statement-ecj-ruling-genome-editing.

42. LEDs are tiny, no larger than a fleck of pepper, and mixing primary-colored LEDs, i.e., red, green, and blue, results in white light. "LED Lighting," Energy Saver, https://www.energy.gov/energysaver/save-electricity-and-fuel/lighting-choices -save-you-money/led-lighting.

43. In 2016, the City of Angels achieved an 11 percent emissions cut (the same as 737,000 cars coming off the streets) via an improved public transport system and solar energy investments, while creating "30,000 new green jobs," wrote Matt Simon in *Wired*. M. Simon, "Emissions Have Already Peaked in 27 Cities—and Keep Falling," *Wired*, September 13, 2018, https://www.wired.com/story/emis sions-have-already-peaked-in-27-cities-and-keep-falling/.

44. It turned out that the source of the pollution was "problematic connections between the pipes that carried sewage and those that were meant to carry clean rainwater out to the river," according to a Citylab story by Stephanie Garlock. When a rainstorm hit, "everything in the pipes, sewage and all, would flush directly into the Charles and its tributaries through older drainage pipes." Renovations of the sewage systems have all but eradicated the problem. S. Garlock, "After 50 Years, Boston's Charles River Just Became Swimmable Again," Citylab, July 19, 2013, https://www.citylab .com/life/2013/07/after-50-years-bostons-charles-river-just-became-swimmable -again/6216/.

45. The farm cost 200 million Australian dollars to construct and has a solar plant made up of 23,000 mirrors that reflects the heat of the sun to a solar tower. Rather than soil, the tomato plants grow in "a watery solution fed by nutrient-rich co-conut husks." E. Bryce, "These Farms Use Sun and Seawater to Grow Crops in the Arid Australian Desert," *Wired*, February 14, 2017, https://www.wired.co.uk /article/sundrop-farms-australian-desert. See also Sundrop Farms, http://www.sun dropfarms.com.

46. Letter from Joseph Wharton, December 6, 1880, https://giving.wharton.upenn .edu/wharton-fund/letter-joseph-wharton/.

47. P. Sopher, "Where the Five-Day Workweek Came From," *Atlantic*, August 21, 2014, https://www.theatlantic.com/business/archive/2014/08/where-the-five-day-work week-came-from/378870/.

CONCLUSION

1. E. Pesheva, "Rewinding the Clock," Harvard Medical School, March 22, 2018, https://hms.harvard.edu/news/rewinding-clock; see also A. Das, G. X. Huang, M. S. Bonkowski, et al., "Impairment of an Endothelial NAD+-H2S Signaling Network Is a Reversible Cause of Vascular Aging," *Cell* 173, no. 1 (March 2018): 74–89, https://www.sciencedirect.com/science/article/pii/S0092867418301521.

2. J. Li, M. S. Bonkowski, S. Moniot, et al., "A conserved NAD+ Binding Pocket That Regulates Protein-Protein Interactions During Aging," *Science* 355, no. 6331 (March 24, 2017): 1312–17, https://www.ncbi.nlm.nih.gov/pmc/articles/PMC5456119/.

3. President's Council on Bioethics, *Beyond Therapy: Biotechnology and the Pursuit of Happiness* (New York: HarperCollins, 2003), 190.

4. Ibid., 192.

5. Ibid., 200.

6. "ICD-11 for Mortality and Morbidity Statistics: MG2A Old Age," World Health Organization, December 2018, https://icd.who.int/browse11/l-m/en#/http://id.who.int/icd/entity/835503193.

7. Bravo Probiotic Yogurt, https://www.bravo-probiotic-yogurt.com/.

8. Y. Guan, S.-R. Wang, X.-Z. Huang, et al., "Nicotinamide Mononucleotide, an NAD+ Precursor, Rescues Age-Associated Susceptibility to AKI in a Sirtuin 1–Dependent Manner," *Journal of the American Society of Nephrology* 28, no. 8 (August 2017): 2337–52, https://jasn.asnjournals.org/content/28/8/2337; see also S. Wakino, K. Hasegawa, and H. Itoh, "Sirtuin and Metabolic Kidney Disease," *Kidney International* 88, no. 4 (June 17, 2015): 691–98, https://www.ncbi.nlm.nih.gov/pmc/articles/PMC4593995/.

Sinclair Disclosure

Dr. Sinclair is committed to turning key discoveries into medicines and technologies that help the world. He is involved in a variety of activities beyond being an academic including being a founder, equity owner, adviser, member of the board of directors, consultant, investor, collaborator with, and inventor on patents licesnsed to companies working to improve the human condition. These include Vium; CohBar; Galileo Bioscience; Wellomics; EdenRoc Sciences and its affiliates Arc Bio, Dovetail Genomics, Claret Medical, Revere Biosciences, UpRNA, MetroBiotech, and Liberty Biosecurity; and Life Biosciences and its affiliates Selphagy Therapeutics, Senolytic Therapeutics, Spotlight Therapeutics, Lua, Animal Biosciences, Iduna, Continuum Innovation, Prana (now Alterity); and Jumpstart Fertility. He is an inventor on over forty patents, most of which are licensed to industry or have been filed by companies, including a patent application filed by Mayo Clinic and Harvard Medical School and licensed to Elysium Health, of which any proceeds to him are donated to research. He gives lectures at conferences, museums, not-for-profit events, and occasionally at companies, and he sits on the boards of not-for-profit organizations, including the American Federation for Aging Research. He also serves as an adviser to the Lorraine Cross Award. For an updated list of activities, see https://genetics.med.harvard.edu/sinclair/.

The Scale of Things

1 grain of sand = 10 skin cells	0.5 millimeter
1 skin cell = 5 blood cells	50 micrometers
1 blood cell = 2 X chromosomes or ~2 yeast cells	10 micrometers
1 X chromosome = 1 yeast cell = 10 *E. coli*	5 micrometers
1 *E. coli* or mitochondrion = 2 *M. superstes*	0.5 micrometer
1 *M. superstes* = 4 ribosomes	0.25 micrometer
1 ribosome = 6 catalase enzymes	30 nanometers
1 catalase enzyme = 5 glucose molecules	5 nanomaters
1 glucose molecule or amino acid = approximately 4–6 water molecules	1 nanometer
1 water molecule = 275,000 atomic nuclei	0.275 nanometer
1 atomic nucleus	1 picometer

1 inch = 25.4 millimeters
1 foot (12 inches) = 0.3048 meter
1 yard (3 feet) = 0.9144 meter
1 mile = 1.6093 kilometers

1 million = 10^6 (1 with 6 zeros)
1 billion = 10^9 (1 with 9 zeros)
1 trillion = 10^{12} (1 with 12 zeros)
milli = 10^{-3} (1 thousandth)
micro = 10^{-6} (1 millionth)
nano = 10^{-9} (1 billionth)
pico = 10^{-12} (1 1,000 billionth, or a trillionth)

32°F = 0°C
212°F = 100°C

Cast of Characters

JOSEPH BANKS (February 24, 1743-June 19, 1820): English naturalist, botanist, and former president of the Royal Society who accompanied Captain James Cook on his voyage round the world. With Lord Sydney a staunch advocate of starting a colony in Australia at Botany Bay on Cape Banks. Namesake of the flower called the *Banksia*.

NIR BARZILAI (December 23, 1955-): Israeli-born American endocrinologist and professor at the Albert Einstein College of Medicine in New York best known for his work to elucidate genes that enable members of Ashkenazi families to live over 100, hormones that control lifespan, and the effects of metformin on lifespan.

ELIZABETH BLACKBURN (November 26, 1948-): an Australian American Nobel laureate who, with Carol W. Greider and Jack W. Szostak, discovered telomerase, the enzyme that extends telomeres. In 2004, she was controversially dismissed from the Bush administration's President's Council on Bioethics, allegedly for her advocacy of stem cell research and politics-free scientific enquiry.

ARTHUR C. CLARKE (December 16, 1917-March 19, 2008): British science fiction writer and futurist known as the "Prophet of the Space Age." Spent most of his adult life in Sri Lanka foreseeing the advent of space travel and satellites. Advocate for protection of gorillas. Polio in 1962 led to postpolio syndrome.

ALVISE (LUIGI) CORNARO (1464 or 1467-May 8, 1566): Venetian nobleman and patron of arts who wrote four books of *Discorsi* about the path to health and longevity that included fasting and sobriety.

EILEEN M. CRIMMINS: American demographer at the University of Southern California who was the first to combine indicators of disability, disease, and mortality to predict healthy life expectancy. She showed that the prevalence of dementia in women stems largely from their longer life.

RAFAEL DE CABO (January 20, 1968-): Spanish-born scientist at the National Institutes of Health, an expert in the study of the effects of diet on health and lifespan in rodents and primates.

BENJAMIN GOMPERTZ (March 5, 1779-July 14, 1865): British self-educated mathematician who is best known for the Gompertz-Makeham Law of Human Mortality, a demographic model (1825). He became a Fellow of the Royal Society and then an actuary at Alliance Assurance company, founded by his brother-in-law Sir Moses Montefiore with his relative Nathan Mayer Rothschild.

LEONARD P. GUARENTE (June 6, 1952-): American molecular biologist and professor at MIT, best known for codiscovering the role of the sirtuins in aging and the necessity of NAD⁺ for sirtuin activity, linking energy metabolism to longevity.

ALEXANDRE GUÉNIOT (1832-1935): Centenarian and French physician who wrote the book *Pour vivre cent ans. L'Art de prolonger ses jours* (To Live a Century). He attributed great significance to the "hereditary vital force" that he suggested determines the natural duration of human life at no less than 100 years.

JOHN B. GURDON (October 2, 1933-): British biologist who in 1958 cloned a frog using a nucleus from an adult tadpole's cell, demonstrating that aging can be reset, for which he shared the Nobel Prize with Shinya Yamanaka in 2012.

DENHAM HARMAN (February 14, 1916-November 25, 2014): American chemist who formulated the "Free Radical Theory of Aging" and the "Mitochondrial Theory of Aging." Harman was a founder of the American Aging Association, ran two miles a day until he was 82, and eventually died at the age of 98.

LEONARD HAYFLICK (May 20, 1928-): American biologist who invented the inverted microscope; best known for his 1962 discovery that normal mammalian cells have a limited capacity for replication. The Hayflick limit on cell division overturned a long-held belief promulgated by the French surgeon and biologist Alexis Carrel in the early twentieth century that normal cells in culture would proliferate continuously.

STEVE HORVATH (October 25, 1967-): Austrian-born American professor at the University of California at Los Angeles known for his pioneering work on epigenetics and aging and for codeveloping algorithms that predict the age of organisms based on DNA methylation patterns, known as the Horvath aging clock.

SHIN-ICHIRO IMAI (December 9, 1964-): Japanese-born American biologist known for his Heterochromatin Hypothesis of Aging, his work on mammalian sirtuins, and the discovery with Lenny Guarente that sirtuins need NAD⁺ for their activity.

CYNTHIA J. KENYON (February 21, 1954-): American geneticist who showed that Daf-2 mutations double nematode worm lifespan, after studying under Nobel Prize winner Sydney Brenner using nematodes as a model organism. Kenyon is a professor at the University of California, San Francisco, and vice president of aging research at Calico.

JAMES L. KIRKLAND: American physician and biologist working at the Mayo Clinic in Rochester, New York; a pioneer in the study of senescent "zombie" cells and the development of drugs called senolytics that kill them.

THOMAS B. L. KIRKWOOD (July 6, 1951-): South African–born biologist and associate dean for ageing at Newcastle University, UK. Proposed the Disposable Soma hypothesis, the idea that species aim to balance energy and resources between reproduction and building a robust, long-lasting body.

PIERRE LECOMTE DU NOÜY (December 20, 1883-September 22, 1947): French biophysicist and philosopher who noticed that the wounds of older soldiers healed more slowly than those of younger ones. His "telefinalist" hypothesis that God directs evolution was criticized as unscientific.

CLIVE M. McCAY (March 21, 1898-June 8, 1967): American nutritionist and biochemist who spent decades at Cornell University researching the soybean and flour. Best known for his early work confirming that calorie restriction extends the lifespan of rats. In 1955, he and his wife published "You Can Make Cornell Bread."

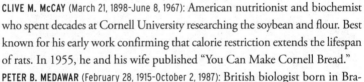

PETER B. MEDAWAR (February 28, 1915-October 2, 1987): British biologist born in Brazil whose work on graft rejection and the discovery of acquired immune tolerance was fundamental to the practice of tissue and organ transplants. Realized the force of natural selection declines with age due to reduced "reproductive value."

ARTHUR PHILLIP (October 11, 1738-August 31, 1814): British admiral of the Royal Navy and first governor of New South Wales who sailed to Australia to establish the British penal colony in Botany Bay that later, after moving one harbor north, became the city of Sydney, Australia.

CLAUDE E. SHANNON (April 30, 1916-February 24, 2001): American mathematician and engineer who worked at MIT and is known as the "father of information theory." His paper "A Mathematical Theory of Communication" (1948) solved problems of information loss and its restoration, concepts that laid the foundation for the TCP/IP protocols that run the internet. His hero was Thomas Edison, who he later learned was his relative.

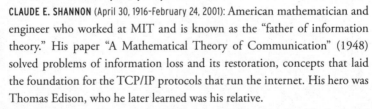

JOHN SNOW (March 15, 1813-June 16, 1858): English anesthesiologist and leader in the adoption of anesthesia and medical hygiene; best known for his work tracing the source of a cholera outbreak arising from the Broad Street pump in Soho, London, in 1854.

LEO SZILARD (February 11, 1898-May 30, 1964): Hungarian-born American physicist and humanist who proposed the DNA Damage Hypothesis of Aging. Wrote the letter that resulted in the Manhattan Project. Conceived of the nuclear chain reaction, nuclear power, chemostat, electron microscopes, enzyme feedback inhibition, and cloning of a human cell.

CONRAD H. WADDINGTON (November 8, 1905-September 26, 1975): British geneticist and philosopher who laid the foundations of systems biology and epigenetics. His Waddington Landscape was proposed to help understand how a cell can divide to become the hundreds of different cell types in the body.

ROY L. WALFORD (June 29, 1924-April 27, 2004): American biologist who rejuvenated the field of caloric restriction. One of eight crew members inside Arizona's Biosphere 2 from 1991 to 1993. In medical school, reportedly used statistical analysis to predict the results of a roulette wheel in Reno, Nevada, to pay for medical school and a yacht, and sailed the Caribbean for over a year.

H. G. WELLS (September 21, 1866-August 13, 1946): British science fiction writer who foresaw air raids in World War II, tanks, nuclear weapons, satellite television, and the internet. Best known for *The War of the Worlds*, *The Shape of Things to Come*, and *The Time Machine*. His epitaph is from *A War in the Air*: "I told you so. You damned fools."

GEORGE C. WILLIAMS (May 12, 1926-September 8, 2010): American evolutionary biologist at the State University of New York, Stony Brook, known for developing a gene-centric view of evolution and "Antagonistic Pleiotropy," a leading theory about why we age; essentially that a gene that helps young individuals survive can come back to bite them when they are older.

SHINYA YAMANAKA (September 4, 1962-): Japanese biologist who discovered reprogramming genes that turn regular cells into stem cells, for which he shared the Nobel Prize in Physiology or Medicine with John Gurdon in 2012.

Glossary

ALLELE: One of several possible versions of a gene. Each one contains a distinct variation in its DNA sequence. For example, a "deleterious allele" is a form of a gene that leads to disease.

AMINO ACID: The chemical building block of proteins. During translation, different amino acids are strung together to form a chain that folds into a protein.

ANTAGONISTIC PLEIOTROPY: A theory proposed by George C. Williams as an evolutionary explanation for aging: a gene that reduces lifespan in late life can be selected for if its early benefits outweigh its late costs. An example of this is the survival circuit.

BASE: The four "letters" of the genetic code, A, C, T, and G, are chemical groups called bases or nucleobases. A= adenine, C = cytosine, T = thymine, and G = guanine. Instead of thymine, RNA contains a base called uracil (U).

BASE PAIR: "Teeth" on the twisted "zipper" of DNA. Chemicals known as bases make up a DNA strand, each strand runs in the opposite direction, and bases attract their opposite partner to make a base pair: C pairs with G, A pairs with T (except for in RNA, where it's a U).

BIOTRACKING/BIOHACKING: The use of devices and lab tests to monitor the body to make decisions about food, exercise, and other lifestyle choices to optimize the body. Not to be confused with biohacking, which is do-it-yourself body enhancement.

CANCER: A disease caused by uncontrolled growth of cells. Cancerous cells may form clumps or masses known as tumors and can spread to other parts of the body through a process known as metastasis.

CELL: The basic unit of life. The number of cells in a living organism ranges from one (e.g., in yeast) to quadrillions (e.g., in a blue whale). A cell is composed of four key macromolecules that allow it to function: protein, lipids, carbohydrates, and nucleic acids. Among other things, cells can build and break down molecules, move, grow, divide, and die.

CELLULAR REPROGRAMMING: The changing of cells from one type of tissue to a prior stage of development.

CELLULAR SENESCENCE: The process that occurs when normal cells stop dividing and start to release inflammatory molecules, sometimes caused by telomere shortening, damage to DNA, or epigenomic noise. Despite their seeming "zombie" state, senescent cells remain alive, damaging nearby cells with their inflamatory secretions.

CHROMATIN: Strands of DNA wound around protein scaffolds known as histones. Euchromatin is open chromatin that allows genes to be switched on. Heterochromatin is closed chromatin that prevents the cell from reading a gene, also known as gene silencing.

CHROMOSOME: The compact structure into which a cell's DNA is organized, held together by proteins. The genomes of different organisms are arranged into varying numbers of chromosomes. Human cells have 23 pairs.

COMPLEMENTARY: Describes any two DNA or RNA sequences that can form a series of base pairs with each other. Each base forms a bond with a complementary partner: T (in DNA) and U (in RNA) bond with A, and C bonds with G.

CRISPR: Pronounced "crisper." An immune system found in bacteria and archaea, co-opted as a genome-engineering tool to cut DNA at precise places in a genome. CRISPR, which stands for "clustered regularly interspaced short palindromic repeats," is a section of the host genome containing alternating repetitive sequences and snippets of foreign DNA. CRISPR proteins such as Cas9, a DNA-cutting enzyme, use these as molecular "mug shots" as they seek out and destroy viral DNA.

DAF-16/FOXO: An ally of sirtuins, DAF-16/FOXO is a gene control protein called a transcription factor that activates cell defense genes, upregulation of which extends lifespan in worms, flies, mice, and perhaps humans; required for Daf-2 to extend lifespan in worms.

DEACETYLATION: The enzymatic removal of acetyl tags from proteins. Removal of acetyls from histones by histone deacetylases (HDACs) causes them to be more tightly packed, switching off a gene. Sirtuins are NAD-dependent deacetylases. Deacylation is a catchall term that includes deacetylation and the removal of other, more exotic tags such as butyryls and succinyls.

DEMETHYLATION: Demethylation is the removal of methyls and is carried out by enzymes called histone demethylases (KDMs) and DNA demethylases (TETs). Attachment of methyls is achieved by a histone or DNA methyltransferases (DMTs).

DISPOSABLE SOMA: A hypothesis proposed by Tom Kirkwood to explain aging. Species evolve to grow and multiply quickly or build a long-lasting body, but not both; limited resources in the wild don't allow for both.

DNA: Abbreviation of deoxyribonucleic acid, the molecule that encodes the information needed for a cell to function or a virus to replicate. Forms a double-helix shape that resembles a twisted ladder, similar to a zipper. Bases, abbreviated as A, C, T, and G, are found on each side of the ladder, or strand, which run in opposite directions. The bases have an attraction for each other, making A stick to T and C stick to G. The sequence of these letters is called the genetic code.

DNA DOUBLE-STRAND BREAK (DSB): What happens when both strands of DNA are broken and two free ends are created. May be done intentionally with an enzyme such as Cas9 or I-*Ppo*I. Cells repair their DNA to prevent cell death, sometimes changing the DNA sequence at the site of the break. Initiating or controlling this process with the intent to alter a DNA sequence is known as genome engineering.

DNA METHYLATION CLOCK: Changes in the number and sites of DNA methylation tags on DNA can be used to predict lifespan, marking time from birth. During epigenomic reprogramming or cloning of an organism, methyl marks are removed, reversing the age of the cell.

ENZYME: A protein made up of strings of amino acids that folds into a ball that can carry out chemical reactions that would normally take much longer or otherwise never happen. Sirtuins, for example, are enzymes that use NAD to remove acetyl chemical groups from histones.

EPIGENETIC: Refers to changes to a cell's gene expression that do not involve altering its DNA code. Instead the DNA and the histones that the DNA is wrapped around are "tagged" with removable chemical signals (see Demethylation and deacetylation). Epigenetic marks tell other proteins where and when to read the DNA, comparable to sticking a note that says "Skip" onto a page of a book. A reader will ignore the page, but the book itself has not been changed.

EPIGENETIC DRIFT AND EPIGENETIC NOISE: Alterations to the epigenome that take place with age due to changes in methylation, often related to an individual's exposure to environmental factors. Epigenomic drift and noise may be a key driver of aging in all species. Damage to DNA, especially DNA breaks, is a driver of this process.

EXDIFFERENTIATION: The loss of cell identity due to epigenetic noise. Exdifferentiation may be a major cause of aging (see Epigenomic Noise).

EXTRACHROMOSOMAL RIBOSOMAL DNA CIRCLE (ERC): The generation of extrachromosomal ribosomal DNA circles leads to the breaking apart of the nucleolus in old cells, and in yeast they distract the sirtuins and cause aging.

GENE: A segment of DNA that encodes the information used to make a protein. Each gene is a set of instructions for making a particular molecular machine that helps a cell, organism, or virus function.

GENE EXPRESSION: A product based on a gene; can refer to either RNA or protein. When a gene is turned on, cellular machines express this by transcribing the DNA into RNA and/or translating the RNA into a chain of amino acids. For example, a highly expressed gene will have many RNA copies produced, and its protein product is likely to be abundant in the cell.

GENE THERAPY: The delivery of corrective DNA to human cells as a medical treatment. Certain diseases can be treated or even cured by adding a healthy DNA sequence into the genomes of particular cells. Scientists and doctors typically use a harmless virus to shuttle genes into targeted cells or tissues, where the DNA is incorporated somewhere within the cells' existing DNA. CRISPR genome editing is sometimes referred to as a gene therapy technique.

GENETICALLY MODIFIED ORGANISM (GMO): An organism that has had its DNA altered intentionally using scientific tools. Any organism can be engineered in this manner, including microbes, plants, and animals.

GENOME: The entire DNA sequence of an organism or virus. The genome is essentially a huge set of instructions for making individual parts of a cell and directing how everything should run.

GENOMICS: The study of the genome, all the DNA of a given organism. Involves a genome's DNA sequence, the organization and control of genes, the molecules that interact with the DNA, and the way in which these different components affect the growth and function of cells.

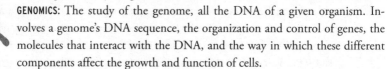

GERM CELLS: The cells involved in sexual reproduction: eggs, sperm, and precursor cells that develop into eggs or sperm. The DNA in germ cells, including any mutations or intentional genetic edits, may be passed down to the next generation. Genome editing in an early embryo is considered to be germline editing since any DNA changes will likely end up in all cells of the organism that is eventually born.

HISTONES: The proteins that form the core of DNA packaging in the chromosome and the reason three feet of DNA can fit inside a cell. DNA wraps around each histone almost two times, like beads on a string. The packaging of histones is controlled by enzymes such as the sirtuins that add and subtract chemical groups. Tight packaging forms "silent" heterochromatin, while loose packaging forms open euchromatin, where genes are turned on.

HORMESIS: The idea that whatever doesn't kill you makes you stronger. A level of biological damage or adversity that stimulates repair processes that provide cell survival and health benefits. Originally discovered when plants were sprayed with diluted herbicide and afterward grew faster.

INFORMATION THEORY OF AGING: The idea that aging is due to the loss of information over time, primarily epigenetic information, much of which can be recovered.

 METFORMIN: A molecule derived from the French hellebore used to treat type 2 (age-associated) diabetes that may be a longevity medicine.

 MITOCHONDRIA: Often called the cell's powerhouse, mitochondria break down nutrients to create energy in a process called cellular respiration. They contain their own circular genome.

MUTATION: A change from one genetic letter (nucleotide) to another. Variation in the DNA sequence gives rise to the incredible diversity of species among different organisms of the same genus. Though some mutations have no consequences at all, others can directly cause disease. Mutations may be caused by DNA-damaging agents such as ultraviolet light, cosmic radiation, or DNA copying by enzymes. They can also be created deliberately via genome-engineering methods.

NAD: Nicotinamide adenine nucleotide, a chemical used for more than five hundred chemical reactions and for sirtuins to remove acetyl groups of other proteins such as histones to turn genes off or give them cell protective functions. A healthy diet and exercise raise NAD levels. The "+" sign you sometimes see, as in NAD^+, indicates that it is not carrying a hydrogen atom.

NUCLEASE: An enzyme that breaks apart the backbone of RNA or DNA. Breaking one strand generates a nick, and breaking both strands generates a double-strand break. An endonuclease cuts in the middle of RNA or DNA, while an exonuclease cuts from the end of the strand. Genome engineering tools such as Cas9 and I-*Ppo*I are endonucleases.

NUCLEIC ACIDS OR NUCLEOTIDES: The basic chemical units that are strung together to make DNA or RNA. They consist of a base, a sugar, and a phosphate group. The phosphates link with sugars to form the DNA/RNA backbone, while the bases bind to their complementary partners to form base pairs.

NUCLEOLUS: Located inside the nucleus of eukaryotic cells, the nucleolus is a region where the ribosomal DNA (rDNA) genes are situated and where the cellular machines for stitching together amino acids to form proteins are assembled.

PATHOGEN: A microbe that causes illness. Most microorganisms are not pathogenic to humans, but some strains or species are.

PROTEIN: A string of amino acids folded into a three-dimensional structure. Each protein is specialized to perform a specific role to help cells grow, divide, and function. Proteins are one of the four macromolecules that make up all living things (proteins, lipids, carbohydrates, and nucleic acids).

 RAPAMYCIN: Also known as sirolimus, rapamycin is a compound with immunosuppressant functions in humans. It inhibits activation of T cells and B cells by reducing their sensitivity to the signaling molecule interleukin-2. Extends lifespan by inhibiting mTOR.

REDIFFERENTIATION: The reversal of epigenetic changes that occur during aging.

RIBOSOMAL DNA (rDNA): A key component of the manufacture of new proteins within cells; the source of the genetic code for ribosomal RNA, which is the building block of the ribosome. These molecules knit together amino acids that become new proteins.

RNA: Abbreviation of ribonucleic acid. Transcribed from a DNA template and typically used to direct the synthesis of proteins. CRISPR-associated proteins use RNAs as guides to find matching target sequences in DNA.

SENOLYTICS: Pharmaceuticals currently under development that are hoped to kill senescent cells in order to slow down or even reverse aging-related issues.

SIRTUINS: Enzymes that control longevity; they are found in organisms from yeast to humans and need NAD⁺ to function. They remove acetyl and acyl groups from proteins to instruct them to protect cells from adversity, disease, and death. During fasting or exercise, sirtuin and NAD⁺ levels increase, which may explain why those activities are healthy. Named after the yeast *SIR2* longevity gene, *SIRT1–7* (Sir2 homologs 1 to 7) genes in mammals play key roles in protecting against disease and deterioration.

SOMATIC CELLS: All the cells in a multicellular organism except for germ cells (eggs or sperm). Mutations or changes to the DNA in the soma will not be inherited by subsequent generations unless cloning takes place.

STEM CELLS: Cells with the potential to turn into a specialized type of cell or divide to make more stem cells. Most cells in your body are differentiated; that is, their fate has already been decided and they cannot morph into a different kind of cell. For example, a cell in your brain cannot suddenly transform into a skin cell. Adult stem cells replenish the body as it becomes damaged over time.

STRAND: A string of connected nucleotides; can be DNA or RNA. Two strands of DNA can zip together when complementary; bases match up to form base pairs. DNA typically exists in this double-stranded form, which takes the shape of a twisted ladder or double helix. RNA is typically composed of just a single strand, though it can fold up into complex shapes.

SURVIVAL CIRCUIT: An ancient control system in cells that may have evolved to shift energy away from growth and reproduction toward cellular repair during times of adversity. After response to adversity, the system may not fully reset, which, over time, leads to a disruption of the epigenome and loss of cell identity leading to aging (see Antagonistic Pleiotropy).

TELOMERES/TELOMERE LOSS: A telomere is a cap that protects the end of the chromosome from attrition, analogous to the aglet at the end of a shoelace or a burned end of a rope to stop it fraying. As we age, telomeres erode to the point where the cell reaches the Hayflick limit. This is when the cell regards the telo-mere as a DNA break, stops dividing, and becomes senescent.

 TRANSCRIPTION: The process by which genetic information is copied into a strand of RNA; performed by an enzyme called RNA polymerase.

 TRANSLATION: The process by which proteins are made based on instructions encoded in an RNA molecule. Performed by a molecular machine called the ribosome, which links together a series of amino acid building blocks. The resulting polypeptide chain folds up into a particular 3D object, known as a protein.

 VIRUS: An infectious entity that can persist only by hijacking a host organism to replicate itself in. Has its own genome but is technically not considered a living organism. Viruses infect all organisms, from humans to plants to microbes. Multicellular organisms have sophisticated immune systems that combat vi-ruses, while CRISPR systems evolved to stop viral infection in bacteria and archaea.

WADDINGTON'S LANDSCAPE: A biological metaphor for how cells are endowed with an identity during embryonic development in the form of a 3D relief map. Marbles representing stem cells roll down into bifurcating valleys, each of which marks a different developmental pathway for the cells.

XENOHORMESIS HYPOTHESIS: The idea that our bodies evolved to sense the stress cues of other species, such as plants, in order to protect themselves during times of impending adversity. Explains why so many medicines come from plants.

Index

NOTE: Bold page numbers refer to picture captions.

ABOUT THE AUTHORS

DAVID A. SINCLAIR, PHD, AO, is one of the world's most famous scientists and entrepreneurs, best known for understanding why we age and how to reverse it. He is a tenured professor of genetics, Blavatnik Institute, Harvard Medical School; co-director of the Paul F. Glenn Center for the Biology of Aging Research at Harvard; co-joint professor and head of the Aging Labs at the University of New South Wales in Sydney, Australia; and an honorary professor at the University of Sydney. His work is regularly featured in print, podcasts, TV, and books, including *60 Minutes*, a Barbara Walters special, *NOVA*, and Morgan Freeman's *Through the Wormhole*. He is best known for his work on genes and small molecules that delay aging, including the sirtuin genes, resveratrol, and NAD precursors. He has published more than 170 scientific papers, is a coinventor on more than 50 patents, and has cofounded 14 biotechnology companies in the areas of aging, vaccines, diabetes, fertility, cancer, and biodefense. He serves as co–chief editor of the scientific journal *Aging*, works with national defense agencies and with NASA, and has received 35 honors, including being one of Australia's leading scientists under

45, the Australian Medical Research Medal, the NIH Director's Pioneer Award, *Time* magazine's list of the "100 Most Influential People in the World" (2014) and the "Top 50 People in Healthcare" (2018). In 2018, he became an Officer of the Order of Australia for his work in medicine and national security.

MATTHEW D. LAPLANTE is an associate professor of journalistic writing at Utah State University. A selection of his work as a journalist, radio host, author, and cowriter can be found at www.mdlaplante.com.

Connect with Dr. David A. Sinclair on

———————

lifespanbook.com

✉ info@lifespanbook.com

🐦 @davidasinclair / @mdlaplante

📷 @davidsinclairphd